Quality and Profit in Building Design

Quality and Profit in Building Design

Edited by

P.S. Brandon
Head of Department of Surveying
Portsmouth Polytechnic

and

J.A. Powell
School of Architecture
Portsmouth Polytechnic

LONDON NEW YORK
E. & F.N. SPON

First published 1984 by
E. &. F.N. Spon Ltd
11 New Fetter Lane, London EC4P 4EE

Published in the USA by
E. &. F.N. Spon
733 Third Avenue, New York NY10017

© 1984 This collection:
P.S. Brandon and J.A. Powell

Printed in Great Britain at the
University Press, Cambridge

ISBN 0 419 13390 9

British Library Cataloguing in Publication Data

Quality and profit in building design.
1. Building — Estimates
I. Brandon, Peter S.   II. Powell, James A.
690'.068'1   TH435

ISBN 0-419-13390-9

TRANSACTIONS

OF

DESIGN : QUALITY : COST : PROFIT

A Conference held at

PORTSMOUTH POLYTECHNIC

(School of Architecture)

(Department of Surveying)

29th NOVEMBER — 1st DECEMBER 1984

CONTENTS

PAGE

(v)

(vii)

(viii)

FOREWORD

Design:Quality:Cost:Profit was the title for a conference bringing together those interested in exploring the relationship between these factors in building design. This book of its proceedings, carrying the title "Quality and Profit in Building Design", attempts to reveal the essence of the debate at that conference through its papers, and more particularly to articulate those discussions which must face all involved in the building industry, namely, how to design a quality building at a reasonable profit to enable future development. Forecasts for the next decade all point to the fact that Britain will only survive as an advanced nation if it becomes more effective and inventive at utilising its scarce resources. This is especially so in the building industry where there is truly a need to learn how to "Design for Profit" - the title of a campaign led by the Department of Industry to bring home the importance of design to our manufacturing industries. While it is agreed that profit is a necessary precursor for economic growth, it is clearly not sufficient in itself. Profit at all costs is not what Government or any of us envisage as the future context for design. At the very least we in Britain wish to keep our existing standard of life and, if possible improve it.

The idea for the present conference was stimulated as a result of rewarding feedback from two previous conferences held under the auspices of the Faculty of Environmental Studies at Portsmouth Polytechnic:

In 1982, the Department of Surveying and the Association of Heads of Surveying in Polytechnics, initiated the first building cost research conference, entitled "Building Cost Techniques: New Directions" ; its proceedings (Brandon, 1982), now in second printing, has provided those involved in the construction industry with an essential reference to the latest developments and techniques concerned with building costs.

In 1983, supported by the Design Research Society and the Science and Engineering Research Council's Building Sub-Committee, the School of Architecture organised a conference entitled "Designing for Building Utilisation". The aim of that conference was to produce guidelines (Powell,Lera and Cooper,1984) to aid designers' considerations of human factors when trying to produce a fuller utilisation of resources in future building designs.

Both conferences focussed their attention on the development of techniques concerned with specific aspects of building assessment, design and construction; as one would expect from their titles the former conference dealt with cost/profit issues while the latter with considerations of quality assessment relating to the needs of users and designers. As a result of these conferences and a burgeoning

xi

interest in general building evaluation techniques, due to greater resource restrictions, there are now many more well substantiated techniques available to those who want to design for improved building resource utilisation; unfortunately the techniques so far developed are restricted to sub problem considerations rather than dealing with the building design:quality:cost:profit problem as a whole. This may be because, neither of the above mentioned conferences nor the ensuing debate has realistically dealt with the middle ground between design and profit. It appeared to the present conference organisers that, at the very least, attention ought to be focussed on some strategic exploration of the issues that lay between Design and Profit and on developing practical tools for bridging the gap between traditional architectural designers and their cost consultants. How might this relationship become more symbiotic and how can all take a more active and constructive part in designing with quality and good value for a reasonable cost ? These, and related questions, came readily to mind when we began to think about topics for future discussion. This conference, and its proceedings, naturally grew out of those early ideas thanks largely to the financial and moral support of the Specially Promoted Programme in Construction Management of the Science and Engineering Research Council. Our primary object was to bring together leading practitioners and academics to discuss the benefits derived from building projects and the resources required to achieve them. Over thirty key individuals, concerned with the building industry, gathered together in Portsmouth during the latter part of 1984 to discuss their views and practices with respect to this important topic. Many of their discussions concerned the impact of design decision making on the quality or degree of excellence with which the objectives of the client may be satisfied; there was also much discussion on such topics as cost planning, value analysis and cost effectiveness. Current research featured strongly in the debate and hence forms a major part in the proceedings. Furthermore, to establish the present 'state of the art' at the leading edge of practice, we were pleased to invite and now include in this book extremely valuable case study material from several building firms.

Our hope was that the fruitful interchange between those involved in what are traditionally differentiated aspects of the same problem would yield new understandings that will improve the quality, efficiency and effectiveness of building design and construction. We also hope that the papers presented here will form the context for setting priorities and guiding future interdisciplinary research and development in construction and allied industries.

Peter Brandon and James Powell
Portsmouth, June 1984

BRANDON P.S.,   Building Cost Techniques: New Direction   Spon , 1982

POWELL J.A.,LERA S. AND COOPER I., Designing for Building Utilisation
        Spon , 1984.

**Section I
Editorial**

AN EDITORIAL CONJECTURE CONCERNING BUILDING DESIGN, QUALITY, COST
AND PROFIT

JAMES A POWELL, School of Architecture, Portsmouth Polytechnic
PETER S BRANDON, Department of Surveying, Portsmouth Polytechnic

1. INTRODUCTION

Pressure for Change in Design

At its most general level design is an adaptive process, a means
whereby members of the human race are able to cope with their
surroundings and deal with difficulties and change. Traditionally
designers were craft based people who had learned to develop and
produce simple physical objects well; design-by-drawing was the
method they used to externalise and then hone their initial concepts. However, over the past decade there has been a change in
the designers' role in society, a change in the nature of the problems they deal with, and a gradually rising vision of design itself. At every level of British Society people are becoming generally more design aware and now insist design matters are their
concern; at the highest level, Government is pursuing a campaign
for more profitable but higher quality design (especially in the
building industry); the two appropriate Research Councils (the
SERC and ESRC) are preparing major changes of strategy in the way
they handle design research; the CNAA and the Engineering Boards
are leading new design orientated changes in relevant degreee
courses; there are now several B. Tec, '0' and 'A' levels offered
in design based topics. The preceeding statements are a clear
testimony to the increasing interest and urgency with which design
is being considered but furthermore there is now a general recognition that design problems are more complex and less tangible
than ever before; that designs now have to be more reflective of
context, whole-life-costs, the need to produce profit, designers'
time spans and the design's own useful life spans. The designers
of today and tomorrow will have to design in space, and time, for
people, with people.

The problem is how do they cope with this new era of complex socio-
technical decision making ? Furthermore, how will they be able to
compose, in the time available and to a strict cost limit, building

3

design patterns which have a sense of beauty and which are truly reflective and responsive to a changing cultural context. These and related questions formed the framework for a conference from which this proceedings has emerged.

The Context for Change in the Building Industry

In recent years the attention of the world has been drawn to the dwindling stocks of its non-renewable resources. The construction industry has been close to the centre of that concern producing as it does commodities which are major users of energy, mineral resources and land. As a result, conservation has become the order of the day and, for the first time, the building industry as a whole has had to focus its attention on the performance of the buildings it designs. The greatest waste of resources undoubtedly occurs when a building, designed to a so-called "tight performance specification", fails to match the functions required of it either when new or in the future. The sacrifice of such resources, painstakingly acquired by existing and usually other members of society, is a bitter legacy for future generations. Furthermore, in order to raise the resources needed to construct and run the built environment, it is clear that society, and especially its financial community, is now requiring greater confidence that future building will bring forward reward in terms of profit, efficiency and/or welfare. The building industry will have to learn quickly from its past mistakes if it is to remain competitive (particularly with its European rivals) and not be a drain on society. Hopefully, it will respond positively to the challenge and go beyond by developing better and more rewarding design, construction and management strategies; it was with this in mind that those presenting papers herein were brought together.

Problem Definition

The problem restated becomes; "how to understand and thereby learn to control better the complex 'balancing' act between those factors and forces which constitute the title of this conference Design: Quality:Cost:Profit." The purpose of the papers in this proceedings is manifold: to describe in detail the precise nature and role of each factor (mentioned in the title) in the building design process; to explain why various sectors of the building industry may hold the views they do about these factors and the existing relationship between them; to investigate the procedures by which that balance has been derived; to suggest the sort of balance needed to improve efficiency and effectiveness; to indicate the kinds of understanding and strategy needed to enable improvement; and to suggest the sort of research which would aid our industry in this task. This introductory working paper is an attempt to provide a contextual backdrop against which to see the rest of the proceedings. It is a questioning framework - a conjecture to aid thought and discussion about Design:Quality:Cost:Profit; whether that con-

4

jecture will be refuted, or supported, only time will tell, but
hopefully the debate at least will be of value to the research and
professional community.

A Paradox and Caveat

There are, of course, great dangers in trying to achieve a balance
between such diverse elements as design, quality, cost and profit,
because the very word 'balance' suggests that all aspects can be
quantified and presumably measured with a common unit of measurement. For instance, most economic decision-making has, and still
does, attempt to quantify in this way and there is the danger that
this process will distort the higher values of life which lead to
the concept of quality and which by their very nature cannot be
measured or priced. Some of the gravest warnings in this respect
have come from Schumacher (1). For example:

> ".... out of the large number of aspects which in
> real life have to be seen and judged together be-
> fore a decision can be taken, economics supplies
> only one - whether a thing yields a money profit
> to those who undertake it or not."

and

> "To press non-economic values into the framework
> of the economic calculus, economists use the meth-
> od of cost/benefit analysis. This is generally
> thought to be an enlightened and progressive de-
> velopment, as it at least an attempt to take acc-
> ount of costs and benefits which might otherwise
> be disregarded altogether. In fact, however, it
> is a procedure by which the higher is reduced to
> the level of the lower and the priceless is given
> a price...to undertake to measure the immeasurable
> is absurd and constitutes but an elaborate method
> of moving from preconceived notions to foregone
> conclusions."

It is extremely important for a conference whose emphasis is on
the relationship between building cost and utility to have its
assumptions clear on these fundamental issues of value and preference. The theoretical explorations of economist Ken Arrow (2,3)
provide, perhaps, the most elegant, succinct and academically
rigorous proof of the fallacy of the common unit concept for measuring and assessing social preference - appearing in economics as
the notion of "good" or "utility" (where economists seem to believe
that utility is what people choose or "prefer" and where "preference" and "utility" have become almost interchangeable terms) or
to the followers of Bentham, the utilitarian principles of "collective happiness" (leading to the greatest happiness for the greatest number). Arrow endeavoured to design a theoretical "machine"

which would receive all citizens' preferences and attempt to aggregate them in a sort of electronic referenda (perhaps now actually possible through teletext); the hope was that his "machine" would somehow yield the best possible choice on each social issue, including quality and opportunity cost, in order to aid decision makers. Unfortunately, for the supporters of the unitary measure concept, Arrow showed that "if we exclude the possibility of interpersonal comparisons of utility, then the only method of passing from individual tastes to social preferences, which will be satisfactory and which will be defined for a wide range of sets of individual orderings, are either imposed or dictatorial." Arrow's treatment of the problem is highly mathematical and itself might obscure the importance of his views in the present argument. However, Douglas (4), whose recent concern has been with "how political processes cope with the paradoxes of social change", has neatly encapsulated the essence of Arrow's proof (or Theorem as it has come to be known) as follows:

"Arrow postulates that a rational method of arriving at a social or collective preference should satisfy the following four conditions:

i) Unrestricted choice. Whatever the order in which each member of a collectivity (of three or more members) places his preferences among three or more alternatives(always assuming each individual ordering is, of course, logical and transitive), it should be possible to aggregate them.

ii) If any member of the collectivity prefers one alternative to another, the social preferance should reflect this - usually called the Pareto Principle.

iii)The preference of one individual in the collectivity should not automatically become the social preference regardless of the preferences of all other individuals. Arrow called this "non-dictatorship".

iv) Independence of irrelevant alternatives. This requires that the social ordering of a given set of alternatives depend only on the individual members' preference orderings of those alternatives. Amongst other things, this condition excludes the possibility of interpersonal comparisons of utility.

Arrow's proof demonstrates the surprising fact that these four apparently commonsensical conditions are mutually incompatible. Furthermore, while several thousand papers have attempted to disprove or dismiss his theorem, it is now commonly accepted that

Arrow has provided a formally valid proof concerning aggregation of social preferences.

A summary of the Theorem is that "it is impossible to devise an aggregation device that can be sure of producing a single unitary set of social preferences under Arrow's four conditions of rationality"; this should make those of us involved in long term Building Cost/Design Research think very carefully before aggregating individual preferences into a summary statistical representation of social utility. Political systems have clearly understood Arrow's paradox and responded accordingly with suitable electoral mechanisms; however, those researching on such topics as <u>quality</u>, <u>utility</u> and <u>value</u>, especially in the area of building cost-utility, seem to have evaded or dismissed Arrow's work – perhaps because it is economically too difficult to handle with the traditional tools available to the economist. It appears that researchers in this topic area seem set to engage, at all costs, in the rather narrow and 'autistic' exercise of producing pragmatic and parsimonious tools simply to make the utility-cost decision making process more *efficient* . But the *costs* are *high* especially in terms of *lost opportunity* . For, while we are the first to welcome an appropriate application of Occam's razor to this topic area, we cannot condone its use to the detriment of the overall *effectiveness* of the decision making process. There is, therefore, a need for future building cost utility research to understand the implications of Arrow's paradox and to provide strategies and tools which are both context dependant and reflective of Arrow's general concerns.

The above discussion may lead the reader to believe that any attempt to derive a useful indication of social preference for the purposes of improving building design is pointless. However, Arrow himself indicates one solution when he says that "it is <u>not</u> impossible to devise some sort of aggregation framework that would enable us to elicit an idea of social preference from individual preferences" provided that framework evades the constraints implicit in the four conditions of rationality.

It is our contention that at least one way forward lies in negotiating a consensus for a particular preference rather than assuming it is a 'given', waiting to be revealed. For, as Douglas(4) concludes, "consensus, of course, infringes one of Arrow's conditions. If we have to put value on something other than revealed preferences of citizens, unanimity, the desire to be all of one mind, seems to be something on which we could well place positive value".

2. A CULTURAL/NEGOTIATIVE FRAMEWORK FOR CONSIDERING DESIGN:QUALITY: COST:PROFIT

We have noted the difficulty of trying to treat everyone as though they were the same. Another equally problematic alternative might be to negotiate consensus from the starting point that every individual is a separate and special case; a situation where needs would be infinite and their ordering – the sequence in which they come to be satisfied – would be brought about solely by the finiteness of resources; in the real world it is easy to see the endless and impossible nature of such a task of trying to negotiate total consensus from such an extreme. However, it may not be necessary to treat individuals in such an infinitely polarised way since most people are social beings as well as psychological individuals. When people operate within social networks and contexts, or as part of a social group, their behaviour seems largely mediated by those around them. For, as Thompson's (5, 6, 7 & 8) well substantiated research into industrial cultures and their related problems – poverty and housing, risk assessment and management, energy conservation policy, smoking and building utilisation – has shown, coercive and deepseated properties of social groups operate to limit some forms of behaviour/attitude (or at least does not make them worthwhile) and promote others. It might be useful to look at Thompson's model and propositions in greater depth with the aid of diagram 1 which depicts the three dimensional nature of the way people manage their needs and resources. Thompson identifies these degrees of freedom (or its inverse constraint) when it comes to citizens coping strategies for managing needs and resources – scope to manage needs, scope to manage resources and scope to manage the overlap between them. In theory, a person could reside almost anywhere in this conceptual management space depending upon his/her socio-psychological strategy. In practice, Thompson has shown that people's actual strategies are confined to the five corners of the diagram since there appear to be only five possible ways of viewing human social experience. Thompson's assertion, now backed by a wealth of cultural survey, is that "these five perceptions/strategies are "given" to us, or witheld from us, according to the ways in which we are caught up in the process of social life. So long as human social life exists these five paradigms will also exist". Both Thompson and his colleague, Mary Douglas (9) suggest that such strategies are substained by people's idea of nature and by certain socio-logics; these socio-logics are based upon two main orthogonal dimensions 'grid' or people's desired relationships with each other in networks (either controlling their own network or at the edge of someone elses) and 'group' or people's desire or otherwise to be a member of a group from which they can seek succour. Furthermore, according to Thomson, three of these five socially induced strategies are adopted by those who are purposefully engaged in the professional/business milieu, whereas the other two strategies relate more to those who have either totally opted out or have been so constrained by circumstances that they feel they are no longer in control; for brevity the former will henceforth be referred to as professional strategies while the latter as non-professional strategies. Starting with the three professional strategies they will now be detailed in a non-abstract way to give the reader a feel for the cultural framework.

8

DIAGRAM 1 (after Thompson, 1984). A Cultural Framework for Personal Management showing the five logically possible coping strategies

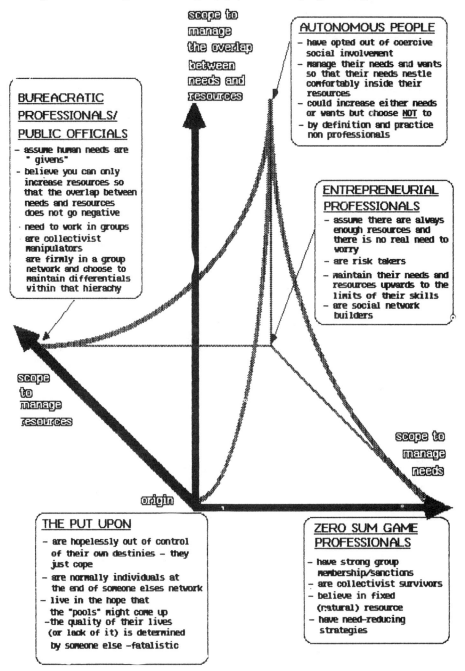

scope to manage the overlap between needs and resources

AUTONOMOUS PEOPLE
- have opted out of coercive social involvement
- manage their needs and wants so that their needs nestle comfortably inside their resources
- could increase either needs or wants but choose NOT to
- by definition and practice non professionals

BUREACRATIC PROFESSIONALS/ PUBLIC OFFICIALS
- assume human needs are " givens"
- believe you can only increase resources so that the overlap between needs and resources does not go negative
- need to work in groups
- are collectivist manipulators
- are firmly in a group network and choose to maintain differentials within that hierachy

ENTREPRENEURIAL PROFESSIONALS
- assume there are always enough resources and there is no real need to worry
- are risk takers
- maintain their needs and resources upwards to the limits of their skills
- are social network builders

scope to manage resources

scope to manage needs

origin

THE PUT UPON
- are hopelessly out of control of their own destinies - they just cope
- are normally individuals at the end of someone elses network
- live in the hope that the "pools" might come up
- the quality of their lives (or lack of it) is determined by someone else -fatalistic

ZERO SUM GAME PROFESSIONALS
- have strong group membership/sanctions
- are collectivist survivors
- believe in fixed (natural) resource
- have need-reducing strategies

Bureaucratic Professional/Public Officials

These people _need_ to work as a member of a large group network. As collectivist manipulators they are prepared to work as part of a strong heirarchy since they need the social/emotional support of others; members of such groups recognise that they can only improve their "share of the cake" through "collective opportunity"; to some extent they also realise that an improvement in their own lives is only possible so long as they do not "speak out of turn" or deliberately try to overtake people above them in the heirarchy. This is the group of people whose strategy is concerned with resource utilisation and needs imposition - usually by the use of standards or common units of assessment. Their view is that "since you ca<u>nnot</u> do anything about your needs" then the only viable strategy to adopt is to increase your resources so as to make sure that the overlap between needs and resources does not go negative. However, these professionals do not go to too much trouble to increase "the size of this overlap because to them needs are fixed - almost 'givens'". They would not agree in any sense to give people control over their own needs or resources because people cannot be trusted; these officials are the "professionals" or "experts" who "know" best what people really need and they want to control people so that in specifying needs they also have control over people's utilising of resources; furthermore, they feel those who do not act in a standard or normal sense (taken both literally and statistically) are abnormal and actually need controlling. This group do not use any more resources than absolutely necessary and since they are not prepared to question the decisions of those above them or "rock the boat" they are perfect controllers/guardians of the country's "standards".

Zero Sum Game Professionals

These professionals, as with bureaucrats and public officials, are strongly motivated to work in groups. Their view is that the world's resources are fixed and nothing can be done about physical "givens". As Thompson points out this group feels that since resources are finite "then one man's gain must be another man's loss". For this group, therefore, the only option is a 'need-reducing strategy' - a decrease of needs only to a level so as to ensure a comfortable overlap between the needs and resources." Such a need-reducing strategy, to be effective, has to be followed by everyone. These strongly collectivised survivors can all see the advantages of such behaviour and in the background, of course, there is always the threat of strong group sanctions to encourage those individuals who may be dragging their altruistic feet".

10

Entrepreneurial Professionals

This is the first professional who would prefer to work on his/
her own. He/she is a network builder and an individualist
characterised by Thompson as the "exhuberant business person who
is wont to say 'I am not able to reconcile my net income with
my gross habits'". These professionals are predominantly risk
takers who attempt to manage their needs and resources. They
reject the simple overlap management that we shall see is adopted by autonomous individuals; rather they maintain their needs
and resources upwards to the very limit of their entrepreneurial
skills and physical abilities.

For our three professionals, therefore, we have one group who
regard the needs as fixed and manage their resources; another
group who feel the resources are fixed and they manage their needs
and the final group of individuals who maximise their needs and
their resources. The two non-professionals manage their needs and
resources in slightly different ways.

The Put Upons

These people do not seem to be able to work together in a group;
they have no management strategy - no scope to manage either their
needs or their resources; they just cope as best they can with
their existing situation. These individuals feel they have no
control over their circumstances which appear to be determined,
and to a large extent controlled, by others, usually by one of
the kindhearted officials mentioned earlier. These "put upons"
are controlled by such officials and not in their own control.
According to Thompson they spend much of their pastime engaged
in chance related activities, such as bingo, in the hope of
"getting out" - they clearly feel they need luck to improve
their lot. For these people to be otherwise would mean public
officials expanding their needs for them to reach a, so called,
correct level; this level may be just above the poverty line
or it may be above the limit set by socio-political/economic
consideration (often these levels are assessed for society by
human environment scientists who themselves see human beings
as "ball bearings" rather than as interacting negotiators taking
a full part in life's "game"). Officials therefore manage the
resources on behalf of the "put upons" - extending a bit here
and trimming a little there so as to fill the rather rigid
existing resource framework.

Autonomous Individuals

Many yearn for this category yet few succeed. This is the
realm of the truly individualistic person who has opted out of
society. This individual is a simple resource needs overlap
manager whose needs are made to nestle comfortably inside his/

her resources. Such individuals clearly have the scope to manage the overlap between their resources and needs, and while they could easily increase either, they choose not to - "for that way lies coercive involvement and they have had enough of that".

For the present purposes it might be better to take on board, not the detail of Thompson's model, for that would need reference to his original texts, but rather the idea that there are only a small number of socially induced strategies for maintaining and managing people's needs and resources. For it is our contention that such a model of the social process, if applied within the building industry, would yield useful strategic and tactical understandings of the "problem space" that lies between building design and building profit; in this fundamental "space" reside a number of socio-technical considerations such as quality, value and utility, which must be understood in human terms if we are to have any chance of improving the effectiveness of our building design process. In this respect Thompson's analytical framework of cultural bias suitably disaggregates social actions and preferences in a most constructive way to provide a mechanism which:

i)   reflects the fact that human strategies for coping with needs
       and resources have social malleability; strategies which are
       slowly and continuously changing, so that people remain in
       control of their own circumstances rather than the circum-
       stances controlling them, but which appear static because
       they are sustained and explained by and through people's idea
       of nature.

ii)   is a valuable base from which future strategies for negotiat-
       ive consensus might be achieved for the building industry.

iii)   should enable any one member of society (social context) to
       recognise, and have some reciprocity with, the perspectives and
       actions of others; for it would appear that there are legitim-
       ate and constructive alternatives for viewing the world. Hope-
       fully the above framework will suggest the need to listen more
       to others before rejecting their views; otherwise consensus
       can never hope to be more than partial.

iv)   should enable us to interpret the sort of debate that occurs
       when different parties are discussing issues such as build-
       ing quality, building value and/or building utility; such
       an interpretation, in Thompson's terms "allows us to uncover
       the social processes that, working sometimes towards stability
       and sometimes towards instability, cause the debate to develop
       in one way rather than another".

When it comes to understanding the relationships between people, groups, institutions and government we now have a framework which enables us to identify five distinct kinds of social individual (or, rather, four, if we excluded the autonomous individual who has deliberately opted out of social involvement). Therefore any attempt to develop a methodology for negotiative concensus with respect to the building industry ought to recognise these distinctions.

Furthermore, we feel we ought to use these distinctions in conjunction with the highly developed techniques from the world of management decision making to provide a strategy and procedures which would, at least:

i)   produce consensus where it ought to be possible - within each distinct social grouping; a consensus which should indicate that grouping's constructive alternative.

ii)   initiate and support a debate between those who represent each distinct viewpoint; or as Thompson would say when it comes to the matter of building quality, value or utility":, who gets to talk about it and, out of those who get to talk about it, who gets to be listened to ?

iii) attempt to produce a global "preference ordering" through the building design process itself; for as O'Brien (10) has shown, design can be used to encompass differences in sub-group preference orderings which cannot be sensibly processed by any traditional, non-design, decision making.

Our particular suggestions would be to develop a transactional evaluatory process (see Powell and Sime, 11) which would attempt to understand and allow for social differences. This should lead to the development of designs which are more reflective of those differences; ideally "good" design should reduce the gap between different groups (and hopefully remove them altogether) so that each can agree that a final design is "good sense" but from their own perspective.

The above are important issues to raise within a conference on this topic although they tend to fit rather uncomfortably within the commercial world in which we all work. Unfortunately, much of the preeeding discussion may seem rather remote, or perhaps even irrelevant, to those who have come to listen to (or read) papers on pragmatic building-cost techniques. We clearly understand why that might be the case for those from whom figures are demanded at all stages of the design-construction process. Needless to say we feel that such an analytical framework of social interaction is important and would commend its use to you when reading the rest of this paper and indeed the rest of the proceedings. It is often very difficult to find words and visual images to counteract the weight of numbers and for this reason we felt the need to give over so much space in this paper to a full development of such an argument. However, the issue is, as already stated, a conjecture to promote discussion; we introduce it here as a kind of caveat to avoid "blinkered" focussing on mechanistic solutions to the "balancing" problem. Numerical techniques clearly have their place

but the underlying assumptions should be challenged and the wider
issues not discounted.

3. THE TITLE TOPICS ELABORATED, COMBINED AND COMPARED

In this section, using the developed framework, an attempt is made
to describe the important roles and existing relationships between
certain of the topic factors appearing in the title of this conference.

Quality in the Context of Time

Building quality is not something which is judged at a single
point in time and thereafter accepted without challenge. It is
continually under review throughout a building's life from a variety of sources and viewpoints - owners, tenants, managers, public
etc. Neither is it something which is limited to the performance
of materials or the correctness of the detailing, important as
these are. It embraces such diverse issues as spatial arrangement,
general ambience, aesthetics and flexibility. To encompass all
these things (and many more) for every unit of time in a statement
about building quality is extremely difficult indeed.

What assumptions of the future will we accept therefore when we
make our judgements as to quality? Will social values and patterns
of behaviour continue as they are today or will they vary? Will
there continue to be economic growth? Will there be a solving of
the energy problem or will it get worse? In the past there has
been a tendancy to expect stability in these and the other issues
relating to the future. Forecasting is now however a national
pastime and despite the monumental errors in prediction it would
now be considered irresponsible to assume a stable future.

These issues are of major importance to all concerned with the
construction process for they provide the context within which the
quality judgements can be made. More than this they are directly
related to cost budgeting for they affect the decision as to how
much the client should spend now and what materials, spatial arrangement, building form etc he should specify.

Scenarios for the Future - Time Constraints and Design Costs

We have just touched on issues on the key issue of future forecasting. In this respect it is interesting to note, from Thompson's
cultural bias framework that our three professional experts have
socially induced time perceptions with respect to future resource
scenarios.

   The Bureaucrats or Public Officials are moderates who have a balanced discrimination between the short and the long term. They

14

believe that a technical fix will solve their current problems because the future, while different, is a far way off.

The Zero Sum Game Professionals are conservative and believe in radical change now - for them the long-term implications must dominate the short-term.

The Entrepreneurial Individuals are optimistic and see "business as usual". For them the short-term dominates the long-term since they feel the future must be a simple extrapolation of the recent trends - as Thompson says "they feel, to get to the future, all we have to do is to carry on as we've been doing - innovate with skill and confidence".

It is clear that these different perceptions of the future will colour the cost information designers accept and use while designing.

Design Philosophy and Cost Planning

The complexity of the problem is so great that some commentators have decided that it is not worth considering. However there is a growing body of opinion arising from the suggestion by Professor Robert McLeod (taken up by Alex Gordon) that design philosophy should develop from the principles of long life, loose fit, low energy. These are issues which are directly related to the future and which will influence current expenditure and our present view of building quality.

This conference is concerned largely with the relationship between cost and quality and in particular the cost planning process and its effect on design quality. It is inevitable that a discussion on this topic will bring in the concept of value. Value was defined by Burt (12) as the relationship between quality and cost. If Quality is defined as the 'degree of excellence' by which he reaches his objective then the relationship between quality and value can be clearly seen.

Quality can of course be considered at a number of different levels.

i) Quality related to the workmanship employed in a project and which will largely depend on the skill and supervision of the workforce.

ii) Quality related to the level of specification and which will depend on the performance of the components and materials chosen.

iii) Quality related to spatial arrangement, circulation, function, aesthetic and so on which will have its origins in the skill of the designer.

15

In conventional cost planning terms the major considerations of quality are found in (ii) and (iii) above and then largely constrained to specification levels and the area/spatial needs of the building.

Those qualitative aspects related to workmanship and detailing are assumed to be satisfactory and outside the control of, or too early in, the design process for the design team to consider. However it is realised that too tight a budget may well lead to economies in these areas which could be to the detriment of the overall quality.

The Department of Education & Science Building Bulletin No 4 (13) describes cost planning as - "a term used to describe a method of allocating a pre-determined total sum of money between the elements of a building, using data from cost analysis, with the object of providing

i)   a proper balance between the area provided and the cost per square metre

ii)   a proper balance of cost between the constituent elements."

The key words in this definition are <u>total sum of money</u> and <u>proper balance.</u> The concept is one of 'designing to a cost' and as such is more concerned with value than quality. Although the definition appears to provide a rational strategy for design, in practice the operation tends to be largely one of accountancy and accountability based on whatever total sum is thought appropriate. In the years following the general implementation of cost planning techniques it is noticeable that the 'total sum' gradually became relatively smaller and areas and specification levels became more constrained. Dr Alan Spedding (14) commented from his research into educational buildings that ' ... falls in the value of the cost limit particularly in the 1970s caused many schools to be built to lower than desirable space and other quality standards'.

Unfortunately because cost planning techniques are based on what has been built before i.e. historic data, and as the incentive is to prove you can do better than the 'norm' a downward spiral develops. Each design team uses a previous project as its reference point and is inclined to cut the cost slightly. That project then becomes the norm and so the process continues. Eventually the lack of quality becomes evident, at considerable cost in putting it right, and a long battle develops to stop the spiral and improve the situation. Meanwhile a building stereotype has developed to which design teams retreat for safety in getting within the cost limit and because of the long term nature of building it is some years before this is found to be inadequate. Public sector building is littered with examples of this type of problem and although it cannot be laid entirely at the feet of the cost constraint there is no doubt that its influence was considerable.

16

Cost Constraints

It might be fairly argued that this is not the fault of cost planning but of the resource providers. This is true, but it must
also be recognised that it is in the nature of the technique to
constrain, and therefore if benefits (possibly in terms of quality)
are not also considered then an unsatisfactory situation may develop. This is not to deny the important contribution that elemental cost planning has made to the thinking of the design team but
to warn of the possible consequences of incorrect implementation.

At the other extreme where cost limits have been loosely applied
then the picture is not significantly better. In economic terms
research appears to suggest that the higher the initial cost then
the higher the maintenance cost. The reason for this anomaly may
lie in the fact that the extra latitude in the cost constraint
allowed for more experimentation with new untried materials or new
combinations of materials which did not match up to expectation.
Novelty and fashion took over, very often overthrowing the conventional wisdom of the past and resulting in rather large monuments
to failure.

Failure and Fashion

For example the removal of cills, copings, arches and so forth in
order to provide a clean, sharp image has resulted in widespread
weathering and damp problems due to jointing problems aggravated
by the water not being thrown from the buildings.

Perhaps the political fashion for system buildings has given
rise to the most public of these failures (see Russell, 15). Structural, jointing, condensation and roof problems have pervaded most
large systems on such a scale that many are having to be pulled
down or refurbished within two decades of their construction.
There is little evidence to show that system building is appreciably
cheaper in initial cost than traditional building (except in freak
economic conditions) and indeed considerable evidence to show that
they are more expensive in future cost.

It is therefore obvious that design quality is not directly
related to building cost. Only at the low cost margin can a strong
conflict between quality and cost be considered to be true. If a
cost limit is so tight that the design team are constrained to an
inadequate solution, then a marginal increase in the amount of
money available may well result in more than a proportional increase
in quality. The question therefore remains, where there is a reasonable sum of money provided, why do we not get better buildings ?

Design and Cost Education

Part of the problem may lie in the somewhat polarised philosophies behind current thinking within and between UK architectural and surveying education. Several architectural courses still assume that artistic values should provide all the major core programme of design work for architectural students whereas other courses, developed since the 1958 Oxford conference on Architectural Education, have laid greater emphasis on trying to integrate so called "necessary specialisms" - including economics, technology and the social sciences - into the more professional and design aspects of architectural education. Both course styles have led to problems, as has the difference between the ways in which architects and surveyors are taught.

In <u>artistically based Schools of Architecture</u> there is still a strong tradition of the designer as an artist. Like the painter he starts with a clean canvas for each new project and "... there is not need to go back to previous works to establish the new design ... it is a creative act and should not be tampered with ... it is of course influenced by outside pressures, standards, fashions, etc. but essentially it is an individual statement ..." These words, or words like them, are often heard in schools of architecture and elsewhere and are deeply ingrained in architectural education where the 'suck it and see' approach is still the major way of teaching design. There is therefore a tendency for the lessons of history to be forgotten until they become all too painfully obvious. In such schools art is divorced from technology to the detriment of both.

This is an exaggeration because no designer, or artist come to that, can completely ignore the lessons from the past. However, the 'crit' system of assessment in many schools of architecture encourages novelty and flair, quite rightly, but sometimes at the expense of competence in detailing. The conventional wisdom of the past can be overthrown for something new and replaced by systems which are largely untried but have featured quite significantly in the recent editions of the architectural press ! Whilst few of us would wish to see a straight jacket placed on the creative ideas of the designer, particularly at this early stage, it must also be recognised that too much experimentation can interfere with the essential training experience which can lead to a sound foundation for solving construction problems.

On the other hand, in the <u>Science-integrated-into-design Schools of Architecture</u>, the implicit assumption is that scientific understanding is necessary in order to produce better design or at least avoid major mistakes. While these aims are laudable, a pilot study by Cardona et al (16) revealed that very few schools of architecture can show how these specialist subjects have been fully integrated into the architectural design aspects of the course. As Broadbent (17) pointed out "integration has to be demonstrated in the working methods of an architectural school" and not just appear in the formal course documents. The conclus-

18

ions of Cardona and her colleagues are confirmed by the stated worries of several architectural educators. Specialist architectural lecturers from a number of disparate areas, from cost planning through to history, have all indicated concern over the unsatisfactory nature of existing attempts at information transfer and integration of their specialist disciplines within generalist design. Many authors have documented their individual, and often unsuccessful, attempts to correct what appear to them to be the underlying problems (see review by Powell, 18) Clearly, end-of-year examinations in schools of architecture do indicate that some information has been transferred from specialist to student, but most importantly final portfolios often reveal the limited extent to which detailed technical and cost understandings have made the way through to the student's design skills. Perhaps the wrong information is being taught in the wrong way. Over the years specialists have tried a range of educational techniques including laboratories, simulations, gaming and computer aids, with varying degrees of success, but the content of their message has remained very much the same. It may be this that is causing a failure of total information transfer for exploitation. It is our view that specialists may be putting undue emphasis on teaching the 'know-that' information of their discipline, rather than attempting to provide an educational context which will promote the architectural design skills of 'knowhow'. It is easy to blame this situation on the specialist, but one has to remember that few specialist teachers have ever practised as architectural designers. As a result they may not have the confidence to educate from a design base. And they may lack the understanding to know how information actually fits into, and is used, in practice.

<u>Quantity Surveyors</u> and to a lesser extent <u>property managers</u> receive a somewhat different engineering type of education which can often place too much emphasis on the mechanics of the process rather than the end result.

The conflicts between the approaches of the architect and surveyor were brought to a head in the early post war period when members of the quantity surveying profession were appointed direct by the client for the first time and one or two immediately saw their role as cutting the Architect down to size. A rash of buildings of inferior quality and expensive running cost appeared to be the end result although the architecture profession's adherence to the modern movement could be equally to blame. Fortunately, the education of surveyors has been more sensitive to client needs and design objectives since the 1970's and the profession as a whole now has a wider view of community need.

Despite the growing awareness of each other's contribution to a successful building there may still be strong argument for all the design professions to be educated in common, at least in the early years, before specialising. This may provide the flexibility which each profession will need to survive the technological developments, (the edges between disciplines are already becoming blurred), and more importantly produce a more balanced view of the response to the client's problem. Perhaps if both groups were to consider the implications of our earlier stated cultural/negotiative framework they might more easily see the dilemma facing each other when they come to approach their own contribution to the design process.

Cost Evaluation

Education, however, can only take us part of the way and much more needs to be done to provide the information upon which the balance between the factors featured in the title to this conference can be achieved. One aid to providing some of the information required would be if a systematic 'cost exploration' of the solution space could be undertaken prior to putting pen to paper. Brandon (20) has outlined a possible approach which identified the boundaries to what is economically feasible and specified the flexibility in terms of specification and area within the cost parameters by means of a cost map. If the possible solutions, expressed in very general terms, could be made explicit alongside the cost evaluation then the trade off between the major objectives can be undertaken subjectively in order to obtain a better 'quality balance' in terms of the overall design.

This may provide a step in the right direction but it will still be difficult to perceive it as something other than a constraining influence unless some more generally acceptable qualitative descriptors are found. In fact measures of utility may be more appropriate than quality as it is the degree of satisfaction which is more important than the excellence of the product. The balance which the design team are seeking in attempting to achieve value will be when the marginal increase in expenditure equates with the increase in utility (see Brandon, 21).

Problems with Cost Techniques

The point has already been made that quality is not judged at a single point in time. Indeed with buildings it is the future performance of the construction which is generally in the mind of those who make the judgement. Our standards and certification procedures are geared to performance over time. In some cases the time element may be the whole of the building life.

In cost terms this raises a problem because of the opportunity cost of money. If I have to pay for something in fifteen years' time it is not so critical to me as if I had to pay for it now. In theory I can put aside a smaller sum of money which will accrue with interest to the actual sum of money I have to invest now.

The concept of discounting future costs is well established but it could lead to a devaluation of the role of quality in design decision making.

The following table gives an indication of the effect of different interest rates on the present value of future expenditure.

Present Value of £100 to be paid
end of each time period

Time Period in Years	1%	5%	10%	20%
4	95.1	78.4	62.1	40.2
10	90.5	61.4	38.6	14.9
15	86.1	48.1	23.9	5.7
20	82.0	37.7	14.9	2.2
40	67.2	14.2	2.2	-

There is something rather odd about the replacement of a heating system in ten years' time, being worth only 38.6% (at 10% discount rate - a common figure) of its current value. If tax savings are taken into account then its present worth is even less. What incentive is there to consider any future performance beyond about five years if this method of accountancy is to be pursued.

4. FUTURE DESIGN COSTS

In dealing with long term assets it is our view that the problem of future cost and discount rates themselves should be viewed in a slightly different way.

i)   Future costs should be treated quite separately from initial costs rather than being lumped together as they are in current costin-use techniques, this is because the clients objectives for each are quite different. It may be more difficult to obtain initial capital or indeed it may be fixed by a yardstick. Future costs will depend on a much wider range of assumptions about how the building will be treated, what the world political/financial situation will be, the maintenance policy of the firm, and the patterns of revenue expectations, etc.

     By bringing the two costs together we can fudge the issue by combining different objectives, different degrees of uncertainty and different abilities to control events.

ii) The discount rate chosen should not necessarily be the internal rate of return of the firm or the borrowing rate on a similar type of investment but should reflect the real cost of borrowing money over and above the inflation rate. In Victorian

21

times with no inflation money was loaned at 2½% interest. Marginal changes around this figure to reflect risk would paint a very different picture of the future cost of a building than the use of 10% discount rate mentioned previously.

Here, then, is a strong case where in current methods there is conflict between cost and quality. High interest rates discourage good quality and the combining of initial and future costs tend to distort the view the client gains of his future commitment.

Towards a Yardstick for Assessing Overall Design Quality - Economy, Function and Form

Caudill, Rowlett and Scott (CRS, 19) have develped and successfully used, for sometime, a methodology for assessing total building quality - a crude but effective measuring yardstick which allow for accurate comparisons of buildings (or projects). Interestingly, for consideration of the present reader, their method is based upon three important factors - form, function and economy - linked together in a symbolic and extremely pragmatic triad. CRS get juries to judge a project (building) against each of the three elements in this this triad using a rating scale from zero to ten - where "zero represents no quality, ten is perfection and five is just so-so". By triangulating the three forces of the greatest magnitude, each valued at ten, we would arrive at a "triangle of perfection" - see Figure 1. A building judged to be 5-5-5 would still be a very well balanced entity but its performance

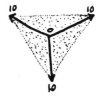

FIGURE 1 (after CRS;19)

By triangulating the forces of the greatest magnitude we get the **triangle of perfection**

FIGURE 2 (AFTER CRS;19)

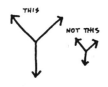

THIS
NOT THIS

dUT EQUILIBRIUM IS NOT ENOUGH. MAGNITUDE, TOO!

FIGURE 3 (after CRS ; 19)
The area of a 10-10-10 triangle is:

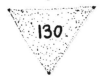

130

This area is called the QUALITY QUOTIENT

on the perfection scales of form, function and economy would clearly not be as high - see Figure 2 - and therefore not a desirable. While the triangle itself is a perfectly good symbolic representation of quality, its area can be used much more easily as a comparative yardstick of the same. Therefore, CRS have designated the triangles areas as the Quality Quotients - see Figure 3; the 10-10-10 triangle would have a quality quotient of approximately 130 while the 5-5-5 triangle would have only 32.

In practice most buildings are neither totally balanced, nor perfect with respect to any elemental quality; a typical score might be something like Figure 4 – with function 2, form 8 and economy 3; this "triangle of reality" has a quality quotient of only 20. Furthermore, the location of its centroid shows it to have a tendency towards the formal.

CRS have assessed many hundreds of buildings using this technique and it has now become one of their many mechanisms for negotiating their own within office designs; designs at least reflective of their considerations of form, function and economy. According to CRS, allowing for the weakness of such evaluatory judgements the technique is surprisingly robust and the results are accurate enough to enable repeated (in a scientific sense) comparisons. Experience has shown them that when the quality quotient for any project rises above 90, that building may eventually be perceived to be "great architecture". Wright's Guggenheim building in New York, USA, for instance, was rated at 92 by the CRS jury.

FIGURE 4 (after CRS ; 19)

Triangulation of the
2-8-3 triangle and
its area

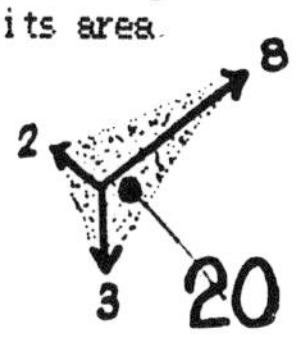

FIGURE 5 (after CRS ; 19)

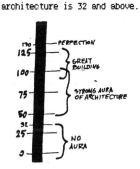

The location of the centroid
of the triangle indicates
the tendency to functionalism
formalism or expressionism

FIGURE 6 (after CRS ; 19)

If one accepts the premise that
there is no "bad" architecture,
only bad buildings; then perhaps
the numerical definition of
architecture is 32 and above.

Like 32 on the thermometer –
the point of freezing – the
Quality Quotient of 32 makes
the beginning of architecture.
the place where mere building
stops and architecture emerges

Figure 6 summarises the general relationship CRS found between their Quality Quotient and what architectural critics and historians would define as quality in architecture.

It is easy to see how a methodology, such as the one described above, could be developed to form the basis for negotiative consensus on issues of design quality, value and utility, providing that the distinct kinds of social individual are involved in the debate.

5. RESEARCH NEEDS AND FUTURE DIRECTIONS

So where does this leave us ? It appears that design quality is not so dependent on initial cost except where the cost limit is exceptionally tight. Therefore research requirements must focus on providing the designer, clients, managers, users with the tools which will make clear the opportunities that are available within a cost limit or cost band both in terms of quality of specification and shape/area considerations. The object of these new models should be to widen the horizon of the design team in order to provide a greater chance of providing better quality. Statutory and non-statutory standards and regulations coupled with site control will ensure minimum quality levels at the component level whereas training in good detailing, and education in aesthetic and spatial arrangement will encourage better solutions to the design problem.

Research into Design, Evaluation and User Control

If you accept the major conjecture of this paper then it follows that there is a need to study the human related mechanisms for improving building utilisation. Here building utilisation is taken to mean those studies concerned with trying to create PLEASING and FUNCTIONALLY REINFORCING (and sometimes cancelling) aspects of total building design which permit SELECTIVE CONTROL on DEMAND of users without RESOURCE WASTAGE and hopefully for a profit; by functional we mean both the ergonomic and symbolic function of the space; by selective control we are suggesting the selection for control should be in the hands of users; resource wastage is taken to mean both the capital costs and the long term life cycle costs of the design. As we have suggested before, different groups will have different needs. We clearly can and should design and manage buildings in such a way as to improve building utilisation for all parties. We therefore need research on all aspects of design for building utilisation (Powell, Cooper & Lera, 22) if architecture is once again to be celebrated by all. With this in mind, research is clearly needed in at least the following areas:

i)   methodology and techniques for promoting a negotiative consensus about building projects. For instance, the CRS total design quality yardstick might be explored and developed. Alternatively one could extend and elaborate the Post-Occupancy Evaluatory Techniques developed by Kernohan and his colleagues (23) in New Zealand or Zeisel and Mills (24) in Canada. It is possible to conceive of a "transdisciplinary walk through" approach being used as a precursor to a negotiation on aspects of quality, value and utility by our four distinct kinds of social individual.

ii)   psycho-sociological studies concerned with determining the functionally reinforcing aspects of many different user-context couples.

24

iii) studies concerned with trying to understand the kind of selective control wanted by users in different situations.

iv) design orientated studies to improve the interface between
research and design practice. It is considered that a far more
active role should be taken in support of the design activity.
This is seen as a major area of enquiry whose goals would be
to increase the impact of new and existing research findings
on design by the dissemination of information and the development of appropriate aids. In tackling this area it is envisaged that a number of related activities will need to be undertaken. These are at least to:

a) evaluate the usefulness and encourage the exploitation of existing information, possibly by employing developments in applied
Information Technology

b) help to meet the knowledge gaps identified by the professional
institutions by carrying out specific and limited studies in
order to meet the currently perceived information needs of
their members.

Future Building Cost Research

Research into the factors influencing future cost is also wide
open for development. The following are suggestions which may
well assist in deciding whether there is conflict between cost
and quality.

i) An improvement in technique which will provide a clearer picture to a client of the true quality of his building in future
cost terms.

ii) An improvement in building documentation which will set standards for building operation and maintenance. The GLC experiments with handbooks (similar to a car servicing logbook)
could be adopted on a wide scale and would provide a useful
reference point for obtaining performance data.

iii) A computerised feedback system for maintenance to identify
and remedy design errors.

iv) The development of management systems and techniques for monitoring maintenance, e.g. inspection frequency.

v) The setting of future performance standards for buildings to
which the design team must comply.

It is true to say that some work has been undertaken in all these
fields in recent years, much of it unreported. There is a need to
persuade clients of the usefulness of such work and to implement

the ideas wherever possible in order to test their viability.

6. CONFLICT OR COMPATABILITY ?

The preceding discussion has suggested that quality and cost only conflict when cost limits are too tight or where the use of the discounting technique reduces the impact of poor quality in the decision making process.

Poor design quality will nearly always result in a cost penalty in the future. One task must to be predict that poor quality will exist, to provide an alternative to a poor solution and to evaluate the economic consequences of a design in such a way that it more clearly reflects the client's view of the problem.

It is unlikely that conflicts will entirely disappear, design is a series of compromises, but with the right tools and the right approach good quality may be more compatible with a realistic cost constraint.

7. ACKNOWLEDGEMENTS

We are indebted to Michael Thompson with whom many of our ideas on the negotiative cultural framework have been developing during the past five years; when not specifically noted, ideas in that section of our paper within quotes are due to him. Furthermore, we would like to thank Barry Russell for introducing us to the work of Cardill Rowlett and Scott on Total Design Duality Assessment

REFERENCES

1. Schumacher, E.F., Small is Beautiful, Abacus, pp.35-38, 1973

2. Arrow, K. Social Choice and Individual Values, New York, Wiley 1951

3. Arrow, K. Social Choice and Individual Values, Yale University Press, 1963

4. Douglas, J. How actual Political systems cope with the paradoxes of social change in Social Choice and Cultural Bias, Ed. Thompson, M., IAASA, Laxenburg, 1984

5. Thompson, M. Information is Power, Ignorance is Bliss, Information Information Reject Styles and the Viability of Organisations, IAASA, Laxenburg, 1984

6. Thompson, M. A Cultural Framework for the Analysis and Design of Energy Policy, SERC, March 1984

7. Thompson, M. <u>Among the Energy Tribes: The Anthropology of the Current Policy Debate</u>, IAASA Working Paper, 1984

8. Thompson, M. <u>Rubbish Theory</u>, Oxford University Press, 1979

9. Douglas, M. <u>Essays in the Sociology of Perception</u>, Routledge & Kegan Paul, 1982

10. O'Brien, D.D. Design and Evaluation Methods: variations on a theme. In <u>Design:Science:Method</u>, Ed. Jacques & Powell, IPC,

11. Powell, J.A. & Sime, J. <u>A Systematic Field Research Strategy for Study of Energy Use</u>, SERC, 1984

12. Burt, M., <u>A Survey of Quality and Value in Building</u>, UK Building Research Establishment Report, 1976

13. Department of Education & Science Building Bulletin No. 4

14. Spedding, A.

15. Russell, <u>Systems Building</u>, Wiley 1982

16. Cardona-Aparicis, C, Powell, J.A., Thompson,M., Weaver, M.J.and Carden, J., 'An introductory essay concerning the integration of special subject disciplines into architectural education' in Jacques, R., and Powell, J.A.,(eds) <u>Design:Science:Method</u>, IPC, Science and Technology Press, Wesbury House, Bury St., Guildford GU2 5BH, UK, (1981)

17. Broadbent, G., <u>Design in Architecture</u>, Wiley, 1963

18. Powell, J.A., Science Research Council Grant, March 1976, B/RG/4161 (1976)

19. Caudill, W., Rowlett, S., and Scott, C., <u>Architecture by Team</u>, Van Nostrand, 1974

20. Brandon, P.S., A Framework for Cost Exploration and Strategic Cost Planning in Design, <u>Chartered Surveyor, Building and Quantity Surveying Quarterly</u>, Vol.5, No.4, pp.60-63, 1978

21. Brandon, P.S., Cost v Quality : A Zero-sum Game ? Transaction of Conference on <u>Building Utilisation</u>, Portsmouth Polytechnic, September, 1983

22. Powell, J.A., Cooper, I. and Lera, S., <u>Designing for Building Utilisation</u>, E. & F. Spons, 1984

27

Section II
Context

COST AND VALUE - A RESEARCH PERSPECTIVE

BERNARD JUPP, Bernard C. Jupp and Partners

1. PREAMBLE

I recall as a very young surveyor remarking to an equally young
architect that every line he drew cost money. His immediate
reaction was to rub some out, and at least for the rest of the
afternoon he was noticeably more deliberate in what he drew.

If the point needs to be made at all, this anecdote demonstrates
that design involves (or should involve) continuous decision-making,
balancing expectations of achievement with expectations of cost, or
in the terms of an economist, balancing outputs with inputs. In our
case, outputs are represented by what is expected of the building,
the shelter it gives, the facility for the people or processes
within it, its aesthetic appeal and so on, in other words its
expected benefits. Similarly, inputs are represented by its various
costs, capital, maintenance, running and operational.

In this latter respect, it is important to regard **all** costs,
rather than just capital costs, as inputs. I have seen maintenance,
running and operational costs represented as an expression of benefit, but this is clearly wrong if one considers that these costs all
represent expenditure. The balance between present and future costs
to maintain the required level of benefit is of course of vital
importance during design as recent studies in life cycle costing so
amply demonstrate, and clearly the better choice is the one by
which, for a given benefit, the combination of present and future
costs is at a minimum, but the point I make here is that costs, of
whichever type, only represent the media by which benefit is
achieved, and should not be confused with benefit itself.

To summarise then, building design is concerned with balancing
building cost with building achievement or benefit, or to revert to
economic terms, balancing inputs with outputs. If inputs exceed
outputs, the building is not justifiable.

31

2. DETERMINATION AND MEASUREMENT OF INPUTS AND OUTPUTS

With regard to inputs, their determination is a relatively simple matter, since they are represented, as we have already seen, by the various costs associated with the building throughout its life. These embrace all capital expenditure necessary to implement the project (the question of site cost perhaps requires examination, as there appear to be good arguments both for and against its inclusion), also future costs necessary to sustain the function for which the project was implemented. The measurement of these costs has mercifully been made simple by the invention of the concept of money, by which the value of otherwise incomparable commodities such as labour, materials, plant, energy and so on can be expressed by a single and readily understood means. Another important concept is present worth, by which future costs can be commuted to their present-day equivalent.

The determination of output (achievement or benefit) is infinitely more complex, and this and the associated problems of output measurement form the crux of the value-for-money problem. Not the least of difficulties lies in deciding the level at which output should be measured, since what is highly desirable to some may bring misery to others, as those living adjacent to fly-overs and busy urban roads will testify. Our valuation surveyor colleagues have an expertise in certain aspects of this problem, as do many designers, but for a building project to represent maximum value-for-money, every decision requires value statements comparable in detail and reliability to corresponding cost statements in order that a reasoned selection from the available design options may be made.

Whilst costs are generally available with varying degrees of reliability, data on benefit are by comparison very sparse indeed except in the broadest sense. For example, whilst a valuation surveyor will have reliable data on probable rental levels for say air-conditioned offices in a certain location, it is unlikely that comparable information will be available on the effect of say circular columns instead of rectangular, or of one room shape instead of another, or of the proximity of one department to another.

The problem becomes even more severe in non-commercial buildings, and whilst the funding agencies of buildings for health, education and housing have done much valuable research as to benefit in their own areas of interest, much remains to be done elsewhere.

3. FORECASTING OF INPUTS AND OUTPUTS

Since building design is essentially a process of forecasting, the means of forecasting cost and benefit also require examination, since without them, value-for-money during design could not be achieved.

Only a moment's study is necessary to discover that the availability of a large volume of cost data is matched to a large extent
by the means of using the data for cost forecasting during design.
Capital and running cost data are much more available and reliable
than maintenance data for reasons that, although interesting, need
not be listed here, but even in this area of cost forecasting,
inroads have been made by the formation some years ago of the
Building Maintenance Cost Information Service.

Unhappily, the situation with regard to benefit is quite the
reverse and the difficulty already mentioned of determining and
measuring benefit is responsible for similar difficulties in
forecasting during design.

This preamble has been necessary to set the scene for describing
the approach that research might take to the subject of value-for-
money in building. The remainder of the paper outlines a line of
enquiry that could usefully be pursued in understanding and
hopefully eventual overcoming the differences that exist in dealing
with cost and benefit forecasting during building design.

4. CURRENT PRACTICE

In embarking on a piece of research, it is customary and prudent to
examine the current position in the field under study, with regard
both to practice and to other research, thus avoiding wasteful
duplication.

With regard to practice, the concept of value is part of the
fabric of existence and certainly pre-dates man. However, only
fairly recently has it been systematised in construction, and one of
the earlier examples of this was the development of cost/benefit
analysis by the Tennessee Valley Authority. The history of this
development is well-known, and since then, cost/benefit analysis has
been used in varying forms for a number of major projects, including
the M1 motorway, the Victoria Line, the third London Airport to
mention only a few.

The analysis in these examples has usually been used to decide
whether and where to build, however, but in the building context,
this decision will usually (but not always) have been taken before
the design/construct team is appointed, and it is necessary to probe
deeper into current practice if any significant development is to be
made.

And so one examines the requirements imposed upon the design team
before and during design, since it is these requirements that may be
regarded as the benefit to be aimed for during design. The list is
formidable and includes

(a) the client's brief. This will range from the vaguest statements to the most precise set of requirements. It is interesting

that the distinction between briefing and design is difficult to define, and in practice this is recognised in that designers are well-versed in assisting clients with the formulation of briefs as part of the design process.

(b) the various "pro bono publico" requirements. These include planning requirements, building regulation or London Building Act requirements, Fire Officer requirements or recommen-mendations and a whole host of other requirements that may be imposed by statutory bodies such as electricity, gas, water, tele-phone and highway authorities.

(c) the various published recommendations as to good practice. Whilst not binding upon the designer, he would be unwise to ignore them owing to the authority and expertise that has produced them. They include British Standard Specifications, British Codes of Practice, standard text-books on building construction and other matters, manufacturers' instructions or recommendations, and the many articles that appear in the technical press concerning good practice.

(d) the designer's own experience and the corporate experience of his profession as to recognised good practice.

During design, each of these requirements must be catered for if the project is to be successful or even proceed, and each require-ment is an expression of the benefit that the project must provide.

Whilst most projects are more or less successful, there have been a number of spectacular failures. These, coupled with the relative ease with which cost can be dealt with, have caused researchers over the last twenty years or so to apply their minds to the more difficult topic of value, especially its identification and measure-ment.

5. EXISTING RESEARCH

A few years ago, research carried out on behalf of the Building Research Establishment as part of its programme of value-for-money research (1) revealed a considerable research interest world-wide in the subject of value. The most significant outcome, however, was not in the volume of value research, but the finding that, unlike the cost research undertaken earlier, so little has filtered through into practice. The conclusion was that, although cost statements are available with varying degrees of reliability for virtually any level of design detail, corresponding statements as to benefit are not. Thus comparison of competing design options is not possible except on a cost basis, and true economy (defined as being achieved when benefit most outweighs cost) is unlikely to result except by intuitive means.

If the renewed interest in value research now being experienced in this country is not to meet a similar fate, the failure of previous research to influence practice requires investigation. Furthermore, the investigation should receive some priority since its outcome could influence the direction of current research and the means by which the research outcomes are transmitted to practice.

6. FURTHER RESEARCH

The investigation that has just been proposed in the previous paragraph would examine the earlier research outcomes and reaction of practitioners to them.   It would also examine the attitude of designers to the suggestion that it is feasible to systematize the consideration of benefit during design.

Possible reasons for the poor take-up of value research outcomes might include

(a) the research outcomes may not be understood by practitioners, who are understandably impatient with reports that require expert knowledge before their application is apparent. Research titles such as "Scenario-matrix method as a systematic evaluation method for building production" are not enthusiastically received by most practitioners, and it may be that more development work is necessary before the research already completed becomes acceptable. Or perhaps researchers should become more familiar with the workings of practice and the minds of practitioners?

(b) practitioners already have to cope with the many design objectives already mentioned, and maybe feel that it is not practical to take on yet more objectives during the often pressurised process of building design. In addition to time constraints, the pressures exerted by fee scales cannot be ignored, and the recent introduction of fee competition will not improve matters.

(c) in view of the many objectives already coped with during design, the application of Pareto's law (2) suggests that only marginal improvements in the present position could be achieved without a massive increase in design effort not normally available owing to time and fee scale constraints.

(d) the attitude of designers may be such that they resent measurement of benefit, including as it must the sensitive subject of aesthetics, and prefer, in the interest of artistic freedom, an idiosyncratic approach to design rather than a calculated approach. The susceptibility of building designers to "fashion" does not dilute this hypothesis.

(e) even if designers were willing to incorporate better consideration of benefit during design, it may be (almost certainly will be) that they will experience very considerable difficulty in

35

locating suitable data except when working with a continuous client who is able to collect data himself and make it available to designers in the form of design guides. Even in procedures requiring designers to undertake user surveys, they are often not in fact carried out because of the dispersal of the design team and the pressures of current work.

(f) it might be argued that, in spite of the opinion that the construction industry has of many of the buildings it produces, users are relatively insensitive and could just as well function in such a variety of designs that too much attention to detail is wasteful. The concept of "loose fit" design tends to support this argument, as does human flexibility in adapting to what is available, and the argument for closer consideration of benefit is thereby eroded.

7. DESIGN GUIDANCE

Mention was made a moment ago of design guides. Good examples are those produced by the government departments concerned with building (for example, health, housing, education) and commercial concerns in close contact with their customers (for example, retail outlets, speculative house builders). The common thread in these examples is that they are all continuous clients, and it may be more realistic to look to such organisations for design guidance rather than to expect designers to provide their own benefit analysis. The reasons for this are not difficult to imagine and have already been discussed; the break-up of design teams before user reaction can be collected; the prolonged and widespread experience of a substantial and continuous client, ensuring the identification and ranking of benefits with real significance; the motivation such a client has for ensuring satisfactory design.

But what of buildings that are not designed for continuous clients; where is their design guidance to be obtained? In spite of the difficulties already outlined with regard to the evaluation of buildings for "one-off" clients, it may be possible to encourage feed-back of performance from the users of the buildings. Even aesthetics are amenable to numerical measurement, in spite of views to the contrary, but this at once raises the question as to the weight to be attached to uneducated reaction. The anguish experienced by designers at the hands of lay planning committees is too well known to dismiss, and an environment ordained by those untrained in design would be disastrous, as events in this country and more especially in some developing countries have shown. There appears to be a sound case for a fusion of elitism and laity, and the work carried out at Strathclyde in user participation in the design of nursery schools (3) is of particular interest.

To return momentarily to Pareto, it is accepted by all that the most crucial decisions in the whole design process are those that occur in the initial stages. Bearing in mind the difficulty already mentioned in being able to distinguish briefing from design, it might be that the most significant advances in attaining value-for-money will occur in helping clients and their advisors to produce more thoroughly considered briefs, rather than in helping designers to refine yet further their already heavily-burdened design processes.

8. SUMMARY

Summarising, this paper suggests that

(a) further research is necessary to discover why existing research on value has not been absorbed into practice in the manner in which cost research has been absorbed.

(b) it may be unrealistic to expect practitioners to take the initiative in routine building evaluation, but at the same time, reliance on lay user reaction may not be sufficient without expert vetting.

(c) it may be more realistic to encourage and help clients, especially continuous clients, to give more expert consideration to their needs and more expert guidance to their designers as to how their needs should be accommodated.

9. CODA

The topic of value is fraught with difficulties. However, one is encouraged by the words of the American psychologist E.L. Thorndyke:

"Whatever exists, exists in some quantity, and can in principle be measured".

REFERENCES

1. Bernard C. Jupp and Partners (1980) 'Cost data and estimating processes in strategic design of buildings', research report commissioned by Building Research Establishment, Garston.

2. Broadly similar to the law of diminishing returns, in that the major part of an objective can be achieved with only a minor part of the effort required to achieve all the objective.

3. Julian Watts and Morven Hirst (1982) 'User participation in the early stages of building design', Design Studies, Vol. 3 No. 1, pp.11-18.

ARCHITECTURAL QUALITY

GEOFFREY BROADBENT, Portsmouth Polytechnic

Since this Conference was organised jointly by a School of Architecture and a Department of Surveying, it is entirely appropriate
that it should deal with Design : Cost, Quality and Profit.
Each of these, does, or should concern us all, but two of them
obviously: Cost and Profit are specific concerns of the Surveyor,
they are his very stock-in-trade. Which set me wondering why,
whilst we have Quantity Surveyors, we do not also have Quality
Surveyors.

We do have architectural critics, and most of them make qualitative judgments. But, as Attoe shows (1978) they do many other
things from simply describing certain buildings - or their architects, to acting as propagandists for specific architectural
styles to writing rhetoric, or even poetry, inspired by architecture which has moved them.

Having considered various ways of carving up the field of criticism into usable categories, Attoe concludes that for architecture the most appropriate categories are:

1.   Normative criticism relating clearly to a 'norm', in which
      the critic has a clear view of how things should be and
      tests a given building against those views.

2.   Interpretative criticism, in which the critic draws people's
      attention to certain buildings and tries to make them see
      these as he does, thus putting them into a new perspective.

3.   Descriptive criticism, in which the critic tries to help
      others see what is actually there, clearly and factually,
      avoiding as far as possible any personal judgment or inter-
      pretation.

We need not go into detail here except to say that each of
Attoe's groupings is broken down into further and further detail.
A descriptive criticism for instance might consist of a more or
less objective record - in words, drawings, photographs, diagrams

38

and so on of the building itself, of what it actually is: or the critic might describe the social, physical, the political, economic artistic or other context into which it was built, the process by which it was designed, observations of the building in use, including the responses of the users. It might even be a biography of the architect, the client, or someone otherwise involved.

Such criticism ought to be objective but since it starts with the selection of a particular building, or a particular architect, it can be so only very rarely. Even photographs are bound to be angled to show the building off in the best - or, conceivably, the worst - possible light.

Even the most experienced reporter - possibly because he is experienced ; will start with preconceptions based on his own perceptions. These will have developed as Bartlett once suggested (1932) according to his appetites, his enthusiasms, his instincts and ideals. They will help determine what the critic actually perceived - the transaction which took place between the physical reality of the building when it stimulated his senses and his personal 'schemata' those residues of experience on which his personal values are based. This will certainly be true if he presents his criticism in words; hardly less so if he presents drawings, with details subconsciously suppressed or 'brought out' to make the building fit his personal conceptions.

Even so, some critics will try to be objective but others will have no such illusions. As Attoe points out - and here I am blurring his categories - the critic may believe in certain doctrines. Indeed it could be argued that there is not much point in his being a critic if he does not! His doctrines may be so strongly held that they force him to advocate certain kinds of architecture at the expense of others. That is fine when one knows what the critic is doing. Such "advocatory" criticism has its place alongside "evocative" criticism - in which the critic uses someone else's building as a vehicle for presenting his own emotional response, or "impressionistic" criticism, in which he uses it as a vehicle for his own written, photographed or even painted work of art.

The problems come when critics such as Giedion (1941) or Pevsner (1936), convinced that to 'express' the 20th Century, buildings should be simple, rectangular, flat roofed and preferably glass-walled presented their personal advocacy as if it were objective history.

We are all faced with similar temptations but Attoe suggests a number of ways in which we might mitigate their worst effects. He describes "normative" criticism which includes:

39

(i)   Systematic criticism in which the critic presents a series of interrelated doctrines rather than a single advocatory one. In the First Century AD for instance Vitruvius felt that buildings should possess convenience, durability and commodiousness or, in Sir Henry Wootton's memorable translation (1624) Commoditie, Firmness and Delight.

(ii) Typal (others would call it Typological) criticism, in which, having established the "essence" of a particular building type one can test any building of that type against the standards thus established to see how well it conforms to 'type'.

(iii) Measured criticism, in which, having established standards for the dimensions, say, of staircases, ceiling heights, corridor widths, structural strength, lighting levels temperatures, noise levels, costs and so on one can use them to test a particular building's performance.

Attoe's pioneering analysis of this field is most interesting and useful. What follows is an attempt to set down conditions, as it were, for what he would call Normative: systematic criticism.

But first of all we shall have to look at what it is about buildings which actually can be criticised. One can discuss a building's form or its shape, in visual or other terms (see my Design in Architecture, 1973). One can discuss its planning, structure and construction, the materials and components of which it is built, servicing systems and so on. One can discuss its performance in human or other terms. Most of these are parts of Attoe's scheme, but he groups them differently.

We are exploring a "system" but there are many such systems from Vitruvius's onwards. Over the last fifty years, notable contributions have been made to this field by Roth (1939), Hillier, Musgrove and O'Sullivan (1973) and Caudill (1971) amongst others. My own version, which has been published many times, (1979, 1979, 1982) derives from Hillier, et al, and it is an attempt to answer their fundamental questions as to what it is that a building will do, irrespective of the architect's or anyone else's intentions. Since it is available in so many other forms, I shall merely summarise it here.

Bill Hillier asked the question: "What, in spite of anyone's intentions, is a building going to do anyway?" He published his version of the answer (1973). My answer to that question is:

FIRSTLY, ANY BUILDING IS A CONTAINER FOR HUMAN ACTIVITIES: there is no point in building if the things we want to do can be carried out, conveniently and in comfort, on an open

site or in some existing building. The building provides
interior spaces for human activities, which in size and
shape are appropriate for the things we want to do. Relationships between spaces may be more important than the
spaces themselves because they enccourage, or inhibit, patterns of movement, social interaction and so on.

SECONDLY, THE BUILDING IS AN ENVIRONMENTAL FILTER: the
building, as we have seen, is a filter between the users of
the building and the external environment. The building
should work effectively in terms of thermal, lighting and
acoustic control; we can define, crudely, the limits of
what it should do with reference to human physiology and
psychology.

THIRDLY, THE BUILDING IS A CULTURAL SYMBOL: Concern for the
building's function as an environmental filter implies concern for the satisfaction and - hopefully - the delight of
the human senses. It would be ludicrous to design, say for
thermal pleasure unless we designed also for visual pleasure, and whether we like it or not, this, finally is a
cultural matter. People will read meanings into our buildings, based on their past experience. We can no longer
pretend that the building is merely 'functional' in the
utilitarian sense.

FOURTHLY, THE BUILDING IS A CAPITAL INVESTMENT WHICH CHANGES
THE VALUE OF MATERIALS AND SITE: building involves the use
of materials, which have to be located, extracted, transported, transported again, and assembled into position. Each
operation consumes resources in the form of labour and
plant charges. The site itself will be changed in value
because it has been built on. Buildings also cost money to
run and maintain.

FIFTHLY, THE BUILDING HAS AN IMPACT ON THE ENVIRONMENT:
Every time we put a building on the ground we change many
things. We change the appearance of the place, we change
the micro-climate, we may attract more traffic than has
been attracted in the past and we shall certainly cause
pollution of some kind. If it is a very bad building, then
this may be physical and the more any building causes the
consumption of energy - say for heating and/or cooling,
then the more it causes pollution where that energy is
being generated!

My intention was that this could be used as a check list, and
so it can. One can test the plans of a building to see if it
has the right number of rooms, in the right sizes, shapes and
relationships, for the activities it is supposed to house.
One can check the actual structure: walls (internal and exter-

41

nal), floors, roof and so on to see the extent to which they do keep out unwanted solar energy and noise whilst permitting views out as required, sufficient thermal capacity, thermal insulation and so on. Attoe would describe this as 'measured criticism'.

But clearly there is more to quality than the mere satisfaction of measurable criteria which, in any reasonable kind of architectural world, we would have a right to expect in any case! For the fact is that many buildings - including some recognised as great - by no means satisfy such criteria, at least not in any balanced way.

Nor indeed is measure enough. We can measure room sizes and shapes, lighting levels, temperatures, air movements and humidity. And, of course, we can measure quantities and costs. Since others here are concentrating on such things, I should like to explore, as coherently as I can, how we go beyond the measurable, how architectural judgments can be made.

A former tutor of mine, Peter Collins, was quite clear (1971) that architectural judgments are, or should be, very much like legal judgments, based on precedents one knows to be good. One compares the new and untried against the old and accepted, the 'known to work' as a baseline against which the new can be judged. But Collins used that, essentially, as an argument for architectural conservatism.

I would like my system to accommodate the new and the old. So we shall have to go further into what it is about a building about which judgments can be made.

As van de Ven shows (1978) there is an enormous literature, starting in China with Lao-Tsu (6 BC) and later in Greece with Plato (427-348 BC) to the effect that architecture is space. A sculptor makes solid objects - by modelling or carving - and we contemplate them from the outside. We walk around them, view them from various directions, and then we make our judgments about them.

Van de Ven stops his survey in the 30s and the most prolific writer on the subject since has been Bruno Zevi, especially in his Architecture as Space (1957).

According to the "as space" tradition architecture is a form of 'inside out' sculpture. We walk into it, move around the interiors - each of which is a space - and then make our judgments in much the same way as we make our judgments of sculptures.

Scruton (1979) will have none of that. As he says (p 43):

If space were all that interested us, then not only must a large part of the architect's activity seem like so much useless decoration, but it is even difficult to see why he should bother to build at all. If I stand in an open field, then I can have a full experience of all the separate spaces that are enclosed in St Peters in Rome. The only difference is that here the shell which Bramante and Michelangelo constructed around those spaces does not exist, and so does not interfere with the pure, unmediated contemplation of the spaces as they are in themselves.

Scruton argues instead that the essence of architecture lies in its detail. For a start it is the details which tell us what kinds of architecture we have in view. If it has columns, cornices, and pediments then probably it is Classical architecture of some kind. If it has pointed roofs, pointed arches, ribbed vaults and pointed windows, it is probably Gothic and so on.

What is more, according to Scruton, the details must relate to each other, and to the whole, in a coherent way. If you are putting words together to form sentences, then you have to choose the words which convey your meanings, and put them together in a sequence which is a grammatically correct. You have to get the syntax right! Scruton does not find it useful to make such comparisons with language, but I do (see my 1977, 1980a, 1980b, 1981). And, what is more, as Scruton himself says (p 172).

> Details have implications, and cannot be combined in just any way without producing nonsense. The details themselves impose a possibility of organisation.

So, he goes on (p 173):

> It will never be possible to separate the quality of a building from that of its detailing ... a disposition of badly worked details, however finely patterned, may well be meaningless, in comparison to the same disposition when the details are properly executed.

So, for Scruton (p 205):

> The central operation in all aesthetic taste, whether in its primitive or intellectual form, is the sense of detail.

There are times, as Scruton points out (p 211) when detail is the only thing the architect can enforce. The shape of the site, the height permitted by the planners, the needs of the client may all constrain the overall form to such an extent that the architect can do little more than work within the constraints. But even under these circumstances, he still has control over detail.

43

Conversely the finest proportions will be destroyed if the detail
is crude, weak, inappropriate or ill-executed!

All this is crucial to Scruton's central point that if detail is
so important - and of course it is - then the details themselves
and the ways in which they relate to each other, must be so
appropriate, so obviously 'right' for their purpose, so 'natural'
in the way they fit together that they transcend aesthetic discussions of the "I like this and you like that" kind. His aim is
to go on to argue that the only kind of detail which satisfies
these criteria is age-old Classical detail. So like Collins,
Scruton too uses his doctrine to justify an innate conservatism.

But, of course, he is right in arguing that architecture is much
more than space. There is force too in his argument that the
essence of architecture lies in detail which, however, leaves us
with some further problems.

For details have to be on something, such as surfaces, and these
surfaces in their turn must be of something, the surfaces of
solid materials, from which the building is constructed: the
walls, floors, roofs, columns and other constructional elements.

We may see these elements themselves: the bricks, steel columns,
concrete panels, wood boards from which they are made. Or they
may be covered with something else: a finish. You can cover a
concrete wall with marble panels and the marble surfaces will
look different, feel different, be diffferent from the original
concrete. The one will look rich and expensive, the other may
look cheap and nasty.

Of course, you can plaster the concrete and paint it to look
like marble, but since Ruskin (1849) most of us have had the
uneasy feelings that such things are patently deceptive, not to
say dishonest and immoral. When Ruskin knew that the surface
he saw was a thin, applique skin, such as gold leaf, then he was
quite happy in that knowledge. So he was with coloured decoration. It was the cheating, the immorality involved in making
one thing to look like another that worried him.

As for judging the quality of a building merely by the richness
of its material, others, such as Muthesius suggested (1911)
that this really is quite irrelevant. If the basic design is
good enough in form, proportion and so on then these will speak
for themselves, whatever the cost of the materials. Nor am I
worried about how expensive materials look, but I am worried
about their durability.

Materials are exposed to the weather. In all but the most benign
of climates they will streak and stain as they age. They may
corrode, grow lichens or whatever. Certain materials, such as
brick and stone, if they are carefully detailed, can look better

as they age. Others, such as glass, metals, ceramics, plastics, and certain concrete and smooth-plastered surfaces can only look worse. If the aesthetic of the building depends on the clean precision of simple, smooth, machine made surfaces and their joints, then what of quality is left when that clean precision disappears?

What is more the kinds of details that Scruton advocates: cornices, string courses, mouldings and so on enhance the 'ageing gracefully' process. Classical buildings also streak and stain, but the pattern of streaking and staining is controlled by the very detail itself. Which makes it look as if I am joining Collins and Scruton to use my critical system as a way of advocating their kind of conservatism.

Not so, by any means. I can think of many architects, unequivocally 'modern' who have used surface treatment and detail to enhance the capacity of their buildings to age with dignity and grace. The Czech 'cubists' for instance, such as Chochol and Gocar, used the plain white surfaces of the 20s. But they were facetted like Picasso paintings, and the facets direct the streaks and stains. Even where patches of plaster have fallen away, their elevations are complex enough, and robust enough, to contain these localised calamities.

Utzon's Sydney Opera House has a robust skin of tiles, intricate enough to contain the streaks and again certain works by the Catalan Bofill - even though they be of painted plaster - have modelling of a kind that directs the weathering.

Conversely when I think of the 'greats' of the Modern Movement in Europe, such as Le Corbusier, Gropius and Mies van der Rohe, I find that without exception, their early buildings, such as the Tugendhat House (Mies) or the Villa Savoye (Le Corbusier) simply have not aged as they should. Their plain, white, smooth abstract surfaces streaked and stained in a matter of weeks. The paint flaked, the plaster cracked and fell off leaving scars. They have to be rebuilt from time to time.

What is more every other building I know in which architecture has been reduced to pure abstract geometry has equivalent, or worse problems.

It is true of high rise flats formed of simple, geometric concrete panels. It is true of steel-and-glass office towers where, in addition to the problems of cracks, streaks and stains, there are terrible problems of solar overheating, heat loss, noise, glare distractions and so on. Such buildings are the worst environmental filters there have ever been in history, but similar problems are inherent in the current fashion for metal-panelled 'Hi-Tech' buildings.

Which brings me to the heart of the matter. Such buildings are recognised as 'modern' because of the simple, geometric abstraction of their forms. Once Giedion, Pevsner and others had taught architects that 'modern' buildings should be simple and abstract in this way, we could recognise them, easily. For critics of that kind, of architectural quality lay simply, directly and unequivocally in the degree of abstraction!

Once this was accepted all one had to do was recognise such a building when one saw it. All buildings of these kinds have basic environmental and other problems, but their protagonists made fine distinctions between the work of a 'Master' such as Mies van der Rohe and the work of his his second-rate followers.

Let me illustrate the heart of their arguments with reference to the work of the painter, rather than the work of an architect, the painter, who, more than any other, introduced architects to abstract geometry: the Dutchman Piet Mondrian.

I have learned, over the years, to recognise Mondrian paintings just as I have learned to 'read' Rembrandts, Picassos and so on.

There is nothing aesthetic about that recognition. I have simply learned to "read" the idea of Mondrian off a painting just as I have learned to read the word "Mondrian" off a page.

And once I have "read" it in this way, I can then start to look at the Mondrian as an aesthetic object.

I can think about each line and each rectangle of colour, making judgments of the kind he must have made, as to whether this line is exactly right, should be two or three millimetres higher, lower, to the right, to the left and so on.

Mondrian, P. Compositions, unfinished, 1939-44. We "read" or recognise them as Mondrian paintings and they make aesthetic decisions about them. The first one shows Mondrian 'trying out' lines in various positions, the second has "too many" lines. These are aesthetic decisions.

46

We know he made judgments of this kind because Mondrian left various paintings unfinished. In one of the cases illustrated he tried a vertical line in no less than 16 positions. Judgments of that kind are aesthetic, and just as we can apply them to Mondrian paintings, so we can apply them to buildings. Once we have recognised the overall forms and surface detailing of a building, the size, shape and 'fitness', we can go onto make aesthetic judgments the proportions of on a room, the surface detailing of its walls, floor and ceiling. And, given access to the plans, we can also apply such judgments to the sizes, shapes and arrangements of the solid materials from which it is built. Indeed if we are skilfull enough, we may be able to read such things from what we perceive of it in three-dimensions.

Given the three essential parts of architecture therefore, the spaces, the surfaces and the solids, I submit each one, willy-nilly to a two part process of decision: recognition that it is of such and a kind (reading the signs) followed by my judgment that of that kind it is, or is not, a good example.

The two stages of the reading I have described, of course, put different demands on the reader. The first stage: recognition, demands a basic knowledge in the first place. You cannot read "Mondrian" into a painting if you have never seen a Mondrian before.

If you have then your reaction will depend on many things of which the most important by far will be the sum total of social pressures acting on you to form opinions of paintings such as family background, educational background and, most particular-ly, peer-group influences.

Like many other artists of high moral tone, Mondrian thought that life had gone wrong. Human behaviour could become pure and beautiful again, if the environment became pure and beautiful. His paintings were meant to show designers how to make beautiful environments.

Our cities have become much closer to that state of abstract purity which Mondrian so earnestly desired, thanks to the steel and glass office tower, the high rise apartment block in precast rectangular concrete panels and so on (see my 1982). This is not to say that Mondrian would have approved these results but, in rather more than a metaphoric sense, his paintings are their progenitors.

To appreciate them, of course, you have to have learned what they mean. You have to belong to those social sub-groupings of archi-tects, developers, chairmen of multi-national corporations and so on whose memberships of their groups depend on their liking for such things.

A steady stream of them identified their membership during the Public Enquiry held in 1984 to determine whether a Mr Peter Palumbo should be permitted to build a seventeen-year old design for an abstract, steel-and-glass Tower by Mies van der Rohe on a site near the Mansion House in London.

In each case they stated a commitment, of which the most intense appear to have been:

Sir John Summerson: "Mies's design is of irreproachable excellence" (Proof of Evidence, 13).

Richard Rogers: "(Mies's) building will be unsurpassed in elegance and economy of form. It will be the culmination of a master architect's life work" (Proof of Evidence 6.11).

Peter Carter: "The building would be among the world's most technologically and environmentally advanced office buildings". (Proof of Evidence 8.15).

Not a word, other than dismissing them as irrelevant – about the climatic problems – both internal and external – to which such buildings are kknown to give rise, the logistic problems of moving people vertically, the need for constant maintenance, the bleak inhumanity and so on.

But of course, it requires an act of faith to accept such building forms in the first place and once one has experienced 'conversion' one becomes blind to such faults. Those who raise such things are seen as enemies of the faith and treated with appropriate contempt.

For the fact is that, whether we like it or not, our first readings of such things as building forms are based on the ideologies social, political, economic, aesthetic, religious and so on, by which we identify ourselves with particular social subgroups. These are what determine the things we deem to be important in our later and deeper 'readings' of these things, including our aesthetic readings.

So the best we can do, if we hope to make sensible judgments of quality, is to recognise that this is how things are and try to hold them in some sort of balance, preferably by the use of some kind of critical system of the kind which Attoe approves and which I have tried to develop.

REFERENCES

Attoe, W. (1978), Architecture and Critical Imagination, John Wiley and Sons Ltd., London.

Bartlett, F. C. (1932), Remembering : A Study in Experimental Social Psychology, Cambridge University Press, Cambridge.

Broadbent, G. (1973), Design in Architecture,, John Wiley & Sons Ltd., London.

Broadbent, G. (1977) 'A Plain Man's Guide to the Theory of signs in Architecture' in Architectural Design, July/August, 1977, pp 478-483.

Broadbent, G. (1979) 'Recent Developments in Design Method Studies' in Open House, Vol. 4., No. 3., 1979, Stichtinng Architecten Research Group, Eindhoven, Holland.

Broadbent, G. (1980a), 'Architeccctural Objects and their Design as a Subject for Semiotic Studies' in Design Studies, Vol. 1., No. 4., April, 1980, IPC Business Press, Guildford, pp 207-216.

Broadbent, G. (1980b), 'An Interview with G. Broadbent on Meaning in Archiitecture' in METU : Journal of the Faculty of Architecture, Vol. 6., No. 1., Spring 1980, Middle East Technical University, Ankara, Turkey, pp 7-30.

Broadbent, G. (1980c), 'Towards Post-Functional Design' in Industrial Design and Human Development, Proceedings of the XI Congress and Assembly of the International Council of the Societies of Industrial Design, held in Mexico City, October 14-19, 1979. Edited by Pedro Ramirez and Alejandro Lazo Margain, Excerpta-Medica, Amsterdam-Oxford-Princeton.

Broadbent, G. (1981), 'Architects and their Symbols' in Built Environment, Vol 6., No. 1.

Broadbent, G. (1982) 'Design, Economics and Quality' in Building Cost Techniques : New Directions, Edited by P. S. Brandon, E. & F. N. Spon. Proceedings of Building Cost Research Conference held at Portsmouth Polytechnic 23-25 September 1982.

Carter, P. (1984), Proof of Evidence, Mansion House Square Public Inquiry.

Caudill R. S. (1971), Architecture by Team, Van Nostrand Reinhold Company, New York.

Collins, P. (1971), Architectural Judgment, Faber & Faber, London.

Giedion, S. (1941), Space, Time and Architecture, (1962 Edn. consulted), Harvard University Press, Cambridge, Mass.

Hillier, W. R. G., Musgrove, J. & O'Sullivan P., (1972), 'Knowedge and Design, in Edra 3, The Proceedings of the Environmental

Research Association Conference No. 3. (Ed. W. Mitchell).

Lao Tsu : Tao Ten Ching (Asian Institute Trans. 1960), St Johns
University Press, New York.

Mutthesius, S. (1911), Who Stehen Wir?. Speech to Deutsch Werk-
bund, Berlin, quoted in Banham, R. (1960), Theory and design
in the First Machine Age, London, Architectural Press.

Pevsner, N. (1936) Pioneers of the Modern Movement, Faber and
Faber, London.

Pevsner, N. (1943), An Outline of European Architecture (1973
Edn. consulted), Penguin, Harmondsworth.

Plato, Timaeus (and Critias). Trans. Lee, H. D. P.. (1965) Penguin, Harmondsworth.

Rogers, R. (1984), Proof of Evidence, Mansion House Square Public
Inquiry

Roth, A. (1940), La Nouvelle Architecture, also in English and
German. 1947 Edn. consulted. Les Editions d'Architecture,

Ruskin (1949), The Seven Lamps of Architecture (1901 Edn. consulted)

Scruton, R. (1979) The Aesthetics of Architecture, Methuen,
London.

Summerson, Sir J. (1984), Proof of Evidence, Mansion House Square
Public Inquiry.

Van der Ven C. (1978) Space in Architecture, Van Gorcum Assen,
Amsterdam.

Vitruvius, The Ten Books of Architecture (Trans. M. H. Morgan,
1960) Dover Publications, New York.

Wootton, H. (1624), The Elements of Architecture Collected from
the Best Sources and Examples, John Bill, London. Reprinted
1969, Gregg Press, Farnborough.

Zevi, B. (1967), Architecture as Space. How to look at Architecture, New York, Horizon Press.

A SENSE OF ARCHITECTURE

ANDY MacMILLAN, Mackintosh School of Architecture

It is something of a challenge to accept a given title for a
conference paper, and seek to structure an adequate response. The
title must be presumed to contain some preconception of the possible
contribution expected.

'A Sense of Architecture', what could it mean? - surely not literally another sense like sight, hearing, touch, smell or taste; but
perhaps some innately connected perceptual faculty conferring the
ability to recognise architecture. Would this be important? or only
important to architects or admirers of architecture.

I would hope to show that it is vitally important,that not only is
architecture itself an extremely old human activity, appearing at the
very dawn of pre-history, at that threshold of evolution when human-
ity began to emerge from its animal origin and assert its cognitive
difference from the rest of the sentient life on the planet, but that
an associated sense of architecture can be seen to be a fundamental,
human attribute central to the deepest necessities of the species, as
necessary today in the evaluation of (the worth of) contemporary
urban environments as it was in those primaeval times.

This proposition can be seen to be sustained by an examination of
history, which demonstrates an ever widening role for architecture as
the availability of the means to engender and sustain it is progressively extended through society. Formerly the prerogative of kings and
high priests, then from Renaissance through to Victorian, progressively employed by Princes and Dukes, Lords and Landowners, manufacturers and middle class, and now in the 20th Century widely employed
across the entire social spectrum, architecture seems to be necessary.

By 'architecture' I would propose to mean that total synthesis of
end and means, that holistic process by which the human species
modifies its planetary environment to create a cultural and physical
habitat, a man-made or 'built environment' which I believe is the
most effective evolutionary tool in the human survival kit.

In the words of Aldo Rossi[1], "I use the term architecture in a
positive and pragmatic sense, as a creation inseparable from civil-
ised life and the society in which it is manifested. By nature it is
collective.....built with aesthetic intention. Architecture is
deeply rooted in the formation of civilisation and is a permanent,

51

universal, and necessary artifact".

   To properly understand the nature of architecture, in other words, "demands an ecological approach in which the structure of IDEAS and SOCIETY, the mode of gaining a LIVELIHOOD and the (domestic) ARCHITECTURE are interpreted as a single interacting WHOLE in which no one element can be said to determine the other", as Mary Douglas perceptively remarked in another context [2].

   Animals too may demonstrate a capacity to build, but merely for shelter; humans structure a milieu which reflects and sustains their lifestyle and aspirations, they erect a familiar fabric which not only protects their physical being, but strengthens their psychological identity.

   Manifestation of the ordering of that milieu is the evolved societal task of architecture. The related 'sense of architecture', the ability to recognise and respond appropriately to that structuring is engendered in every human as part of the primary environmental conditioning of the species towards efficient operation in the habitat, towards survival.

   Norberg Schultz, has observed[3] that "From remote times man has not only acted in space, existed in space and thought about space, but he has also created space to express the structure of his world. His interest in space stems from a need to grasp vital relations in his own environment, to bring meaning and order into a world of events and actions. It aims at the establishment of a dynamic equilibrium between man and his environment."

   I would assert that this aim determines architecture's primary function, and that the search for order in and understanding of the human environment the reconciling of the local, planetary and cosmic space which man inhabits has in turn led to the elicitation of the fundamental concepts of self, of space, of place; and to the important subsequent ideas of magic, religion, and science, by which man has progressively attempted to gain understanding of and a measure of control over the erratic world of chaos he inherits.

   Kevin Lynch[4] in America, and Enrico Guidoni[5] in Italy, have pointed to the original function of architecture as relating to the structuring of Territory, and to a need for what Lynch called "purposeful mobility". Guidoni had this to say, "the basis of the articulation of space is thus fixed in the elementary realities of the point and the true line which have the specific values of stasis and of movement that are inherent in territory".

   He also elsewhere remarked on the symbolism underlying architectural organisation in relation to the historical and mythical, patrimony of a people, "the line between prevalent myths and natural reality, whose spatial interpretation is a valid instrument for establishing order at the level of architecture as inserted into the environment and at the level of cosmological referent."

   The concept of "cosmological referent" is particularly vital in that it clarifies the probable origin and motive for the elementary transformation of environmental intervention for shelter into "architecture" as an instrument of environmental structuring.

   Leach[6] in his basic introduction to structural anthropology, comments on the striking contrast between human culture and nature,

52

"Visible wild nature is a jumble of random curves, it contains no straight lines and very few regular geometric shapes of any kind. But the tamed, man-made world of Culture is full of straight lines, rectangles, triangles, circles and so on", and adds that the contrast between the man-made "geometrical topography" and "random natural topography" is a metonymic sign for the wider, general contrast between Culture and Nature.

Perceptive though this comment is, it neglects the Cosmic (astro-nomical) aspect of Nature where the geometric properties of circles and straight lines are only too observable to those aware of their existence, as early human society, living under the sky must have been.

Further, the existence of such perceptible order and a desire to comprehend it, to intervene and exert some, however notional control over events, can be seen to further condition human effort towards self-alignment with the idealised eternal universe of the Gods, to lead to the association of 'regular' structures with the higher (cultural) functions of human life, and to the introduction of geometry as a constituent necessity in architecture.

Lynch and Guidoni make it clear that way-finding and purposeful mobility demand or engender an environmental construct or mapping and that therein lies the original motivation for what Guidoni refers to as 'the architecture of the territory'. This construct is a dynamic creation constituted of a cosmic and topographic percep-tion mediated by human intervention in the form of routes, land-marks and locii, a concept of space predicating a concept of place; ie. specific space, memorable and identifiable, a "sensible" space.

Lynch, in his seminal 'Image of the City' suggests that early man was dependent on nature for this structure, hills, rocks, river bends and junctions were his landmarks, but his developing social organisation made possible the modification of intensification of natural features, a cairn on a hill, a menhir on a plain or the blazing of a tree in the forest, both to structure and to lay claim to the territory.

Later man created his own landmarks recognising that their nature "involves the singling out of one element from a host of possibil-ities; the key   physical characteristic of this class of object is singularity..... some aspect that is unique or memorable has a clear form..... some prominence of location".

This description of the role of the landmark in the environment – singular... memorable..., clear form or prominence of location, is also a description of the primary architectural characteristics inherent in building to which the sense of architecture can be attuned.

I would suggest those characteristics therefore not only cover the criteria inherent in the concept of the "architecture of territory", but also that of the "architecture of building" (dwell-ing, as Guidoni has it) ie. the prescription covers the architect-ure of the nomad and the settler.

Memorability through singularity is the key constituent of the structuring process and once comprehended, singularity of design can proliferate.

53

In 'Culture and Communication", Leach made it clear that in human cultures, recognisable singular changes of state or condition - birth, puberty, marriage, death etc., are always recognised and celebrated (singularised) in ritualised responses. These ritual celebrations map social progress in the same way as physical landmarks map territory and it is easy to see how architecture can develop as a social instrument to celebrate and manifest the gradations and purpose of societal space. Aldo Van Eyk has lectured convincingly on the potency attached to thresholds, those tenuous interfaces between conditions of inside and outside, which occur at all levels of environmental ordering. Raglan too has commented on the threshold as the barrier between the man's world and the woman's in primitive society, as a relic of which to this day brides are carried over the doorstep.

The sense of architecture integrates the physical and social structure of the culture serving as an interpretive tool for the undifferentiated users, and a symbolic field of investigation and development for the specialised architect users working towards the development of group memory and tradition.

The sense of architecture relates to this, responding to recognition of a physical, societal structuring to match the psychological.

The aspect of memorability in the environmental construct is vital and it is fascinating to find in the work of Dame Frances Yates on memory[7], clear evidence of the early manifestation of the sense of architecture in relation to pre-literate memory and to the mnemonic devices of Classical oratory.

She shows Cicero to have observed, "....that persons desiring to train this faculty (of memory) must select places and form mental images in the places, so that the order of the places will preserve the order of the things, and the images of the things will denote the things themselves, and we shall employ the places and images respectively as a wax writing-tablet and the letters written on it," and later, "We have to think of the ancient orator as moving in imagination through his memory building whilst he is making his speech, drawing from the memorised places the images he has placed in them". Surely this is a clear exposition of the acknowledged power of the architectural image, and of the expected social ability to easily memorise and recall it.

Again she quotes him as observing, "It is essential that the places should form a series and must be remembered in their order, so that we can start from any locus in the series and move either backwards or forwards from it. If we should see a number of our acquaintances standing in a row, it would not make any difference to us whether we should tell their names beginning with the person standing at the line or at the foot or in the middle."

Thus by Classical times, and by inference even earlier, architectural imagery as a familiar and memorable system of reference was not only well established but also well understood, and the sense of architecture could operate through memorability as a vehicle not only for movement within the environment but for the intelligible structuring of human thought.

The concept of Place follows from the construct of a memorable environment; Place is "specific space", Place is a locus, fixed and memorable, Space is infinite and permeable.

Renfrew's studies of man in relation to the environment quantified in terms of energy flow, show that in prehistoric times man ritualised his occupation of territory in quite specific architectural terms as demonstrated for example, by the distribution of burial cairns in Scotland in terms of the availability of arable land to support extended family groupings.

Similar studies in Wessex of the relationship between barrows, causewayed camps and henges, seem to suggest a hierarchic architecture indicative of a hierarchic, tribal, social structure.

That the earliest monuments relate to the dead rather than the living and that few remains of structures for the living remain would seem to indicate a deep human need for a sense of territory and of continuity; of being and belonging, individuality and community.

These important related characteristics of our social species, the desire for Identity, for a sense of self, and the need for Community, for social intercourse, can be seen to generate the natural human habitat "the Built environment", a built collective that goes beyond the mere need for shelter.

Paradoxically, it can be seen that the individual need for identity is reinforced by a developed sense of community, and equally the desire for social contact, the proximity to others of the same kind, reinforces self-fulfilment, an opportunity which has continually operated throughout man's history in favour of the densification and extension of the built environment, or urbanisation, even under such apparently unfavourable conditions as exist in the present day third world.

Rossi points out in the 'Architecture of the City', "The contrast between particular and universal, between individual and collective, emerges from the city and from its construction, its architecture.", but Renfrew's research would appear to confirm that the need for such contrast is already in existence in pre-urban society.

Inherent in this idea of Identity are the concepts of self, of place and of location; not only 'who are you' but 'where are you' and 'how you got there', while the concept of Proximity not only also embodies the idea of place but polarises the idea of community or the human need for social intercourse, face to face contact, for co-operation, to achieve societal tasks beyond the resources of a single man or even a family.

These concepts concern an idea of space, and an idea of place (specific space) useful in isolating the societal tasks of architecture and identifying and defining the sense of architecture.

Architects can be seen to have an important cultural role as the creators of place, involving their unique ability to offer society an environment which not merely provides shelter but offers it a culturally structured habitat; legible to those possessing a shared culture and a common "sense of architecture", a three dimensional paradigm of its aspirations, "the built environment".

They have another 'technical' role as an organiser of space, a less unique activity shared these days with planners, builders, engineers, developers, administrators and the producers of building and building components.

Thus the irreducable component of architecture and the unique characteristic of the architect's role is concerned with the manifestation of the societal/cultural aspect of the environmental objects and places architects create, their mandate to exercise judgement in those areas where human value systems rather than parameters of force condition design.

Those areas of the environment which are shaped by forces like gravity, hydraulics or speed etc: Dams, Bridges, Roads; or which necessitate the modification of the occupant by training or even these days by physical and chemical adaptation; submarines, space capsules and satellites, properly lie outside architecture in the realm of engineering. Architects deal with people and their values, engineers in forces and their consequences; and many of the problems in our society today are created by an increasing tendency to treat economic parameters as having similar inevitable characteristics of force, a situation not unnaturally resented by the victims.

The societal tasks of today are conditioned by number and by very large scales of operation, our technical mastery of the environment makes possible cities of a size which takes on the nature of a natural force in its capacity to destroy the natural environment, hence our present concern with ecology, and to disenfranchise and dehumanise all but a few of the inhabitants on the other, hence the current questioning of the achievements of the modern movement through industrialisation.

As Metzstein observed in Oxford[9] in I979, "The delicate relation between consumer and producer is a feature of any industrialised society, and very evident at the present time. Building can be seen as possibly the last and most important survivor, of pre-industrialised production. A constituent element of the economic capitalisms of the late I9th and 20th Centuries is the commodification of products, and system building may be the weakest link in the building process in resistance to this basically unpopular process. The unconscious public assessment of these buildings as a serious threat to the quality of life is reasonable, as the popular simplistic model is almost a commodity, virtually consists of commodities, and is sold and bought nationally by catalogue. It is a reasonable surmise that the current disenchantment with urban change, and particularly the hostility to tall buildings, manifests deeper consumer resistance to the education of human values by all industrial processes."

Thus the present architectural re-awakening of interest in the "classical language of modern architecture", as John Sumnerson[10] called it, manifested in post-Modernism, the re-evaluation of the virtues of the Victorian Town, (the last reassuring built environment comprehensible totally through the sense of architecture) by Perez D'Arcy[11] and the Kriers[12], the weaker populist taking of refuge in pseudo vernacular[13] or the specious sub-classic of Quinlan Terry[14], even the world-wide re-interest in Mackintosh[15] can be seen as attempts to re-establish in the human habitat envir-

onmental constructs susceptible to comprehension by all through
the inherited sense of architecture.

It can equally be seen as a restoring of the design of cities
and urban artifacts to the realm of choice, proper to architecture,
a belated recognition of the psychologically destructive potential
inherent in a world designed only as a programmatic response to
simplistic, quantifiable, physical needs in the context of a value
system determined only by economic considerations.

Returning to the realm of choice and the need for architecture
to provide a reassuring legible environment accessible to and
acceptable by the community in general through a shared sense of
architecture - what does this demand?

I hope it has been demonstrated that a basic "territorial
imperative" as Ardrey[16] called it, generates a need for a memorable
mental construct of the social territory of a culture, to permit
purposeful movement in the environment, and that the need for
shelter goes beyond the physical and embraces the psychological,
the reinforcement of self, and that considerations of Identity and
Community act to condition the built environment towards that end.

In turn, this suggests that the sense of architecture demands
of both architecture and society in general a shared perception
which encompasses:

SOCIAL ORDER

Structuring of a built complexity of habitat to correspond to the
actual complexity of society living in concert not compartmentalised,
young/old, rich/poor, infirm/strong; and the complexity of spatial
need, civic, commercial, industrial, dwelling, service etc.,
recognition and revolution of the complex relationships of providers
users and managers in the built environment.

SPATIAL ORDER

A holistic and a particular comprehension of space, permitting
meaningful movement through the habitat and encouraging psychological reassurance through familiarity.

FUNCTIONAL ORDER

Recognition of route, circulation, accessibility of public and
private domain, recognition of appropriateness of location and
comprehension of the causal and dynamic relationship between social
density and social amenity, need for open and closed space for
natural and artificial, sacred and profane, even recognition for
a need for disorder and cycles of decay and renewal.

VISUAL ORDER

Recognition of locus and landmark, of public responsibility in the
public and private domain, and private responsibility in public and
private domain, the resolution and manifestation of societal and
tectonic demands on building towards greater legibility, apprec-

iation of implications of scale, surface, texture in the information (message) conveyed/comprehended intuitively through the "sense of architecture".

The goal is a totally comprehensible environment which not only functions but is seen and understood in its functioning, an environment which places a premium on the holistic resolution of human needs and values rather than on the solution of separate and conflicting self-contained systems of force.

Such an environment reinforces and reassures the sense of self and of community rather than demeaning the individual and degrading the environment while permitting meaningful occupancy, use and even pleasure.

Recognition of the sense of architecture and its social role is the the surest method of achieving such a holistic environment. Architecture after all still is the only environmental discipline which demands holistic involvement in design and recognises both the cultural and the physical needs which must be satisfied.

I hope I have shown that architecture is a vital human activity which has existed throughout the history of humankind, implying a basic irrevokable need for reassuring order in the environment and a manifestation of human values to which we are all responsive through our sense of architecture.

REFERENCES

I.   Rossi, Aldo (I98I) 'The Architecture of the City',
     MIT Press, Chicago.
2.   Douglas, Mary (I9II) 'The Seasonal Life of the Eskimos',
     Maas and Beuchats.
3.   Schultz, Christian Norberg (I97I) 'Existence, Space and
     Architecture', Studio Vista, London.
4.   Lynch, Kevin (I960) 'The Image of the City',
     MIT Press, Harvard.
5.   Guidoni, Enrico (I975) 'Primitive Architecture',
     Electra Editrice, Milan.
     (Translated by Robert Erich Wolf (I978).
6.   Leach, Edmund (I976) 'Culture and Communication',
     Cambridge University Press.
7.   Yates, Frances A. (I966) 'The Art of Memory',
     Routledge & Kegan Paul, (I969) Peregrine Books Penguin.
8.   Renfrew, Colin inaugural address (I973) 'Social Archeology',
     University of Southampton.
9.   Metzstein, I. and MacMillan A. (I974) 'Amenity and Aesthetic
     of Tall Buildings', International Institute of Bridge
     and Structural Engineers/Institute of Structural
     Engineers, Residential Conference St Catherine's College
     England, Oxford.
I0. Sumnerson, John (I980) 'The Classical Language of Architecture'
     Thames and Hudson.
II. Architectural Design (I978) Volume 48 No.4.
I2. Architectural Design (I977) Volume 47 No.3.
     Architectural Design (I983) Profile 49.
I3. (I973) 'A Design Guide for Residential Areas',
     County Council of Essex.
I4. Architectural Design and Academy Editions (I98I) 'Quinlan
     Terry'.
I5. MacMillan, A. (I979) Global Architecture No.49.
I6. Ardrey, Robert (I967) 'The Territorial Imperative',
     Collins.

CAN BUILDINGS GENERATE PROFIT AND, IF SO, FOR WHOM?

THOMAS A MARKUS, University of Strathclyde

In writing a paper many months before a conference is to take place,
one can only predict from presently available information what might
be the dominant themes within which one's paper will have to be
relevant - the brief programme, the title of the conference, the
papers and the knowledge of the authors one is fortunate to have.
    First, then, the title - 'Design: Quality: Cost: Profit'. This
suggests four themes.   The first is the process by which buildings
are created. The second is some desirable property of buildings
which is beyond quantification in economic or other terms and hence
has to be veiled under a label which suggests a contrast between
humdrum, plebeian concerns of money and other constraints and something difficult to define, associated with a select group in society,
and, as the Shorter O.E.D. defines it, "the degree or grade of
excellence etc., possessed by a thing". 'Excellence' is not of
course the result of the operation of a law of nature but the result
of a judgement by some person or group. It can be predicted that
it will be used in the sense which reflects the judgements of those
groups which own the majority of land and property and whose judgements are faithfully reflected by the professions which serve them,
primarily architects. The third word in the title has a hard-edged
ring about it, but also a suggestion that the elements in a cost
analysis are 'facts' and that the models and techniques used for
carrying out the analysis are 'objective' and neutral.
    The fourth, and last, 'profit', is drawn from the economic
language first fully raised to the status of a law of society by
Adam Smith in 1776.   Book One of 'The Wealth of Nations' - "Of
the Causes of Improvement in the Productive Powers of Labour, and
of the Order According to which its Produce is Naturally Distributed
Among the Different Ranks of the People" - opens with the famous
sentence: "The greatest improvement in the productive powers of
labour, and the greater part of the skill, dexterity, and judgement
with which it is anywhere directed, or applied, seem to have been
the effects of the division of labour". From an analysis of the
division of labour, how it involved the development of machinery
(of which he had not yet seen industrialised examples), how it
greatly increased productivity, how money developed to displace
barter of goods or labour, Smith rapidly moves on to price and value

analysis and shows the role of profit in these mechanisms. The whole
process is part of an 'Order' which is natural, to be interfered
with minimally by the State; and so, equally natural, are the 'Ranks'
of people who play different rôles in the economic system. Although
this powerful analysis is over two centuries old, its structure and
rules are inherent in our 'market' view of the economy and society,
and appear to be inherent in the Conference and many of its papers.
Design services are spoken of as having a 'market'; 'developers'
equations' occur; developers will speak of property 'worth';
and throughout the papers titles occur in which the apparent
polarity of 'cost' and 'value' (or 'quality') is highlighted.

This paper sets out first to examine the consequences of the
market as a formative influence on buildings and towns and second,
to challenge the idea of 'quality' as a mysterious property of
objects which is at the same time absolute and beyond the possibility
of inclusion in resource models.

In many areas of building the operation of the market has a long
tradition and investment for profit has roots in classical times;
certainly it was one of the major modes of wealth creation in the
city states of Italy, in seventeenth century London in such projects
as Covent Garden, throughout the eighteenth and nineteenth centuries
in industrial and housing projects, and in our time in office and
factory buildings, speculative housing and shopping development.
Prior to the Industrial Revolution such investment projects were
instigated by Royalty, the Church, guilds or wealthy landowning
families, such as Grosvenors, deriving capital from mercantile as
well as rural investments. Since that time the prime developers
have been industry, commercial enterprises, (mainly banks and
insurance companies), the railway, canal, warehousing, ports and
harbour companies, and developers of speculative housing and shopping. In such development the return on investment is a central
design issue but leads, according to the nature of the market, to
quite different forms of architecture.

In the case of industrial projects the buildings are treated as
normal plant, with initial capital and commissioning costs, a
notionally fixed life, maintenance, repair and refurbishment costs,
and, finally, a demolition cost. The land and reusable materials at
that point have predicted capital value. The total resource cake
is sliced so as to yield the most efficient production tool,
production being largely a process in which machines, and people
attached to the machines, carry out a planned operation which is
predictable, controllable, and systematised, characteristics which
the coming of robotisation and electronic control systems have
increased. Occasionally either the product or the site demand that
more than normal be spent on the image as in a Herman Miller or
IBM factory. Aspects of image which are <u>internal</u> to the workers in
the factory are largely kept to bare simplicity or even crudity;
whilst the functional aspects of spaciousness, controllable and
comfortable environment, sanitary provisions and safety are designed
in accordance with the minimum standards which from time to time
legislation has imposed over the heads of reluctant owners.

Since the processes <u>within</u> the factory are based on Adam Smith's

61

notion of the division of labour, spaces and functions are structured
according to the rules of labour specialisation which determine
spatial relationships and which take precedence over all other
functional needs, other than the need for surveillance and control,
functions which spread across the process divisions. Thus surveyors
and controllers have different spatial worlds and of course
correspondingly different social relationships with the work groups
and the management. So forms, functional programmes and spatial
structures, all related to social structure, arise directly from
market forces.

In building types such as offices the process is less mechanised
and the image projected by the organisation, which is usually
located on a central urban site, is an important part of its product
promotion. Hence more resources will be allocated to the creation
of appropriate images, possibly involving the use of lavish
materials, landscaping and sculpture. A key element in such promotion beyond advertising the actual product (insurance, for instance)
is a visible role of the enterprise as a patron of art and architecture and, hence, its validation in cultural and heritage terms.
The Palumbo scheme is a case in point.

In the case of shopping developments the market forces require
characteristic planning solutions. Shops are grouped round an
arcade or deep internal mall, to which entry is gained from the
street through a single, or a few, 'gates'. This spatial control
also unifies all the tenants under one roof, one management, one
rent and, by selection and rejection of tenants, one marketing
strategy. This ensures that competition between tenants is minimised
and that products or information resources which present any kind
of ideological challenge are excluded.

In the case of buildings which have no ostensible investment
purpose the market operates less obviously. Recently a great deal
of attention has focussed on art galleries - Stirling's Stuttgart
Gallery, the National Gallery extension, Foster's Sainsbury Centre
at the University of East Anglia and the Burrell Gallery in Glasgow.
These were built for a variety of clients - for a University or
city to house great private collections, or for the state or a city
for a national collection. In each case certain objectives had
to be achieved: the commemoration or celebration of private collectors
such as Burrell or Sainsbury: the demonstration of the State's
or the City's committment to a cultural heritage; the particular
artistic or anthropological theories whose classificatory systems
have to be mapped in the arrangement of objects; the role of the
international art market shown in the flaunting of market value as
a major hurdle which only wealthy collectors can overcome; and the
establishment of specific relationships between objects, the public
and the curatorial staff who stand in place of the 'owner'. Market
forces operate here not only through the emphasis on the value of
the unique objects, but through association of private or public
wealth generation to 'culture'.

A few more specific effects of the market on buildings - be they
regarded as machines, sales promotion or cultural symbols - can be
described.

In the case of the factory the difference in the spatial
provision, access from the street, environment and privacy between
the spaces for various grades of management, supervision and workforce are the architectural concretisations of Adam Smith's
natural 'rank'. It is the same principles which result in tight
zoning of cities, including the establishment of industrial estates
in areas of favourable land prices where transport and time limitations ensure that workers are unable to participate in any
communal city life during the working day, and that the areas themselves become isolated, deserted and silent at night and during the
weekend. Beyond this the detailed negative effect of market forces
would have to be teased out for each individual case. In some it
would be evident in the absence of a private social space for
workers with the result that most of the informal communication
network is centred on the toilets! In others the division of
the total cake allows minute sums for landscaping, sports areas,
or more than minimal finishes, even in non-production spaces.
    In the case of offices and other urban commercial development
the effects are similar but clearer. The total system which allows
investment in a Palumbo scheme whilst thousands of Londoners are
homeless or living in squalid, decaying and overcrowded public or
private housing, is symptomatic of the market operating at global
level. At the level of the individual building the debate which
surrounded it brought in every kind of formal issue to the fore -
scale, image, the need to commemorate Mies van der Rohe and conservation of the aesthetic precedents set by the Mansion House.
But not a single issue around the actual functional and spatial programmes of the building itself was raised from the view point of
those who will work in it. It suits the market to reinforce a view
of architecture, which architects, the public and the Schools have
grown accustomed to accepting, as a question of large scale public
sculpture and architects as a species of artist. Thus the function
of the building, initially prescribed in the brief by the client,
and its spatial structure, are taken out of the debate, and the
hands of the architects. The myth is maintained that the brief
is 'neutral' and 'objective' - that what language says, and those
things on which it is silent, can somehow be innocent, and that
therefore the limitations on behaviour, activity and organisation
of people's relationship to each other in space are an inevitable,
unarguable given - again the tradition of Smith's 'Order'.
    Space standards which vary for different categories of office
workers, grouping by function, the increasing reliance on electronic
systems for communication and information handling, the lack of
environmental control for individuals or small groups and the system
of supervision and surveillance, all accepted by staff as part of
this natural 'Order', perpetuate the usefulness of buildings as
major instruments for the maintenance of a compliant, uncritical
workforce. There are spatial obstacles to the development of
solidarities, outside the Unions, which significantly are, in Bill
Hillier's terms, 'transpatial' in that nothing in the building is
visible which would suggest the existence of another institution;

hence they are not only transpatial, but a-spatial – having no
concrete reality and largely disembodied in terms of everyday
experience.

So the market largely determines what is built, on which sites,
much of the imagery, the functional programme, the spatial (and
hence social) structures, the standards of space, environment and
finishes as well as the degree of landscaping. In the private
sector this can be shown to have direct profit motivations; and
the critical question is: "whose profit?". Strangely, in the public
sector, where market forces may be expected to be weaker, the same,
or often even lower standards of design prevail, with no greater
attempt to reduce the alienation caused by buildings. In part this
is the result of the strong bureaucratic institutions under which
hospitals, schools and local authority housing are designed and
built. Electors and elected councillors, of whatever political
colour, usually reproduce the values of the market even though
'profit' in the technical sense is not at stake. Their professional
advisors, especially architects, behave in public service as in the
private sector – that is as artists and technicians whose task does
not include deliberate search for social meaning in buildings
through control of all aspects of design. Moreover, any 'islands'
of councils who may wish to operate in a new way are surrounded
by the market: land values, competitive tendering in the
contracting industry and, today, national government, can easily
swamp such effort. The fate of urban local authorities in a hostile
national environment is no different from that of Allende's Chile
in a hostile international one.

It is time to explore the second main issue of this paper – the
problem of dealing with 'quality' in resource models. First it
needs to be said that even assuming, especially assuming, a new
pattern of building design and construction, based on the goals of
weakening economic and social injustice, there will still be
limited resources for building. So resource models which examine
the consequence of alternative decisions will be as much needed
as today. However the variables appearing in the models, the
values given to various terms (such as interest rates), and the
nature of the models themselves would be very different. This would
be specially true of those models which guide decision makers and
designers on the relative priorities of conflicting demands on a
fixed budget.

The angles of the segments into which a total cake is sliced are
a rough guide to values and political position. In the national
cake, the main differences between the parties is the angles which
are assigned to, say, health, education, industry, defence,
housing and social services. At regional and local level these
angles are an equally good measure of values. In an individual
building project the same applies – the segments being either
building elements – foundations, walls, roofs, finishes, services
etc. – or spatial elements – for instance, in a health centre,
circulation space, consulting rooms, waiting areas, children's
waiting spaces, interview rooms and nursing accommodation. More
allocated to one item results in less to another – but this is only

a crude economic framework for within any fixed allocation, for each element there is still an infinity of solutions which will cost the same amount.

Some of these solutions may be entirely novel, so the designer who invents them has to find guidance on the extent to which these will meet needs. If they do not resemble something which already exists, it can at least be simulated - verbally or in two or three dimensional representations. Where solutions exist preferences can be discovered from analysis of purchase or rental where these have been made with genuine freedom of choice. Unfortunately such data is rare, for the market rarely operates with equal power in the hands of the participants; thus houses have to be bought or rented, not as a preference but as a last resort from the limited range available within a (probably inadequate) family income,from developer or local authority.

In cases where alternatives have to be simulated people can either be asked to evaluate them directly, as if they were free from unjust economic constraints, or to equate different goods against each other. For instance, in housing both the quantity of space and the degree to which it can be made comfortable is controlled by resources. If there is more space it may be necessary to tolerate lower heating standards, or to sacrifice the ability to heat all spaces simultaneously to the same standard. If high heating standards were considered essential, the area of the house will have to be diminished. Similar choices arise with regard to almost any feature - quality of kitchen fittings, amount of garden space, durability of floor finishes or sanitary provisions. By presenting solutions with variable combinations of two or more of these features, an individual's preference and trade-offs can be dis-covered. With two variables only, this can be drawn in the form of an indifference curve, as in Figure 1. With three, it would need

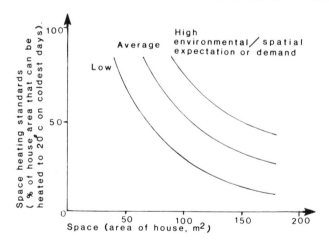

Fig. 1.   Indifference map (three curves) for space heating and space for three levels of expectations or demand

a three-dimensional graph, in which there are indifference planes.
With n-dimensions, one can only represent the choices by means of
matrices in n-l dimensional hyperplanes.

The cost of providing various quantities of two, three or n goods
is known, and therefore the cost of different combinations of these
is also known. These produce so-called 'production' functions.
The indifference curves or planes tend to have a characteristic
shape which indicates that once a basic amount of something is
provided adding more becomes decreasingly attractive; the production
functions take on shapes which indicate not only that the more of
something is made the cheaper the unit (marginal) cost of an item;
but that there may be diseconomies of scale too, where increasing
quantity leads to increase in unit costs. In that case these curves
will have the shape indicated in Figure 2. Then, for a given total

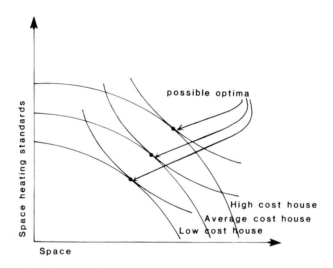

Fig. 2. Production functions for three
         levels of investment and corresponding
         optimum mix of space heating and
         space standards.

cost an individual's best solution is that where his or her
indifference curve is tangential to the production function - that
is the mixture of goods which reaches the highest indifference curve
- any other mixture would be on a lower curve and hence regarded as
less valuable.

Of course there are many difficulties: the data for indifference
curves may not exist, and might be difficult to generate; the
production functions may have the same convexity to the origin as
the indifference curves, and hence there will be no tangential
contact points; the indifference curves of individuals vary with

time, experience and education; in any case different curves for a group of individuals cannot be combined rigorously into a <u>group</u> indifference curve. These are important technical issues but none of them so serious that attempts at this type of formalisation of preferences and cost should not be made. The same problems arise in cost-benefit analysis, which has received a bad press, especially since the '60s, when projects such as the Third London Airport were appraised using crude and slanted data. Nevertheless, analysis needs to be pushed beyond the capital cost appraisal, cost-in-use appraisal, or even cost-effectiveness appraisal - where alternative solutions all meeting the <u>same</u> objectives are pitted against each other in terms of the variations in the running as well as the organisational costs of each.

In some situations one way round is to probe the sensitivity of solutions. For instance, if it is known, or it can be discovered by research and simulation, that alternative solutions are preferred to different degrees, which can be quantified on some preference scale or rank ordered but to which no cash equivalents can be found, it is possible to plot a range of solutions as in Figure 3. In this

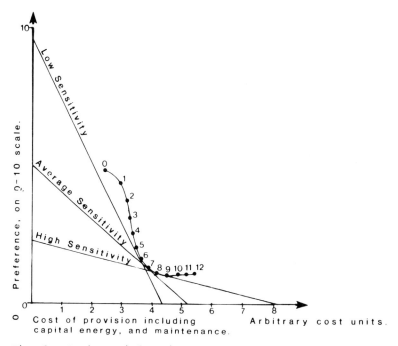

Fig. 3. Optimum window size.

example window design has been chosen - the variable under investigation being size. For each size a preference has been ascertained and scaled on a 10-point scale (say in respect to view) and plotted on the y-axis. For each size also the combined capital, maintenance, energy and other related costs have been ascertained and plotted on the x-axis. Any straight line connecting the two axes represents the equivalence of preference and money for someone - a person relatively insensitive to view will have a steep line, where big shifts of preference account for quite small shifts of money, whereas a sensitive person will have a shallow line, where small shifts of preference equate to large shifts of money. The optimum solution for one individual, with his or her characteristic slope, is where the line is tangential to the base curve. In many cases the base curve, plane or hyperplane has a 'bump' in it, of the kind shown, which results in a whole range of lines of greatly vary- ing slopes all being tangential in the same region of the base curve. Therefore, since the lines represent the evaluations of a range of people varying from highly sensitive to quite thick-skinned, it is not worth worrying about such a decision, and, provided a solution somewhere along the short stretch of the base curve is selected, it will be nearly the best for almost everyone.

It goes without saying that whatever form of economic modelling is used it must encompass all costs, over the whole life cycle of the building. Though this is now commonplace official and profess- ional dogma, everything in the market operates against it: risk, interest rates, inflation and taxation allowances. But the market also creates private and public divisions in resource allocation which are centred on bureaucratic convenience, or the division between investment bodies such as building societies, banks and insurance companies. These divisions reflect, on a corporate scale, 'the division of labour'. They effectively prevent the economic appraisal of problems across institutional barriers, thus operating not only against the general good, but against the greater good of individuals and groups. A case in point is the problem of cold, unheatable houses, where defects of construction and heating system design make it impossible to attain comfortable or even healthy thermal environments in cold weather. It can be shown that there is a high correlation between the onset of a cold winter spell and the incidence of several chest and circulatory diseases which are cold-induced, or brought to an acute stage by cold. The resulting hospital costs and costs of periods off work or school due to sickness can easily be shown to cost more than the cost of improving the house would have been. For instance a typical two- week hospital stay for, say, acute bronchitis, will cost the health service more than it would have cost to insulate and draughtproof a small house, or a flat; an investment which would not only have avoided that hospital incident, but all future ones arising from that house. But the trade-offs between different Government budgets, and those of local government services, make this kind of calculation unusable and thus useless. The expenditure, ill health and discomfort continue. Similar cases can be found in many private investment decisions.

Those designers who argue that 'quality' is something intangible, mysterious and unquantifiable, must at the least come to terms with fields such as the fine arts. Although it may be impossible to quantify the value of aesthetic pleasure from a picture in cash, and although market prices may be grossly distorted, at least an individual intending to purchase a picture from an art gallery, and with a fixed sum to spend, can obtain a price list, so that the adequacy of the sum available can be easily ascertained. The 'qualities' so much discussed in architecture often do not even have an accurate price tag attached to them, so that the purchaser has to buy a pig in a poke and may realise too late that they have been purchased at a cost that is ruinous (say in energy, or maintenance terms).

The inadequacy of models for handling these aspects of buildings can either be used as an excuse for withdrawing such decisions from public debate and treating them as the 'black box' territory of inspired genius, or as a reason for struggling to develop better tools, gather more data, represent more solutions to more people, undertake design education in society and then enter into rational debate.

The critique of the market, and the simultaneous defence of techniques for economic appraisal may seem paradoxical. It is based on a vision where buildings enrich large sections of society; where they are active instruments for co-operation, freedom and creative work. The 'market' forces unleash on society embodiments of the dark side of human nature - greed, lust for power and domination, and exploitation of others as objects. Because of this, societies have struggled to create for themselves those minimal structures called just government, to protect themselves against themselves and against others. The destruction of these defences, and the release of market forces is the economic equivalent of using Darwin's law of natural selection and of the survival of the fittest (literally the law of the jungle) as principles of social organisation. William Morris, in 'News from Nowhere', envisaged a society which had been released from the market, with its corrosive influence on human relationships. His vision was set in the early twentyfirst century - it is doubtful if we can work to such a time-table, or whether, indeed, it can be more than an imaginative goal towards which building designers like others will have to advance step by little step.

INFORMING DESIGN DECISIONS

DOUGLASS WISE, Institute of Advanced Architectural Studies
                University of York

1.   INTRODUCTION

The range of activities and events which could be involved in the
spectrum of 'Informing Design Decisions' is infinite, but for the
purposes of this paper will be confined to the findings of a research
project undertaken at the Institute of Advanced Architectural Studies
between June 1980 and February 1984.

   The interest in the Institute about the way designers behave, and
more precisely, what information they use during the design process
goes back to the early 1970's, with the publication of 'Architects
and Information' by Goodey and Matthew (1). This looked at the form
in which architects preferred their information - trade literature,
technical literature, research reports, digests and the like, rather
than the way in which they used it actively during design development.

   This work had its effect on the form and quality of information
produced for the consumption of the designer, and was enhanced by the
gradual development during the 1970's and early 1980's of more effective literature, particularly by the sector of the building material
and component manufacturers. The Building Centre's Awards for good
technical literature gave an additional impetus, with some impressive
results, and a further piece of work by the Institute, published in
1980 by Margaret Mackinder on 'The Specification of Building Materials and Components' (2) drew attention to the need for trade information which gave details of comparative performance and testing.
The report discussed the way in which both educational establishments
and practice handled the problem of selection and specification, and
found it very patchy, in both sectors. Practice recognised that it
was cautious and conservative in choosing materials, preferring to
select products which had been used before, or which could be seen to
have been used by others for a reasonable length of time, and identified as one main problem the lack of time available for decision
making, which inevitably forced designers back into well tried techniques and practices. The weakness of this, of course, was that the
feedback techniques from buildings in use was found to be equally

patchy, so that there was little assurance that the 'well-tried techniques' had, in fact, been all that successful. Contained in the
research was discussion on the supply and handling of information
within the offices, and facilities were found to be variable and very
much tailored to individual needs.

It was against this background that the Institute was commissioned
in 1980 by the Building Research Establishment to undertake research
into the way in which designers used information in the early stages
of the design process, examining the routes taken by designers to
reach decisions, finding out how these varied from job to job, and to
see how the use of information and experience influenced the decision
making process.

The initial work looked at Stages B-D of the RIBA Plan of Work
(largely) and was published in 1982 (3). A second stage has now been
completed and will be published in Summer 1984.

2.   RESEARCH METHODOLOGY

The work was based on a series of case studies of live design projects in architectural offices of various types in the Yorkshire area.
The projects were varied, and the designers were interviewed as the
project progressed from inception to production drawings. The interviews provided information on the sequence of decision making on the
project and the influences on those decisions.

Six case study offices were chosen in both the private and public
sectors and twelve projects analysed. The projects ranged through
private housing estates, conversions, a primary school and medical
buildings to specialised diocesan work, with clients ranging from
speculative builders to development companies. The projects were
self-selecting in the sense that the research needed to look at buildings which were on the point of starting in the offices, and where
Stages B-D (and into E) could be covered in the time scale of the research.

Development of the projects was assessed through regular interviews between researchers and designers where observed progress could
be analysed, plus self-monitoring by the designers themselves in the
intervening periods.

The research highlighted the following aspects:

3.   THE DESIGN PROCESS

The design process as observed during the case studies followed a
fairly consistent pattern. An initial concept for the building plan
form and general construction developed rapidly, using little information other than the client's brief, the site constraints and the
designer's own experience. This initial concept was then developed
and refined, using more deliberately researched information, and later
modified, as necessary, in response to emerging constraints and

changing requirements. In nearly all cases the initial concept formed the general basis of the final design, only undergoing minor changes. Time factors often seemed to restrict the exploration of alternatives, while requirements of the statutory authorities were generally responsible for the few major changes to design.

This pattern of working contrasted somewhat with simplified models of theoretical design sequences. It showed that aspects of a design might call for the designer's special attention at an earlier stage than generally predicted (by, say, the Plan of Work), while design problems are often short-circuited by the designer on the grounds of his own experience.

4. INFLUENCES ON DESIGN DECISIONS (4)

The case studies showed that there were many outside constraints beyond the designer's control. Factors such as budget, project programming and site constraints were recognised as significant influences on design decision making in all the case studies. The level of input to the design from outside agencies appeared to be causing significant problems of delay and disruption to the design process. These problems involved either general briefing and communication with clients or negotiations for statutory procedures, particularly with local authorities over planning approval. Negotiations, especially over highway matters, involved some designers of the case studies in producing a substantial number of sketch proposals before the application could be lodged. Lack of firm guidance or information from the authorities themselves caused some delays and abortive work.

Regarding the client's influence, incomplete briefing information at an early stage was found to complicate the design and to be a cause of abortive work. In the few cases in which briefing guides developed by the designer or client were used, it was found that these provided a valuable basis for the development of the design and also prevented the communication problems which tended to arise in other projects. Designers also valued meeting the ultimate user whenever possible.

4.1 Experience
Experience was the most often quoted aid to design decision making, accounting for the majority of 'reasons' given for decisions. Its advantage over other forms of information appeared to be its ready availability, it being immediately obtainable from the memory of the designer or his colleagues.

There were three main types of experience:

(a) experience of the decision making process, enabling the architect to predict in advance what problems might arise and to be aware of information sources which might be appropriate,

(b)   experience and general knowledge of building construction gained from both education and practice, enabling design decisions to be made in terms of what would normally be appropriate to the requirements of brief and site, and

(c)   experience of performance of a design decision taken previously.

Use of experience allowed for the prediction of possible problems and helped to avoid abortive work. However, designers rarely kept comprehensive records of the decision making process so offices subsequently lost some of the experience which had been gained in the course of projects.

Use of written and drawn records occurred mainly at the level of detailed design work (outline design concepts are by contrast retained in the memory). Standard details and specifications, where used, needed to be reviewed periodically.

Experience of performance was usually negative, as the designer's attention was more commonly drawn to failures than successes. Positive feedback was most often used to make decisions about visual aspects of the design, and in the preparation of standard details and specifications.

Designers tended to favour the use of design idioms traditional to their offices' output and to the localities in which they worked; a new language of design for each project was unnecessary, and normally impractical. In the case studies, personal choice was seen mainly in the design of minor elements.

Older designers were able to rely far more heavily on all types of experience than less experienced designers, who tended to spend a great deal of time seeking out and consulting a wide range of written information sources. Experience helped to reduce design time, therefore, but was seen as an alternative to the study of current literature, and consequently continuing professional development was largely neglected.

The designers studied seemed to believe that experience was best picked up through the practice of design, although the majority of the offices did not consciously collect feedback from their completed projects, apart from keeping record drawings and specifications. This being the case it seemed likely that what designers most value was the experience of the process of design. This type of experience was seen as a positive advantage in the prediction of problems in a design. The true value of designers' experience of performance, materials and construction depended (as did the value of written data) on such qualities as its being reliable and up to date. Whether the quality of this type of information was any better than that gained from written data has not been proved. In spite of this, designers seemed to prefer to use experience, as it provided a more readily available information source, well suited to the climate in which design decisions were normally taken.

4.2 Recorded design data

The research made note of every written source which the designers
said they consulted over the case study design work. The majority of
references were consulted in order to resolve detailed aspects of the
design. This was usually, but not always, during the latter stages
of the design programmes. These detailed aspects were often delegat-
ed to younger, less experienced designers, who were found to be the
most frequent users of written information. For the outline design
work, independently published data (namely design guides, building
reviews in journals, information handbooks and general architectural
history books) were the main source of reference when any were used.

Written references were mainly used to check points, or to find
solutions to ideas already generated by the designer. They were seldom used to spark off ideas or to predict problems. They were more
frequently used in 'one-off' design projects than in cases where the
office has a history of previous design of a similar building type.

Frequency of use of written references was not related in any way
to the availability of a good library in the office. Use of written
data in the design process was limited. It was seen as a time-
consuming activity. The nature of designers' work schedules, and the
fragmentation of time between a variety of tasks and projects may be
factors in the low levels of information use observed. Designers
tended to refer to a few standard publications only.

Where written data was consulted, technical references were more
frequently used than general design references. There was a tendency
among the case study designers to avoid official literature unless
its use was absolutely necessary, as in the case of the Building Regulations and associated data. The reason given was its generally
complex nature.

The research endorsed previous findings (1) which showed that
designers prefer concise, well illustrated literature such as that
produced by the trade associations, and that they rely strongly on
trade and trade association literature and technical guides published
in current journals.

5. IMPLICATIONS

The research gave a strong indication that designers tended to seek
written information as a 'last' resort, when their own experience or
that available in the office failed to give either an answer to a
problem or the understanding to enable a solution to be pursued. It
showed also that designers needed to know, easily and quickly, the
range of information available, especially when undertaking design of
less common building types - and this had wide implications for office
organisation, library systems and information technology techniques of
both storage and retrieval.

Designers tended to use a narrow range of 'favourite' references. They preferred certain forms of written information (1), and showed an even greater preference for diagrammatic presentation. This re-inforced the findings of previous work which called for information relevant to, and easily understood by, architects in their daily work.

The fact that experience was widely observed to be the preferred source of information seemed to suggest that this should be recognised and exploited more substantially, by designers being encouraged to both increase their experience in areas relevant to their work (ie in depth) and share it with others. The means by which experience was gained, stored and transferred within a practice was tentative, and appeared to be a huge resource, largely ignored by the offices in any formal sense. Systematic and comprehensive collection ofperformance data did not take place in the offices studied.

In the research there was no evidence of continuing education (CPD) playing any part in the upgrading of information about the design process, or the inputs into it - clearly a route which will have to be explored in greater depth as the CPD movement grows.

The question of time available for design decision making was raised continually, with the variety of tasks undertaken every day reducing the time in which actual design decisions are made to very short, unconnected periods. It seemed however that much of this dilution of time might be avoided through both better management and more effective communication documents between the various parties to the project.

The introduction of self-monitoring to the case study designers seemed to be regarded as valuable, suggesting that the recording of design decision making could be helpful to the management of a pro-ject and as an aid to designers' memories (especially when designs were protracted due to delays). It could also prove valuable inputs to subsequent work on the same project or similar projects and thereby maybe save both costs and time.

6. FOLLOW-UP

The research concentrated mainly on the conceptual stages of design development, and suggested that, maybe, during detailed design devel-opment (Stage E) more sources of information might be used than in previous stages. Once the mature sketch design at Stage D was being taken to pieces in Stage E, each element and component designed, detailed, costed, scheduled and the whole building co-ordinated in both design and technical terms, then the need for detailed inform-ation should, theoretically, grow substantially. Further, the work had concentrated on small/medium sized offices undertaking relatively small scale work, Would bigger offices, undertaking bigger work, behave differently?

To find out the Building Research Establishment and the Building Centre Trust commissioned the Institute in October 1981 to develop the research.

It was found that, with minor differences, the process of design decision making was not significantly different between large and small practices, or between large and small jobs, but that more published information was used by both in the later design stages - and this aspect will be analysed.

7. DETAILED DECISION MAKING

Research Methodology
To attempt to analyse the decision making process, and the influences upon it, during detailed design it was decided to take one element only. 'Roofing design' was chosen for a whole range of reasons:

It represented

(a) a universal element in both public and private clients' new building projects;
(b) a frequent element of refurbishment projects;
(c) a topic well covered by trade and trade associations' literature;
(d) an area for which BRE research had produced much design guidance;
(e) a design element where failures were frequent and of significance to users and clients;
(f) a design element where maintenance was a significant consideration during design development;
(g) a significant proportion of the cost of a building;
(h) a visible part of a building.

The research covered the detailed decision making during working drawing and production information stages, which led to selection and specification of roof materials, including detailing necessary for waterproofing and drainage. Design of the main supporting structure was excluded from the study.

Seventeen projects were investigated, ranging through new buildings of £100,000+ in cost, and designed primarily for people, through to refurbishment, including projects where the roof design, for various reasons, could not be an exact repeat of a previous office design.

Host offices were 'recruited' by telephone and post and the designers recorded the data on each project themselves, with support and encouragement from the research team by telephone where necessary.

As the research data was to be provided by individuals who were themselves not researchers, the method of recording the data and questions the designers were to answer had to be considered very carefully. A package of forms and instructions was tested, modified, re-tested and subsequently confirmed for issue to the case study offices.

Each designer received a package containing:

(i)   a form to collect background information on his project and on any decisions about roofing which had already been taken;

(ii)   forms to use on any occasion when he/she considered roofing design. On these forms he recorded what design issue was in question, what printed information needed to be consulted, how the information was used, and certain comments about the presentation of information in the documents, also how the documents consulted were obtained, and whether information from other sources was used (eg in-house design guidance, personal experience);

(iii)   instructions and examples to help with recording.

Towards the end of the research, designers completed three additional forms with some background information about themselves and their practices, and a summary description of the roof designs to date.

8. RESEARCH FINDINGS

As might be expected, more published information was used in detailed design than in conceptual design. The reasons were largely twofold. Firstly decisions taken had an increasing likelihood of being realised and secondly the information required was more detailed, changed with time, and was therefore less memorable. New and larger projects frequently had to be researched, often requiring published information and, importantly, larger projects tended to have the resources available for project related research. But tried and tested solutions were preferred, reducing the need to consult documents and sources of information.

Manufacturers' literature was consulted on nearly every project, and for most was the main source of information. For technical principles designers tended to approach leading firms, known to provide good technical literature, and the use of technical data was often supplemented by discussions with representatives. Manufacturers, however, were criticised for considering their products in isolation, rather than relating them to the overall 'fit' of the building.

Three characteristics of recorded information significantly affected designers' patterns of use of written information:

(a)   information content of the document;
(b)   the presentation of the information;
(c)   accessibility of the information on the document itself.

The information contained in the document consulted was seen as valuable if it

(i)   provided an answer or confirmation of an answer to a problem;

(ii)   provided an understanding which would enable the designer to work out a solution;

(iii)   lead the designer to further sources of information;

(iv)   lead the designer to a redefinition of the problem.

The clouding of technical issues with promotional literature was disliked by designers, who in the main preferred the directness of visual material. The designer's assessment of whether the information in a document would be easy and quick to understand, absorb and transfer to a useable form was a major factor affecting his willingness to consult it.

Accessibility of information was seen as a major problem, and in half the offices questioned, design information of particular importance arriving in the office would by no means be guaranteed to reach the eyes of a designer who might find it of interest, and it was rare for the person servicing the library to take an active role in selecting or suggesting relevant literature for a project.

But experience, as in the first stage of the research, seemed to be the alternative to written information, unless failures had been brought to the designer's attention which called for some 'research'. When projects were pressing, particularly in 'design and build' projects, reading was at a minimum.

The research attempted to identify the kinds of experience which were useful sources of design information. There were broadly three:

o   experience of how a building is put together;
o   experience of performance (tried and tested solutions);
o   experience of practicalities (suppliers, reliability, costs etc).

If therefore experience was the main thing which informed design decision making, then from the point of view of both education and practice, techniques of enhancing experience as a resource was worth investigation.

The work showed that experience was developed in a number of ways:

- on an incremental basis, where designers pick out a few aspects of a design project for particular attention and increase their experience of these aspects through obtaining information and through the exploration and testing of design options. On the next project the designer will build in this experience quickly and concentrate on a new aspect;
- through familiarity with published information sources (often gained in architectural school) and the establishment of a pattern of consultation (keeping up to date);
- by working in multi-disciplinary groups or with outside consultants on larger projects;
- through discussion with manufacturers of building materials and components, particularly where manufacturers take responsibility for sections of design work;
- by undertaking work regularly of a similar order. In some firms there is the 'office expert' in certain fields;
- through feedback, although the technique was very patchy and in many instances did not occur at all and generally, in the case studies, only the larger firms had devised systems for passing feedback information amongst their designers;

- via in-house design guidance. This was only observed in larger practices for reasons which included larger resources, reputations to maintain, need to control quality of design, larger numbers of staff requiring more formal communication techniques and the design of more complex building types on which personal experience and information might be lacking. In-house guides were seen as cost effective, helping the designers to absorb good practice (as seen by the office) quickly, minimising the need for individual updating on each project. But where the guides existed they were often piecemeal and design experience was not automatically recorded and often did not have regard to emerging published information;

- by encouraging individuals to increase their experience by undertaking investigation (or 'research') into design problems and preparing design information for distribution within the practice;

- through CPD, particularly office-centred or in-house.

9. IMPLICATIONS

The actual amounts of <u>written</u> information used in projects was limited, but often because of lack of organisation within the offices to retrieve it, assess its significance, store it, up-date it etc, which had implications for management.

Experience, obviously related to feedback, was casual and if, as seemed clear, it was <u>the</u> main informing agent, then techniques of using it, recording it, ensuring its continuity and enhancing it were crucial. It had to be made more reliable and robust.

There was need also to increase the designer's <u>awareness</u> of the <u>need</u> to consult written information as well as awareness of the information itself. The rate of change of product and technical information with new design factors and constraints was rapid and designers could quickly be out-of-date. The additional incidence of liability claims added to the need to know.

There was also value which could be exploited formally in the 'self-interviewing' techniques used in the research, which although used for research reasons, was felt to have a potential for use in practice in three areas:

as an aid to designing
- as the basis of a search for design ideas or for literature sources at the beginning of a project;
- the process of recording could help to clarify the process of design in lengthy or unfamiliar design problems;
- forming accessible records of collective experience, including feedback data.

as an educational aid
- a source of material for designing in-house CPD events;
- to help structure a design problem;
- for the development of young designers' skills.

for documentation

REFERENCES

1. Goodey, J. and Matthew, K. (1971), 'Architects and Information', RP1, 110pp. Institute of Advanced Architectural Studies.
2. Mackinder, M. (May 1980), 'The Selection and Specification of Building Materials and Components - a study in current practice and educational provision', RP17. Institute of Advanced Architectural Studies.
3. Mackinder, M. and Marvin, H. (April 1982), 'Design Decision Making in Architectural Practice', RP19. Institute of Advanced Architectural Studies.
4. Mackinder, M. and Marvin, H. (July 1982), 'Design Decision Making in Architectural Practice'. BRE Information Paper IP 11/82. Building Research Station.

Section III
Legal/Professional

DESIGN:COST:QUALITY:PROFIT

THE ARCHITECT'S DILEMMA

RAYMOND J CECIL Dip Arch RIBA FRSA

It is most bizarre, that at a conference concerned with building
quality and its relationship with cost and profit, there should
be but one speaker who is an architect in private practice. The
list of 34 eminent speakers includes no less than 14 assorted
academics, five representatives of the public sector, 3 major
private clients, 5 sundry assorted surveyors and estate agents
- but only one private architect.
    Whatever may be the perception of the architect's role in the
building procurement process, it is the architect who is at the
sharp end of the cost/quality equation and also at the blunt end
of the responsibility for the final outcome. If the achievement
of quality results in an overspend of budget, it is the architect
who will be blamed - even if he had no part in establishing that
budget. If the building at some stage in its life does not meet
the expectations of its owner, its tenant, or its occupier or
if it causes damage to some remote passer-by, by reason of inadequate, or wrongly directed, expenditure, then legal action will
ensue. Wherever the main fault lies, one name will for certain
appear on the writ - that of the architect.
    In an increasingly consumerist society, the ways in which an
architect may be held guilty of negligence are almost boundless.
Without limit is the range and assortment of parties to whom he
owes a duty of care, and in many instances the care owed to one
party is directly opposed to the care he owes to another. And so
a situation has been created where architects have become acutely
aware of one further party to whom they owe a duty - themselves
and their families.
    There are those who scorn the 'Victorian virtues' and, in cert-
ain contexts, I would agree with them. But the Victorian concept
of professionalism was that a practitioner would put the interest
of his client before his own. And it is only by applying that
principle that buildings of true quality can be achieved. Thus
arises the dilemma that is the title of this paper. Today, the
architect who disregards his own interest in the pursuit of his
profession, does so at a level of risk that for most is foolhardy
or unacceptable.

BALANCING THE EQUATION

One conclusion that is bound to emerge from this conference is
that any of the decisions relating to cost and quality is a matter
for the fine balancing of conflicting requirements demanding value
judgements of the highest order. You cannot alter any one factor
in the infinite mix of factors in the building formula, without
affecting a whole range of other factors - some to their benefit

and yet others to their detriment.

The ultimate effectiveness of the cost:quality equation for any building is probably decided beyond redemption at the very outset. It is in the initial feasibility studies that the die is cast. That is where the most telling assumptions are made, preconceptions institutionalised, historic experience extrapolated into prophecy and the personal prejudices and predilections of the building owner achieve the legitimacy of considered conclusions. In too many cases, it is only after this stage that the architect for the project will be appointed.

The process is justified by the argument that until the nature of the project is established it is not possible to make a sensible choice of the most appropriate architect to design it. The converse is true. Until an architect has been involved in the problem, it is not possible to establish the most appropriate form for it to take. I would state without fear of valid contradiction that the involvement of an architect at concept stage is essential, and the best guarantee that the resultant building will meet the requirements of the owner and others in terms of cost and quality. I add "and others" because there are few building projects where the needs of only the building owner fall to be considered.

WHAT DO WE MEAN BY "QUALITY"?

Quality in building is a many-faceted concept and to each participant in the procurement process the facets are of differing size, shape and importance.

Principally, we are concerned with:

Functional efficiency
Quality of materials
Durability
Aesthetic delight

Functional efficiency includes not only meeting the physical needs of the occupants so that the building 'works' properly. It means also its effectiveness as a continuing protection from the elements, and from the use or abuse of its occupants.

Quality of materials is concerned not only with the choice of marble as opposed to ceramic tiles or emulsion paint, but also the standards of craftmanship to be employed in their execution. It is also much concerned with durability.

Durability presents a complex problem. Not only are we concerned with the effective life of the entire structure, but also with the individual components. 'Built to last a lifetime' may refer to the shell of a building but mastic seals require regular replacement, building services can be designed to last ten, twenty, or more years, functional requirements change with the passage of time and decor rapidly goes out of fashion.

Aesthetic delight. A subjective area where one man's meat is another's poisson.(sic) Aware of the difficulties, architects have for long put about the idea that good design costs no more than

84

poor design. We have done our own profession a grave disservice.
While the converse may be true, in that a bad design may cost
the same or more than one of excellence, the simple fact is that
good design costs money. There is no building that cannot be
cheapened by removing some of the features that contribute to
the excellence of the design.

THE COST OF BUILDING

There are many speakers at the conference who will seek to define
and differentiate between the various elements in the cost equation. Essentially this is a question of balancing first cost
against later cost. I avoid using the term 'life-cycle costing',
which I consider to be a process limited in its vision and confined to those elements of later cost which appear to be capable
of arithmetic manipulation. There are many elements of 'later
cost' that do not fit into the life-cycle formula at all, including the flexibility of the building to accommodate changing needs,
the impact of maintenance and replacement on the users of the
building, the development of new techniques, components and materials and so on. Life-cycle costing is a delicately fashioned
and lovingly embroidered blunt instrument, to be used with great
care and discretion, together with the incisive scalpels of experience, judgement and foresight.

THE IMPACT OF COST/QUALITY DECISIONS

The impact of the decisions relating to cost and quality extends
far beyond the parties responsible for making them. The social
implications may be very great indeed and we need look no further
than the local authority housing erected as recently as ten years
ago, which is today being massively repaired, altered or dynamited.
Architects have been widely condemned as a profession for the
failure of public housing, but most of the decisions which led
to the disasters were made without or against the advice of prac-
tising architects. The whole system of cost yardsticks and grant
or loan sanction, has within it the perpetuation of this kind
of difficulty.
  Particularly in the public sector, the separation of capital
and revenue budgets militates against truly cost-effective design.
The cost of maintenance is in direct conflict with other running
and staffing costs and in a period of centrally imposed fiscal
constraint, this may lead to the rapid deterioration of buildings.
We need to ask ourselves whether this central authoritarian control of expenditure should have been predicted and allowed for
in the building briefs created in the past.
  Within the public sector also we see conflicts of priorities
in the cost/quality debate. Civic pride, often no more than the
self-glorification of the political incumbents, may take precedence over the true and continuing needs of the local electorate.
In the fifties and sixties we saw the many town centre partnership
schemes where local authorities joined with developers to re-

develop their central areas. Mostly, those partnerships were
doomed to failure as they lacked the one essential ingredient
of all partnerships - identity of interest.

Latterly we have witnessed a new trend: the developer's
design/tender. Here, the local authority with surplus land will
invite developers to submit a money bid accompanied by design
proposals. The theory is that in this way the local authority
can fulfill its obligations to the electorate both in terms of
financial accountability and environmental control. To descend
to the vernacular, this is no more than a cop-out. Potential
developers, faced with no better than one chance in ten, twenty
or fifty will spend little on the exercise. The professionals
work without payment and are also aware of the odds against
success. Seldom does a local authority accept other than the
highest bid - but always stating that it is the best design.
The developer is then stuck with the visual aspects of the design
and in his desire for profit may set a budget that precludes the
creation of a building of true quality.

Political expediency may have its effect on budgets and impact
on quality. And it is appropriate to mention also at the time
this paper is being prepared, the effect of capricious and un-
thinking changes in the tax structure on carefully achieved cost/
quality balance in projects already in the pipeline. I am prepared to forecast a whole chain of building failures as the direct
and clearly attributable result of the way in which the Chancellor
changed the VAT rules in the last budget.

If anything, the private sector presents even greater complexity in arriving at a correct balance between cost and quality.
It is unfortunate that there is at this conference no representative from the development industry, or from private house-
building nor yet from the financial institutions and building
societies. We have speakers from three major private companies
who, by and large, themselves build for their own occupation and
use. They also happen to be users who are very much concerned
with their corporate image and the public perception of it.
Their buildings tend to be paid for from their own internally
generated resources and even if they need recourse to external
finance, their covenant is so good that they will generally be
given a free hand in deciding just what it is that they will build.

However, a great deal of building in the private sector is
commissioned simply as a marketable product, usually with funds
from some external source, often for the satisfaction of the unknown requirements of an unknown user who may purchase or lease
the product for some indeterminate period, taking on the commitment for its future maintenance and operation.

The three main areas of speculative development are housing,
commerce and industry and the cost/quality equation is different
in each of them.

In housing, as you will be aware from the helicopters and
magicians regularly appearing on your television screens, the
first priority is to create a packaging attractive to the consumer.

Wall-to-wall foam-backed needleloom assumes a greater importance than condensation or cold-bridging. Building societies, the last bastion of conservative design, have in recent years forsaken their insistance on georgian windows and mock-tudor embellishment but protected by inflation and an endemic housing shortage seldom have any real fears about the losses they may incur if they have to re-possess. The Defective Premises Act gives the owner protection for six years and the NHBC provides a ten year guarantee. If the owner expects his house to last longer than ten years then he can always look to the architect to compensate him if it does not live up to his expectations.

In commercial development the conflicting requirements of the interested parties assumes the complexity of a Ximenes crossword puzzle. Whether the developer trades as a dealer, selling on his completed projects, or as an investor who seeks to establish a property portfolio, there will be behind him one, or two, financial institutions. If he has sought building finance as a bridge then the institution will be concerned only until the project is complete. It will be concerned that if, due to the failure of the developer, it will have to take over the project and stand in the developer's shoes. It may often insist that the architect acknowledges his responsibility to them by way of a collateral agreement. It may not have the same robust appreciation of risks as the developer.

Institutions providing long-term mortage finance or with agreements to purchase the completed investment are far more concerned with its long-term performance. The institution may have its own quality criteria on whose inclusion it will insist. Seldom will the institution be represented on the development team, but remain a shadowy figure in the background. The building designers cannot fail to be aware of that institution and its requirements which may often conflict with those of their immediate client.

However, neither the developer nor the institutions will actually occupy or use the completed building. Neither will they, if the lawyers have performed properly, have the primary responsibility for its maintenance, servicing or repair. If the building does not meet the expectations of the lessee it is probable that he will seek redress from the building designers and once again the architect will be first in the firing line. It is axiomatic that the landlord/tenant relationship is one of conflict, but the architect has a duty to both.

Lastly, I come to industrial development. Tens of millions of square feet of industrial space have been developed speculatively in the last couple of decades, and I would venture to suggest that most of it has been poor value for money. Further, much of it is unsuitable for the use to which it has been put. The blame for this has to be laid at the door of the institutions, the vagaries of the Use Classes Order and the need for the developer to 'play it safe'. Seldom are the true needs of the eventual user given any real consideration. As a result we have industrial and warehouse buildings in their hundreds which are the wrong size and shape, incorrectly sited, inadequately serviced, that let in the water and are impossibly expensive to heat.

Finally I have to mention the role of the 'development consultant' better known to most of you as the estate agent. Because generally the development consultant receives no payment for his input into the design process - he relies on the commission he will receive for the ultimate sale or letting of the buildng. He is subject to a well-nigh irresistible temptation to influence the design in the direction of marketing features that will look well in his brochure. Thank goodness the architect owes no duty to the development consultant.

I have sought to show the many differing and conflicting interests that have to be served in the design of a building. Of those engaged in the process of establishing the priorities of the various demands there is none better qualified than the architect. The architect by his training and experience is the only professional with the breadth of understanding of the conflicting requirements and the ongoing consequences of the decisions. Is this, I wonder, the reason why his advice is so often not invited until after all the crucial decisions have been made. Is he felt to be an embarrassment to those who have a more limited view and more limited objectives.

THE B.P.F. SYSTEM

Earlier this year the British Property Federation produced its System for the Design and Construction of Buildings. It perpetuated most of the evils in the present system and in its desire to identify 'one backside to kick' sought to make the Design Leader responsible for the whole of the pre-tender design. It went further and made the Design Leader responsible for vetting and sanctioning that part of the design process that the client would decide should be carried out by the builder. Although it states that the architect's 'sanction' would not relieve the builder of his responsibility, we know the view that the courts would take in the case of failure, where third parties sought to recover loss.

The concept of 'contractor's design' is rooted in the belief that the builder knows better than anyone else how to build cheaply and effectively. Certainly, it would be a brave man who asserted that an architect's details are not, on occasions, unnecessarily expensive to construct. But the incentives to the builder to save money within the BPF System are bound to militate against quality. I do no more than warn employers who are concerned with quality in their buildings of the possibility.

THE ARCHITECT'S LEGAL VULNERABILITY

It is not the purpose of this paper to enlarge on the iniquities and inequities of the laws of contract and tort as they affect architects. Much has been written and said elsewhere and it is sufficient to say that in virtually every case concerning defects in buildings, one finger of blame will be pointed at the architect.

Very often, because he is the only participant still in existence and identifiable, he will be the sole target.

Even if, at the end of the day, it is held that the defect was wholly or mainly the fault of some specialist designer, or the defective performance of the builder, the architect will pick up some or all of the bill for damages. If the problem arose mainly because inadequate money was allowed for some particular element, seldom will the client, and never will the quantity surveyor figure on the writ. If the builder has built badly, the architect will be held to have 'supervised' negligently and may, as in the recent case concerning the Westminster Hospital, pick up 80% of the bill (20% was adjudged the responsibility of the Clerk of Works).

Of all the members of the building procurement team, the architect is the most vulnerable - not only to his client, but also to the financial institutions, future owners and occupiers, lessees, adjoining owners and passers-by. With the law in its present state there is not even any certainty that the architect can cover his risk with insurance.

IS THERE A SOLUTION?

Society requires buildings that work, that last, that look well and that keep out the weather. They should be energy conservant, easy to maintain, cheap to run, simple to adapt to changing uses, employ the benefits of developed technology but retain the familiarity of vernacular form. Above all, they should be cheap to construct because it is the capital cost of construction that is the deciding factor, all too often, in the decision whether the building will be commissioned at all.

Within sensible limits, all or most of these objectives can be achieved, given adequate time for resolving the brief and the inevitable conflicts within it. Widely reported building failures notwithstanding, the U K construction industry and its associated professions have an enviable record. But three important factors have to be recognised. Firstly, that any building begins to wear out, even while it is being constructed and its components do not wear out at an even rate. Secondly, every building is a monument to compromise. Money and effort devoted to any one aspect has inevitable and reciprocal effects on other aspects. Thirdly, all buildings are, in effect, prototypes. They cannot be sampled and tested in the manner of nearly every other manufactured artifact in our society.

Even were it possible, it would not be to the advantage of society to insist that every building be 'failure-proof' in all respects. The cost would be quite unacceptable. In every activity there is an acceptable level of risk. The problem with buildings as opposed to washing machines and dry shavers is that when the risk is realised as fact, then the burden to the owner - private, institutional or public - can be very heavy indeed. To attempt to deal with this burden through the law of contract or tort is not only unjust, it provides no certainty that the damaged party will be compensated or that the culpabale party (if there was a truly culpable party) will actually do the compensating. In my view, the correct way to deal with this socially acceptable level of risk is through insurance. Insurance written on a basis that does not allow subrogation or recourse back to the original building creators.

Unfortunately, as the law at present stands, such insurance is not available, nor would it serve to cover some third parties. Even if we achieve a change in the law which would allow 'project guarantee insurance' to be underwritten we have to ensure that it does not act as a dead hand on innovative design and the use and development of new technology. Left to their own devices, the insurers would have but one concern in the design of any building - the protection of their premium fund. Without sufficient safe-guards we could see that protection taking the form of excessive expenditure on foundations and structure to the detriment of other elements in the cost/quality equation. They would, in effect, be charging the building owner twice for the same protection.

The alternative also has to be considered - differential premia. This creates yet another dilemma for the architect who may not wish to be labelled as 'expensive to insure'. If he has confidence in his own judgement and ability he will simply have to spend time convincing his client that the additional premium is money well spent.

CONCLUSION

It would be wrong to conclude this paper without reference to the dilemma faced by clients. In the final analysis, it is the client who has to decide the order of his priorities and the proportion of his building budget that he wishes to allocate to each of the conflicting demands of his building. Inevitably, he will be presented with advice, requirements and demands from all quarters, both lay and expert. If he is an 'expert' client he may have the ability and resources to make the necessary judge-ments and decisions. Many clients, particularly the client with a single building need, are less well-placed. His biggest problem may be to decide to whose voice he pays most heed.

Speaking as an architect, it is perhaps predictable that I should lay claim to the voice that provides the most balanced advice. This is not mere professional arrogance, not yet a further petty blow in the battle for continuing leadership of the building team. It is securely based on the knowledge that of all the pro-fessions involved in the building procurement process, only archi-tects have an education that spans the generality of the building need and its satisfaction. Architecture makes no claim to be an exact science and few architects would claim to be experts, except in a most limited sphere. But they do have the ability to see the whole of the problem and to weigh the balances, the training to make decisions based on compromise and the experience to know the impact of those decisions. It is also a fact that the architect has to 'live' with his buildings longer than any other participant in the procurement process.

Which brings me back to the title of this paper. There is
little evidence that to date the decisions that architects make
are in any way influenced by their own personal interest and the
need to protect themselves. But as an expanding proliferation
of specialists and experts climb upon the construction bandwagon,
there is a real danger that this may change. We cannot produce
neat arithmetical calculations to justify our advice or
decisions, nor have we complex and baffling scientific formulae
defying challenge. If you will not trust us to make these
daunting decisions on the cost/quality equation and accept that
we may sometimes, for someone, be wrong, then we shall retreat
into a carapace of defensive design and protective legalistic
agreements. The built environment and society will be the poorer
for it.

THE ROLE OF THE BUILDING PROFESSIONS IN THE ACHIEVEMENT OF
QUALITY

D.J.O. FERRY, The New South Wales Institute of Technology

The building professions are the only people who can achieve
quality in the built environment and they have to face up to this
responsibility if they are to remain credible. In some respects
a lot of tightening-up has to take place, but in one key area
there is need for more than this as I hope to show.

First, however it is necessary to define terms. I will try to
deal with the easier one first.

What are the building professions? I am interpreting this
narrowly in one sense, to include only those concerned with
implementing the design and build process, and thus excluding
both planners and general practice surveyors.

This does not mean that I believe that these people have no
responsibilities, simply that boundaries have to be drawn around
problems if they are to be dealt with reasonably concisely. If
we don't do this we shall find we are having to consider the
client, government, financing institutions, the community.....the
ripples just go on expanding. In fact, we are well on the way to
compiling a list of those to whom the building professions' responsibilities are owed.

On the other hand the construction manager must be included as
a building professional, even though he is employed in a commercial
environment more often than not.

Mention of this provides an opportunity early in the paper to
dispose of the notion that it is the role of the independent consulting professional or public servant to be concerned about
ethical matters and that the person in a commercial firm is just
there to make money, by cutting corners if necessary.

If somebody, be he architect, engineer, quantity surveyor, or
builder, wishes to be considered as a professional he cannot
escape the consequences of that role by taking shelter within a
commercial ethic. A professional retains professional responsibilities wherever he is employed; they cannot be relinquished
without relinquishing right to the title.

One other person who must be considered is the professional
project manager. Very often he will be drawn from one of the
recognised building professions that have already been listed,

92

and will therefore bring his responsibilities as a building professional to his new role. However, sometimes the project manager may come from a real estate or business background, and responsibility may then fall on the building professionals in his team to see that building quality is maintained, and also perhaps to educate the new breed of project manager in this regard.

The definition of building professional was the easier one. The definition of quality in building is much more difficult. Ireland, in his thesis on managerial actions and cost/time/quality performance was able to measure the first two of his parameters quite easily, but had to resort to subjective assessment of the final one even though his work was restricted to one particular building type.

I doubt whether it will ever be possible to measure quality, although it may be possible to measure some of its attributes. It might even be possible to arrive at some sort of weighted index by assessing some factors and measuring others, but this would contain so many subjectivities both in the assessing and the weighting that its usefulness could be questioned.

An obvious test is to look for the use of high-quality (i.e. expensive) materials, components, and constructions. This is not in itself a valid approach; many of the University buildings of the 1960's, and much of the development around St. Paul's in the City of London, were expensively built on this criterion, but to my mind do not exemplify quality in building.

Fitness for purpose is perhaps a more promising line to pursue - the extent to which a building succeeds in meeting the aesthetic, functional, and cost objectives which were set for it. This does however pose the problem of the quality of those objectives. The satisfying of a brief whose purpose is to make as much money as possible and which calls for the cheapest building that complies with the building and planning regulations would be unlikely to produce a quality product in any ordinary sense of the term, no matter how brilliantly the purpose is achieved.

Whilst not completely turning my back on fitness for purpose I would therefore like to suggest an alternative approach, that quality is simply the end-product of properly qualified people taking care. I am using this phrase not in the too-often employed sense of exercising caution, but in its proper sense of caring about what one does.

This definition has the advantage of encompassing the quality of the design-and-build process itself, as well as that of the finished building, and covers the whole range of professional practice and client service including such matters as communication, keeping to time schedules, and cost control.

If everybody, not just the professionals but also supervisory staff and operatives, is adequately qualified for their job and does it in a caring way I submit that a building of quality is almost certain to result. If, on the other hand, everybody adopts a policy of doing no more than the minimum they are being paid for, and if possible less, then a quality building is unlikely to emerge no matter how lavish the specification or the budget.

This approach to the definition of quality may get over the difficulty of unworthy objectives, since these are unlikely to engender this caring attitude on the part of the people concerned with designing and building the project.

On the other hand it is improbable that a building designed and built with care by competent professionals will not be fit for its purpose, so that I have returned to this particular notion, but seeing it as a result of the quality criteria rather than as the criterion itself.

It is immediately apparent that the responsibility of building professionals for skill and care is absolute. It is what the community has a right to expect from them.

The architect has a primary role in ensuring building quality, but not the exclusive one which is so often assigned to him by other members of the team who see quality as his responsibility rather than theirs. Many years ago the architect and the craftsman (together with the craft-trained master builder) took a mutual pride in achieving quality. Today very little remains of the craft tradition in the building industry, and it would be unrealistic to expect semi-skilled process workers and installers, or financially-trained managers, to have inherited it.

Unfortunately the formal system of relationships between architect and builder in the standard form of contract remains much as it was in earlier days, and takes little account of this quite fundamental change.

On the whole the architect is looking for quality, but is often unable to achieve it. This is sometimes because he does not know how, a topic I will return to. But the virtual collapse of the trade system has left him in a fairly impossible situation on the ordinary price-in-advance contract, where he is required to state all his requirements in detail and check that they have been carried out by a builder who has little duty to do more than he is told, and who was appointed because he cut his price below that of his competitors.

Architects must however take some blame for designing buildings in the recent past (during the anti-human Brutalism craze in particular) that few builders could feel much pride in constructing.

As for quantity surveyors, some members of the architectural profession feel quite strongly that they have exacerbated the situation by encouraging private and public clients to set unrealistically low budgets.

In spite of my earlier refusal to define quality in terms of expensive materials and constructions there is no doubt that quality does cost money in the first instance (although it may lead to economy in the long run). However, nothing like the total initial cost necessarily falls upon the client.

The architect who only prepares 1/100 scale drawings, and sends them out to his favourite sub-contractors to do the detailing,

94

charges his clients no less than the architect who does a proper job. He is however usually able to drive a better car.

Similarly the quantity surveyor whose preferred method of working is to send out some sort of Bill of Quantities as quickly as possible and fix things up during the progress of the contract does not charge a lower fee than the quantity surveyor who takes the trouble to do things properly in the first instance. In fact if he is on an itemised fee scale he may even be able to charge more.

Lastly, the contractor who concentrates on getting the job finished as cheaply as possible, provided that the architect or the clerk of works do not object to what is being done, does not usually intend to share his saving with the client.

Leaving aside contractual responsibilities the architect's prime responsibility compared to other professions should also stem from the fact that he is usually the first to get involved. Care, or alternatively "couldn't-care-less", is extraordinarily infectious and quickly communicates itself down the line. Only psychopaths really want to spoil something that somebody else has devoted care and trouble to (vandals seem able to pick out unerringly the project that nobody loves or loved).

In contrast, a quality building engenders a basic feeling of well-being and users react accordingly. I think of a particularly happy building in this context, the Sydney Opera House - in spite of all the traumas of its construction the designers and builders never gave up trying. As a result the building is as fresh and exciting after ten years as when it was opened, and people just love being there or in its vicinity. This result is not the product of extravagant maintenance or tight security measures.

I can think of other large public buildings which in spite of both these latter attributes have a forlorn graffiti-scarred appearance, and which few people approach with any sense of pleasure. For example, there can be few more hateful environments around a prestige project than the public areas and subways around London's Centre Point, although London does have a few others that can run it pretty close.

The community should have a right to expect of the professionals in the industry that an expensively produced environment will be a pleasant environment.

Quality does not of course show itself only at a macro level. The reverse is more often true and the success of a building may be due to a whole succession of small things done well. To give praise to an architect who has so often been maligned, a Siefert office block may look like just another example of developer's architecture from a distance but a closer inspection usually shows careful detailing as a response to problems that could have been more easily "solved" with a mastic gun.

Quality buildings should have lower running costs than their non-quality equivalents, but in particular they ought to be relatively free of those major unforeseen maintenance costs

95

which arise from faulty detailing, specification, or workmanship, and which earn the building industry and its professions such a bad name.

The professions do not have a good record here. Members of the public find it difficult to understand how a building with rain penetrating its roof can have an architectural award plaque affixed to its entrance. So should we.

This brings me, not before time, to the question of what to do about all this.

Firstly is the need to bring home to building professionals generally their responsibilities in regard to quality, and particularly to those professionals who have always felt that this is someone else's worry. There is no point in being starry-eyed about this, but at the same time there is an underlying responsibility for good building which a professional cannot shrug off on grounds of expediency, profitability, or hierarchical instruction. Those who cannot accept this should stop calling themselves professional.

It is here that the quantity surveying profession has a special responsibility. There is a level of cost in relation to accommodation below which a building of quality just cannot be provided. Not only is the specification cut to the bone, but the low contract price means that even that miserable specification cannot be properly complied with if the builder is to avoid going bankrupt.

The quantity surveying profession in the past has not always been very assiduous in pointing this out. Worse, some would say that through the medium of such devices as the housing cost yardstick it has constructively encouraged this situation to develop.

The counter to this argument of course is that spending more money on a project may be encouraging waste and inefficiency rather than quality, and indeed if there is any truth in the allegations of the previous paragraph it could merely be that quantity surveyors, in helping to eradicate waste and inefficiency from design, have not always known quite when to stop.

Nevertheless I submit that the profession now has sufficient experience, and adequate cost control techniques, to be able to do this, and to ensure that if it advises not going below a certain figure for a project than that money can be spent in the pursuit of adequate quality rather than on feather-bedding of design or construction or on excessive profits. Again this emphasises the need for the professions to pursue quality in their own operations as one contribution to the quality of the final product.

It would be a good thing if the professional bodies would jointly commit themselves and their members to a responsibility for quality in building, perhaps through the medium of one of the national consultative committees.

However, the universities and polytechnics have a job to do, as well. In real terms they are responsible for the basis of

professional education, and the notion of care and quality in all
that is done in the workplace should form a major part of that
education, as well as the teaching of skills.

I am not sure that at present it forms even a minor part - I
have certainly seen few course proposals where responsibility for
total quality permeates the course in the same way as, for
instance, computer usage. Yet it should have an unchanging value
throughout professional life which makes it more important to
long term professional education than training on 16-bit rather
than 8-bit micro-processors. Here, I think, is where the
ultimate responsibility lies.

It is in this context that I would like to return to a most
important matter raised much earlier in this paper, concerning
the ability rather than the will amongst professionals to
produce quality in building.

As a by-product of University level education the situation
has unfortunately arisen where a thorough grounding in the
nitty-gritty of good construction practice is rarely seen as a
strong component of a professional course.

The development of spatial and social awareness and a knowledge of the various architectural movements has increasingly
dominated architectural education. Competence in construction
comes a long way behind spatial and aesthetic design performance
in the teaching, and particularly the assessment, of students.

As quantity surveying education has moved away from an
emphasis on taking off quantities and towards economics and
contractual matters a knowledge of sound building construction
at a level of detail has become less important. The need to be
able to conceptualise constructions in order to measure them
used to ensure that a quantity surveyor's education had a strong
construction component, often stronger than those of the allied
professions. Today this is can no longer be relied on.

Holders of building degrees, who might be expected to be the
obvious answer to this problem, are not. Their education is
equally academic (some would say even more academic) and tends to
emphasise science and management. In so far as construction is
taught it is done with a bias towards the construction process
rather than the performance of the finished building.

In the professional, and particularly the academic, rat-race
'Building Construction' has become the concern of the sub-
professional or the second-rater. To specialise in this and
make it one's main interest is the kiss of death for the
ambitious young practitioner, (or for the young academic unless
it be at an esoteric level). Such persons are most unlikely to
make it to the top in any profession or organisation.

The result of this is that a young architect may neither know
nor care that he is drawing or specifying something that is unlikely to work at all, or almost as bad, that will only work
given an impossible degree of care and luck in carrying it out,
and that nobody else is properly qualified to point this out to
him. The builder may advise him, but his knowledge is much more

concerned with buildability than with the long-term performance of the finished work.

In fact today's designers and constructors have access to a wider range than ever before of written material of high quality on construction topics including analysis of failures, from the Building Research Establishment and elsewhere. It is just that there is often insufficient interest or knowledge to assimilate and use all this.

Side by side therefore with the need to accept professional responsibility for quality is the need for proper training, professional structure and career prospects for the person who is a sound building technologist. All the professional bodies need to take an interest in this, not just the CIOB; in fact it is the traditional role of the Architect in particular that is in question.

However, like producing a rabbit from a hat, I put to you that the type of professional I have been talking about does actually exist. The problem is that he is rarely a member of the design and build team.

This is the Building Surveyor member of the RICS, whose education really does include a good building technology component, but whose services the RICS has never developed or marketed with the same enthusiasm as those, for example, of the quantity surveyor, and whose numbers, perhaps as a result of this, are not very large. This is a branch of the profession that should have a most important role in building procurement but which is rarely called upon. Unless the architectural profession is suddenly going to reverse the emphasis of its education, which I think is most unlikely and not necessarily a good thing anyhow, it is essential to provide a professional role for the Building Surveyors in design and supervision and to encourage an increase in their numbers.

For without the skills of such a person, and an adequate respect for them, all the good intentions and designs are at risk.

Meanwhile the professions generally need to consider whether the teaching of the less exciting aspects of construction technology to architects, builders and surveyors is ever going to fire the enthusiasm of a sufficient number of intellectually ambitious University and Polytechnic staff. If not, perhaps some means must be found to teach this subject in the Technical Education sector, many of whose staff could do it competently and enthusiastically.

REFERENCE

Ireland, V. (1984), *The Role of Managerial Actions in the Cost, Time, and Quality Performance of High-Rise Commercial Building Projects*, University of Sydney.

A LEGAL CONTEXT FOR QUALITY AND COST IN BUILDING DESIGN

PETER HIBBERD, Bristol Polytechnic

INTRODUCTION

It is first necessary to establish what is meant by a legal context
for quality and cost. This in itself present major problems, for
such terms as legal, quality and cost are difficult to define. This
paper concerns itself with how quality and cost maybe effected by
the legal framework upon which the procurement of the building is
based. It will be seen that building design in this context is not
simply a pre-contract stage but the total operation of creating a
building on site.
    Quality is taken to be the standard required for a building proj-
ect and cost is taken to be the amount paid by the building client/
employer for that project.

LEGAL FRAMEWORK

The impact of an adopted legal framework, that is, the building
contract, upon the quality of the building project and upon cost is
dependent upon the following:

    i)   the inherent nature of the framework
    ii)   the ability to judge quality and evaluate cost against a
given set of criteria
    iii) the will and or ability to achieve a particular end
result.

Although, cost and quality are generally related it is necessary
for the purposes of analysis to separate them. It seems logical
that first consideration should be given to quality, for once a
standard has been set it can then be given a cost. Although this
approach may seem logical this paper first considers cost and then
quality and the reasons for doing so will become apparent.

COST

Frequently, the building client will establish a capital cost budget
for a particular project and at the same time communicate his
broad requirements. This means the design team are then required to
design within this budget to provide the general and any particular
requirements specified, achieving whatever standard is possible for
the money available. In this scenario one is not concerned with
predicting cost but rather predicting what quality is achievable
for a particular cost. Great difficulty has been experienced in
approaching the problem this way round and many have resorted to
producing a design with an attributed quality, simply for it to be
costed and trust that it will approximate to the building clients
budget.

COST PREDICTION

The impact of a legal framework is dependent upon the stage at
which cost prediction is being attempted. In turn the stage of
cost prediction is determined by what it is we are endeavouring to
predict. Is it the tender sum, a range within which the tenders
should fall, the contract sum, the final account sum or the costs
in use? In practice many forms of cost prediction are attempted but
logically only two matter, that is the final account figure and the
costs in use.
    Once a contract is let the impact of the legal framework is
established even if not generally apparent for it governs how the
final account is to be calculated. That is, if the strict legal
definition (if known) is adopted as opposed to the reality of a
negotiated compromise.[1]
    The impact of a legal framework upon cost prediction in the
earlier stages, that is, before a contract is let, is not that
obvious and is deserving of much more consideration than has hitherto
been given.
    The legal framework is contained within the building contract and
is reflecting the chosen contractual arrangement. It is the contractual arrangement which should determine the method of tendering and
also the extent and nature of the tendering documents. These
tendering documents may ultimately constitute the contract documents
but this is not invariably the case.
    The tendering documents set out the obligations and risk to be
taken by the parties to the contract. The contract documents
establish the legal obligations and risks and the apportionment of
these obligations and risks will to some extent be reflected in the
price quoted by the contractor. However, the apportionment is not
always expressly stated nor entirely considered[2] and some obligations
and risks may be implied. Unfortunately, agreement as to these
implied matters is seldom forthcoming and will frequently involve
legal argument once a contract has been entered into.
    It is possible and is frequently the case, that cost prediction
can take place by using a model that does not take account of the
contractual arrangement. But, the question is not whether it does

but whether it should. Clearly, if cost is affected by the contract
form then the contract form must be a variable and a component of
any cost model.

The contract form will affect many issues which a good deal of the
cost prediction attempted actually ignores. It cannot be one simply
fails to recognise their significance, it is more likely to be the
nature of the problems and the difficulty of evaluating them. The
contractor however cannot ignore such matters as;

    i)   when and how he gets paid,
    ii) how long is the contract period,
    iii) what is the level of retention, liquidated and ascertained
damages,
    iv) what are the insurance requirements,
    v)   what quality is required,
    vi) is the building occupied,
    vii) are there any restrictions or requirements as to the order
of the works or hours of operation,
    viii) what access is available,
    ix) are fluctuations provided for,
    x)   is a performance bond required,
    xi) nature of the variation clause.

It is quite clear that all these issues have an impact upon the
contractor's tender and therefore if the design team ignores them
in their cost prediction they do so at there peril. The design team
may believe the overall effect of all of these issues is less than
the percentage range one could normally expect to achieve in their
predicted cost, but that range may be generally wider than it need
be simply because of the absence of proper consideration.

An early decision with regard to the type of contractual arrangement is therefore essential. Anyone confronted with the problem of
predicting cost for a project where the contract variables are
known would make some attempt to evaluate them but more frequently
the cost prediction is made in total isolation of such matters with
at best some percentage addition for preliminaries. Such a
percentage addition seldom has any genuine scientific base save
that it is based upon general observation, which constitutes much of
ones professional judgement.

METHODS OF CONTROLLING COST

The control of costs can be viewed from two distinct points, that of the client and the building contractor. Traditionally, the independent design team have not concerned themselves over much with controlling the contractor's costs believing this is really the contractors concern. Although it is clearly understood that a contractor's price is not in itself determined by his costs, it has a very significant part to play and it is only by understanding how these costs are incurred can the design team hope to realistically appraise the costs of a project. One is concerned with controlling the cost to the building client by determining the cost to be incurred by the building contractor. Unfortunately, this still has the inherent problems that a contractor may still have sufficient flexibility to enable him to tender or to recover more than his costs plus profit.

The client's ability to control cost is dependent upon:

(i)   the design team
(ii) the contractual arrangement adopted

The design team is considered first because it will in a majority of cases determine the contractual arrangement. In the choice of the contractual arrangement they are establishing the broad legal framework and as they move towards creating the legal contract they establish the precise legal framework to be eventually adopted. The choice however although having far reaching effects is not generally approached with sufficient consideration. Once again, one is left with the problem referred to earlier, that of not being able to accurately evaluate the effect of certain choices, leads one to choose, if not indiscriminately, without real knowledge, or any degree of certainty.

The design team needs to accept and be fully committed to controlling costs, they must accept the discipline of cost and consider advice given on costs by members of the design team. It is essential to appreciate that the preparation of any project documentation must communicate clearly and unambiguously the project requirements. Even though this goal may be unobtainable it must be sought because without the desire to precisely describe ones intention, progress will not be made. Therefore, it becomes necessary to create a system of procurement which is clear and reduces subjectivity to a minimum by replacing it wherever possible with objective tests.

Nevertheless, the end result even though clearly prescribed may not be achieved because the design team does not possess the desire to achieve it or alternatively the sanctions available are inappropriate. The first is a matter of attitude, the second a matter of contractual terms.

The contractual terms are part of the contractual arrangement and determine by their nature the ability to control cost. In the first instance, that is at pre-contract stage, one is, by selecting the specific terms, creating a framework upon which the cost to the

102

client will be based. As previously stated, this will be significantly related to the contractor's costs but later, that is post
contract stage the control of costs to the client should be solely
dependent upon the legal words used. The fact that this is not
always so has been expressed[3] elsewhere.

The reason why the design team still concern themselves with
predicting the tender sum is an enigma.

The answer may lie with the increased difficulties of predicting
the final account sum but this is in truth an excuse not an answer
for what is the point of prediction even if given with 100%
certainty if the issue predicted has little bearing on what is the
eventual and final outcome. If the tender price equalled the final
account sum there would be no problem, therefore is this something
for which we should strive, that is, to bring the tender price as
close as one can to the final account sum. Or should we accept
that the only real issue upon which we can attach any importance is
the final account.

If fixed price meant what it means to the layman there would be
no need at post contract stage to control cost except for variations.

Regretably, fixed price does not mean what it says for it may
contain;

   (i)   prime cost sums
  (ii) provisional sums
 (iii) provisional quantities
  (iv) dayworks/contingency sums
   (v) errors in documentation
  (vi) fluctuations
 (vii) sundry areas of adjustment
(viii) loss and expense clauses

within its legal framework. Clearly, the prediction of any tender
sum which includes the above provisions is of little use, for it
informs the client of very little.[4]

SPECIFICATION

Once a framework of unit rates has been established it is more
likely that the designer will be able to choose the most effective
material in a given situation, unfortunately, traditional processes
of tendering inhibit this and such a framework is generally not
forthcoming until a contract is about to be let. Not too late to
make adjustments to the design solution but often ensuring that
designs are hastily altered without time for proper consideration of
the overall effect of such a change.

The type of problem where the designer only has historical data
or current data but not related to the project nor the individual
contractor(s) is all too common.

(i) Blockwork is chosen instead of brickwork but the lowest tendering contractor sees it otherwise.

(ii) the use of one material rather than a number, believing that quantity discounts will be created only to find that no additional discounts accrue and the use of another lower cost material for part of the works would have been beneficial.

(iii) the selection of ready mix concrete where the contractor may have otherwise mixed his own.

(iv) the use of only specialist sub-contractors when the main contractor may usefully have tendered.

(v) the historical cost of stud partitions is used yet the designer specifies for the members to be framed.

It is clear that these problems are difficult to avoid in the traditional tendering processes and that to overcome them the contractor is required to become involved in the design process at an earlier stage. Thus creating differing contractual arrangements and which ascertain design responsibility in differing ways.

QUALITY

The most important questions with regard to quality in the legal context, are:

(i) to what extent is the quality of a building prescribed by the contract documents.

(ii) how can one establish whether the prescribed standards etc., have been achieved.

(iii) what sanctions are available to the building client and building designer where prescribed standards have been adjudged not to have been achieved.

TO WHAT EXTENT IS THE QUALITY OF A BUILDING PRESCRIBED BY THE CONTRACT DOCUMENTS?

Quality is generally specified either in the Specification or Bills of Quantities and these documents will constitute a part of the contract documents upon which the project is based. However, where both documents exist it is necessary to establish whether one or both of the documents is to be a contract document. If a Bill of Quantities is to supersede a formal Specification[5] it can and often does create ambiguity if this same Specification becomes available to the contractor. From this ambiguity conflict can arise and that can come about simply because the issue of the Specification in these circumstances was thought to be helpful. Helpful because it contains information in addition to that of quality. Where a Bill of Quantities does supersede a formal Specification[6] it is necessary that the Bills state fully the prescribed quality and this is generally achieved by reading the preambles together with the item descriptions but again this separation can also lead to ambiguity.

In either event, whether using a Bill of Quantities or Specification, is the quality of the building prescribed? The answer is, as a rule, no, for the quality of the building itself is not described, only the component parts are described and this is generally true even where using a performance specification. It can be said that if all the components are adequately described and the standard is achieved then the quality in the whole building is equally specified. This in practice, however, is often not seen to be the case because it frequently comes down to how the components come together and whether a successful interface has been achieved and it is in this area where specification is generally lacking.

It may be that this deficiency is deliberate and recognises that certain aspects of the design process are left to the contractor. However, although this appears to be a logical explanation for what happens, in practice most designers would not acknowledge the contractor's design element in contracts of a traditional nature. Therefore, a situation arises which creates great potential for discussion on site and possibly dispute.

The designer will in many instances simply specify the materials to be used e.g.

Half brick skin of hollow wall in LBC Heather facing bricks in gauged mortar (1.2.9).

and additionally the finish to be achieved e.g.

pointed with a neat recessed joint.

In this instance, nothing else is specified. The question now is will the contractor produce an end product which matches the specification as written and will that in itself be sufficient. The contractor may produce a half brick wall in the specified brick using the appropriate mix of mortar and with the joint as described and produces a wall where:

   (i)   the bed joint is uneven
  (ii)   the perpends are not in line
 (iii)   the bricks have not been mixed
  (iv)   the bricks have chipped arrisses
   (v)   the bricks were laid with frog down
  (vi)   the outer skin of the hollow wall was raised to its full height before the blockwork inner skin was erected
 (vii)   salts have appeared on the mortar joints
(viii)   imported cement which had been incorrectly stored had been used

The workmanship and precise materials have not been specified so what can the designer reasonably expect. Duncan Wallace submits that a contractor undertaking to do work and supply materials implies that he will:

   (i)   do the work with care and skill or, as sometimes expressed, in a workmanlike manner
  (ii)   use materials of good quality. In the case of materials described expressly this will mean good of their expressed kind. (In the case of goods not described or not described in sufficient

detail, it is submitted that there will be reliance on the contractor to that extent

(iii)   that both the work and materials will be reasonably fit for the purpose for which they are required unless the contract excludes such an obligation.[7]

Clearly, it can be seen that the law of contract will imply such terms into a contract and therefore it is not absolutely necessary to expressly include them. This position was also stated in the Defective Premises Act 1972 in respect of dwellings. Nevertheless, would a designer be able to rely upon such implied terms to overcome his objections to the points raised in (i) - (viii). To some extent he would but clearly the burden of establishing such breaches would rest with the designer and where express terms have not been used it will be much more difficult in practice to achieve.

Interestingly, the situation posed raises three distinct aspects to specification, that of

(i)   specifying materials and goods
(ii)   specifying how the materials and goods shall be put together
(iii)   specifying the desired end result.

Each of these aspects of specification may be used individually or in combination, thus creating many legal frameworks in respect of the work. The mere specification of materials and goods has the problems already referred to. The additional clauses which tell the contractor how to use the materials and goods will overcome some of these shortcomings but will inevitably create others. The major issue being that such clauses often impose unnecessary restrictions where acceptable alternatives apply or where contractors are required to comply with a clause generally even though written originally for a specific instance. On balance it would seem that specifying the desired end result has the most to commend itself from a legal point of view. One simply looks at the end product and whether it has achieved the specified requirements. However, in practical terms it will generally be that all three aspects of specification are used on the same project and this may be justified but clearly it requires much greater thought than it is currently getting. It means that the designer in all situations must be able to comprehend the work of the craftsman.

HOW CAN ONE ESTABLISH WHETHER THE PRESCRIBED STANDARDS ETC., HAVE
BEEN ACHIEVED

Obviously, if objective criteria can be established it becomes a
matter of fact whether the standard specified has been achieved but
unfortunately there is much subjectivity left in requirements
contained within specifications. For example, the terms implied by
law, such as, care and skill, workmanlike manner and good quality
all require subjective tests to establish whether they have been
achieved.[8] This subjectivity is also extended further by express
provisions contained in the specification of the works, for
instance;

   i)   the best of their respective kinds,
   ii)   to the architects reasonable satisfaction,
   iii) subject to written approval,
   iv)   or other approved[9]

   Such subjectivity will obviously create problems because the
contractor is required to make an assumption based upon his
interpretation of the Clauses used. This will have some cost
significance and when he establishes that the designers interpretation is different and this is invariably the case, the issue of a
claim manifests itself.
   The use of the term 'best of their respective kinds' does not
necessarily mean best quality available as can be seen from the case
of Cotton v Wallis (1955) where it was decided that the terms have
to be considered in the context of the price paid for the works, thus
creating great potential difficulties in practice.
   The terms to the architects reasonable satisfaction, reasonable
skill and care, and good quality have been said to objective[11] tests
and looked at from a strict legal point of this view this is true
because a standard can be determined. The legal distinction is
between the architects opinion and taste, the latter cannot be
adjudged against objective criteria whereas reasonableness can be.
All manner of problems are however created because the contractor
would automatically interpret that where the work is to the architects
satisfaction for example, his satisfaction must be reasonable but this
does not necessarily follow. Therefore, ones view of what is
objective criteria should be, that which has an absolute standard
which both parties to the contract can perceive at the outset as the
same thing. Conversely, subjectivitity arises when the parties
cannot establish such a test and this situation should be avoided.
However, simply because a test is not established prior to letting
the contract does not mean that the specification is going to be a
matter of subjectivity because the designer may well have the
contractual right to establish objective tests before giving his
approval. Although not a problem of subjectivity it is a problem
nevertheless for should the contractor be expected and indeed should
he price work whereby the standard to be achieved is not absolutely
determined.

An example of this type of issue is the use of a sample panel, whereby the sample panel is prepared after the contract is let. The contractor has priced for work, the standard of which has not been established, it is in his view subjective whether the sample panel he has erected is acceptable. However, once the sample panel has been approved an objective test is made available against which all work of that kind can be judged.

In ones endeavour to establish specification one can always attempt to spell out everything a contractor must do and where this is to be the case one should examine the efficacy of such requirements. For example, the end product although appearing to be the same will in fact last longer because of the way it has been applied, in this instance it can be justified.

When considering the extent of specification it is always worth remembering that even what appears to be precise specification is often ambiguous for instance:

"the hardcore to be blinded with 25mm (consolidated thickness) of sand blinding".

Does the sand fill the interstices to a depth of 25mm or is to be 25mm above the formation of the hardcore. If it is the former how can it be established that 25mm has been achieved.

WHAT SANCTIONS ARE AVAILABLE TO THE BUILDING CLIENT AND BUILDING DESIGNER WHERE PRESCRIBED STANDARDS HAVE BEEN ADJUDGED NOT TO HAVE BEEN ACHIEVED

Any legal framework which is to be effective must not only establish clearly what is required but must also have an effective means by which any breach can be remedied. An effective means by which a breach can be remedied is essential for if sanctions cannot be applied the value of a tight specification is dubious. Conversely, effective sanctions are of little benefit if one is uncertain, because of ambiguity contained within the specification, as to what is required and the significance of what is required.

The independent designer must have a clear authority contained within the contract to enable him to:

i) require compliance with the specifications.
ii) require defective work to be removed and re-executed.

In theory this power should only be given where it can be exercised without creating the conflicts that will inevitably exist if the specification is unsound. This will extend to situations where precise specification is given with regard to operations to be performed but the contractor is unable to see any useful purpose in what is prescribed, having achieved or being able to achieve the desired end result by other means.

However, in practice this power will be given to the independent designer and where it is great care should be used in the exercise of such power.

CONCLUSIONS

No cost predictions should be attempted without first considering and establishing the contractual arrangement to be adopted.

No cost predictions should be attempted without a clear impression with regard to the quality required.

More attention should be paid to final costs as opposed to tender sums.

Quality can sometimes only be effectively established by reference to physical matter and not left to words alone.   All tests should be objective and where a physical specimen is deemed appropriate it should be made available before creating a legal contract.[12]

Detailed observations are required in respect of the effect of contract clauses and the choice of contractual arrangements.[13]

Contractual frameworks cannot in themselves achieve a desired end result, one is still reliant upon individuals and bearing this in mind greater integration of design effort is therefore required.

REFERENCES

1.   The Chartered Institute of Public Finance and Accountancy (1978)
        'A Review of Building Construction Practices in Local
        Authorities (London: CIPFA), p.37.
2.   E.G. JCT Minor Works Agreement 'Standard Form of Contract'
        (London: RIBA)
3.   The Chartered Institute of Public Finance and Accountancy (1978)
        Op cit p.37.
4.   Hibberd P R 'Building Contracts - Variations'
        (UMIST 1982) Tables 26-32.
5.   Wallace I N 'Hudson's Building and Engineering Contracts'
        (10th Ed) (London: Sweet & Maxwell) p. 205
6.   JCT (1980) 'Standard Form of Building Contract'
        (London: RIBA) Clause 8.1.
7.   Wallace I N Op cit pp. 274-275.
8.   NEDO (1976) 'The Professions in the Construction Industry'
        (London: HMSO) p.10.
9.   ICE 'Conditions of Contract for Works of Civil Engineering'
        5th Edition; Clause 13(2). (London ICE)
10. Weekly Law Reports Volume 1 p.1168.
11. Wallace I N Op cit p. 415.
12. Davison, Dr J A (1981) 'Computer Aided Methods in Building
        Design' GMW Computers Ltd. Berkhampsted.
13. Bromilow Dr F J (1969) 'Contract Time Performance,
        Expectations and the Reality' Building Forum, Vol 1 No 3 p.70.

Section IV
Developer's View

THE INFLUENCE OF THE PROPERTY MARKET ON BUILDING DESIGN

ANGUS McINTOSH, Kingston Polytechnic

1.   INTRODUCTION

The structure of the commercial property investment market in the
United Kingdom does not necessarily promote good design. Much modern
building design has become answerable to the investment criteria of
investors particularly financial institutions. In the final analysis
it is often the property's investment market value which is far more
important than a building's cost of construction or it's "architectural qualities".

This paper aims to show firstly how good and bad design may
influence a building's market value, secondly why the property market's
forces and it's investment appraisal methodology may be leading to bad
design and thirdly that there are signs that the commercial property
market is changing which may lead to influences resulting in better
design.

2.   THE COMMERCIAL PROPERTY MARKET

A building is a product and like any product it has a producer and a
consumer who may use the building for a few days or several thousand
years. The consumer may or may not be known to the producer. It may
in fact be an individual or an organization. Whilst houses are
normally produced for unknown individuals, churches are produced for
a known organization. Buildings produced for the commercial property
market have to appeal to a wide range of "uses". Before considering
these it is important to distinguish between a "building" and a
"property". A building is concerned with a physical entity. A property, on the other hand, may or may not contain a building. Agri-
cultural property, for instance, may be devoid of any physical
structure. The main distinguishing feature of property relates to
it's legal tenure, for instance, a property may be "owned" on a freehold, leasehold or licence basis. The type of legal property ownership and the motive for that ownership often have a considerable
influence on a building's design.

The UK commercial property market is principally concerned with
offices, shops and industrial buildings.(1) Within that market there
are a number of building owners.

The freehold owner who also occupies his own building is perhaps the most clearly established. This type of ownership expanded during the industrial revolution and still dominates the industrial property market. This type of ownership has considerable control over the design of the building. This is well illustrated by IBM's building north of Portsmouth.

The system of leasehold tenure by occupiers has expanded considerably in recent decades. Generally speaking this type of ownership has only a residual direct influence on building design; the influence being limited to internal design and decoration for instance in an office building or a shop within a shopping centre. The introduction of mirrors, lights and escalators in the Topman-Topshop chain illustrates this influence.

The third influence in design comes from ownership by developers. Some development companies own a freehold or long leasehold interest in a complete development which enables them to considerably influence the buildings design. The new Ridings Shopping centre in Wakefield, partly owned by Capital and Counties plc., and the Brent Cross Shopping Centre partly owned by Hammerson clearly illustrates this. However many developers own commercial property on a transient basis. They acquire a legal ownership, undertake a development perhaps in partnership with a financier and then sell their interest. Some don't even take a legal interest but work on a fee basis only. These developers are often beholden to the source of their funds or the eventual owner of the investment.

Over the last decade the dominant financial design influence in the UK commercial property market has come from investing institutions.(1) Although in the last 12 months the figures have declined, in recent years £2,000 million per annum has been invested by life insurance companies, pension funds, managed property funds and property unit trusts. A vast number of architects and quantity surveyors have found themselves working directly or indirectly through a third party developer, for these property investors.

Owner occupiers, tenant lessees, developers and institutional investors each "own" buildings in different ways but, perhaps more importantly, with different motivations for that ownership. The design criteria of a building owned as an investment are very different from those of an owner occupied structure which is required as a factor of production process.

Whilst architects are concerned with "design" and quantity surveyors with "cost", an investor is concerned with "value". The irony is that a property's value may have very little to do with it's cost of construction or it's design. What do a market stall and a cramped City office have in common? The answer is a desirable location for their respective consumer markets. The value of an investment property depends far more on it's location than it's design. The tenure of the investment, the nature of the occupying lessee and the actual legal documents which bind the various parties to the building may also be far more important than how it's constructed. However, design can and does play an important part in the investment performance of many investment properties.

3. DESIGN AND MARKET VALUE

Every major town in the United Kingdom now has buildings owned by
investing institutions. Retail investments are found in the high
streets of every regional shopping town. Office investments tend to
be found in the southern half of Great Britain whilst industrial
investments, which comprise primarily warehouse accommodation, are
found predominantly in the south-east of England. The principal
objective of the institutions is to purchase those buildings which
are easily lettable, are flexible in their use and will show good
"rental growth".

To illustrate how bad design can adversely effect the value of
buildings I intend to consider several office buildings which have
been available to let in Kingston upon Thames during the last 12
months. Kingston is a good office market to consider because whilst
there is reasonable demand for office accommodation, that demand is
not totally beyond the level of supply of new buildings. Demand is
not as strong as in the City of London yet it is not as poor as in
towns like Portsmouth and Southampton where some offices have
remained unlet for many years. In other words, because the supply of
and demand for new office accommodation is reasonably in balance,
good design can play an important part in determining a building's
and hence a property's value.

The following is a precis of an investors office design criteria.
Externally the building should be set away from the road with some
landscaping. However both landscaping and the external appearance
needs to be designed for low maintenance. The critical point of the
whole building is the entrance. It should be at street level but
enable a taxi passenger to alight within a few feet of the principal
doors. It continues to amaze me the number of buildings constructed
which forget these basic but important features.

Except for the central area of London every office building should
be constructed with adequate car parking. The objective should be
one car space per 300 sq.ft. (30 m^2) of lettable accommodation.
Planning authorities who only allow one space for every 2000 sq.ft.
have much to answer for although they are rarely blamed when office
buildings continue to remain empty.

Internally the entrance hall is the most important part of any
office building. The catch phrase is that a building never has a
second chance to make a first impression! The quality of the lighting, the finishes and seating for visitors are all very important.

The internal finishes of the lettable area are also important and
must nowadays contain an acoustic ceiling, lighting at 500 lux and
carpeting throughout.

Double glazing and air conditioning are normally only applicable
for buildings in city centres. The additional construction and
running costs make such refinement uneconomic in many instances. In
other words the buildings capital market value is not increased by as
much as the additional cost.

Although a floor should be designed to take an imposed load of 100
lbs per sq. ft. (4.8 KN/m^2) it must contain some form of service
ducting for telephones and computer cables. For this reason raised
floors are becoming increasingly popular.

MILLENNIUM HOUSE
KINGSTON UPON THAMES

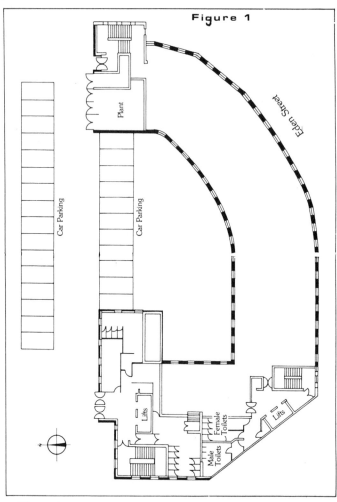

Figure 1

Floor Areas

Ground Floor -	Entrance Hall and Reception Area
First Floor	7600 sq ft
Second Floor	7600 sq ft
Third Floor	LET
	15200 sq ft

Any building which contains more than simply a ground and first floor must contain a lift. Adequate lifts are essential as without them the upper floors have a diminishing value and may in some instances become unlettable. On each floor toilets are needed for both male and female staff.

Perhaps the most important consideration in the whole design process is the actual size of the office module. Generally a 15 metre wide rectangular building is far preferable to a square enabling both sides of the building to enjoy natural daylight. Flexibility is the key word as the building must be capable of being divided up. The position of the window mullions for partitioning, the location of the toilets and the position of heating units especially air conditioning outlets are most important in this respect.

The concept of flexibility underlines the design difference between an owner occupier and an investor. The former may have a specific purpose in mind and will wish to have a building designed accordingly. An investor, on the other hand, will require a building designed to accommodate a variety of different office tenants' needs.

The example (2) illustrated in Figure 1 is the floor plan of an office building which has been to let in Kingston for some years. It does not have direct vehicle access to the main entrance, the entrance is subservient to the main design consideration on the ground floor which is a shopping mall and the banana shape of the building makes it extremely difficult to divide up with partitioning, a factor which is further exacerbated by the location of the structural columns within the lettable area.

Not only has the building taken a long time to let, the asking rent is only £7.50 per sq.ft. pa. Whilst this building has been on the market other buildings have been constructed and let at rents of £10 per sq.ft. pa. and more.

4. DESIGN AND FINANCIAL APPRAISAL

The design of a commercial building falls into three distinct categories; aesthetic, function and cost-in-use.(1) It is the subtle but correct mix of these three factors which may or may not enhance the value of a commercial building. Generally speaking an owner occupier will place greater design emphasis on the aesthetic and cost-in-use aspects than a property investor. To explain why this should be so we need to briefly examine the method employed by the UK property market to value a building.

Valuation 1

Net lettable area	5,000 sq.ft.	
Rack Rent	£8. per sq.ft.	
Rack Rent at Value (net)		£40,000 pa
Years Purchase		
in perpetuity @ 6%		16.6
Gross Capital Value		£664,000

The assumption is that with a bank offering an interest rate of say 10%, the above building should experience rental growth to compensate the investor for the yield gap between 10% and 6% used above. Every 5 years there would be a rent review and the property would normally be let for 25 years with the tenant being fully responsible for repairing, maintaining and insuring the building (a FRI basis). The critical parts of the appraisal are the rents at which the building is let (£8 per sq.ft. - valuers live in an imperial world) and the capitalization yield which the valuer intuitively judges to be correct (6% in the above example). A poor design reduces the rent a tenant is prepared to pay for the accommo-dation but may also colour the valuers judgement when he comes to value the property. The most important design consideration is how well the property has been constructed to meet the function for which it has been built. The efficiency of the heating system, the quality of thermal insulation and the external appearance come lower down the list of priorities. The same would not be so of an owner occupied building.

The above valuation is very straight forward. It is it's very simplicity which makes it attractive and widely used in the invest-ment market. Now imagine the landlord had let the property on a gross basis at say £11 per sq.ft. The calculation to arrive at the net rent might be as follows

Valuation 2

Year One
Gross Rent (per sq.ft)		£11
Less		
Cost of Heating	£1	
Cost of Repairs	£1	
Cost of Insurance	£0.50	
Total Costs		£2.50
Net Rent		£8.50 per sq.ft.

Year Two
Gross Rent		£11
Less		
Cost of Heating (increase of 12%)	£1.12	
Cost of Repairs (increase of 8%)	£1.08	
Cost of Insurance (increase of 5%)	£0.525	
Total Costs		£2.725
Net Rent		£8.275 per sq.ft

Year three and each subsequent year would produce a number of different net rents. In other words letting the property on a gross basis makes valuation difficult. One cannot simply capitalize the initial net rent but a complex discounted cash flow calculation needs to be undertaken taking into account every year's income and all items of expenditure. The net income would be even more erratic if a major item such as a lift or the air conditioning plant needed replacing.

If investors did let buildings on a gross basis they would have every incentive to keep their repairs to a minimum and reduce the heating expenditure by improving the efficiency of the heating system and the level of thermal insulation. The facts are that this is NOT how the property investment market generally works. Landlords normally abdicate all direct financial responsibility for the main-tenance of their own investments. Tenants are normally required to repair and maintain but NOT improve the fabric of the buildings they occupy.

This problem is further exacerbated by the property markets pench-ant for 25 year leases. Investors who purchased offices in the mid 1960s which were let for 25 years, today find themselves still locked into a lease structure and owning a considerably depreciated invest-ment. They are unable to move in and renovate the building except after undertaking the often expensive exercise of buying out the encumbent tenant.

Depreciation may take various forms. There may be aesthetic, functional or locational obsolescence or the building's cost-in-use may increase. The property market takes these aspects of depreciation which often occur as the building ages, into account by increasing the capitalization yield. In the example used, the yield of 6% might be increased to say 8% for a 20 year old building, the concept being that an older building is less likely to experience rental growth.

Indirectly, and over a period of time, tenants have an influence over a building's value. At a rent review, perhaps achieved with the aid of arbitration, a tenant is less likely to pay a full rent for an ageing and poorly designed building.

Thus we have the situation that an owner occupier may be more concerned with a building's aesthetics because it projects his company's image and with it's cost-in-use because he is able to mini-mize it. A property investor, on the other hand, is far less con-cerned with aesthetics and relatively insensitive to the concept of cost-in-use.

5. DESIGNING THE FUTURE

In his address to the RIBA last November Professor J.K.Galbraith (3) observed that "Artistic achievement is now essential to orthodox industrial and commercial development". He stressed that there is an extremely important and much neglected relationship between art and industrial achievement. The problem is that, as I have explained earlier in this paper, the structure of the present financial institution dominated property investment market does not necessarily encourage artistic excellence in building design. There are, however, signs that the property market may and perhaps is changing.

Firstly, it is important to realise the institutional domination of the market is quite recent, in fact less than twenty years old. Institutional investors are becoming increasingly aware of the problem of depreciation. Industrial buildings constructed only 5 years ago are beginning to look obsolete whilst office buildings and purpose built shopping centres are also experiencing similar problems.

Secondly, the letting market has changed. The days of high demand and the unrealistic short supply, often created by planning authorities, of land for new development have temporarily gone. Even in south-east England the level of supply and demand for commercial accommodation is

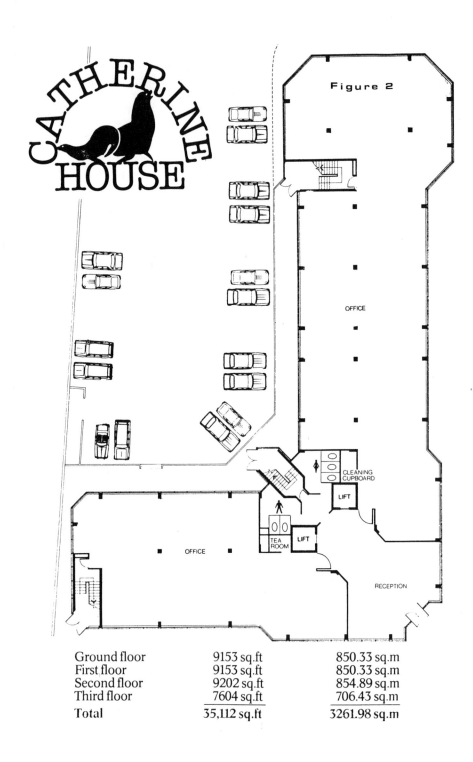

Figure 2

OFFICE

CLEANING CUPBOARD

LIFT

OFFICE

TEA ROOM

LIFT

RECEPTION

Ground floor	9153 sq.ft	850.33 sq.m
First floor	9153 sq.ft	850.33 sq.m
Second floor	9202 sq.ft	854.89 sq.m
Third floor	7604 sq.ft	706.43 sq.m
Total	35,112 sq.ft	3261.98 sq.m

far more in balance. A building constructed in the correct location will not necessarily automatically let; it needs also to be attractively designed. The property investment market, which is different from the property letting market, has I believe noted this change.

Thirdly, man's motivation and ability to avoid paying taxation is beyond exhaustion. The high level of corporation tax, capital gains tax and the low or tax exempt nature of life insurance companies and pension funds has encouraged institutional property investment and leasehold occupational tenure over the last 2 decades. Recent tax changes have begun to redress the balance in favour of private and publicly quoted companies. Firstly, the indexation of capital gains means there is far less of a penalty for a tax paying organisation to owning its own building. Secondly, the removal of tax relief on life insurance contracts may reduce the sums invested via life companies. Thirdly, the lowering of corporation tax eventually to 35% may encourage owner occupation and the expansion of property companies.

The long awaited proposal to make pension schemes portable will, I believe, make pension funds re-assess their investment portfolios. A more aggressive and competitive environment between pension funds may make them reconsider the merits of property as a medium for investment.

Whilst institutional investors will continue to be an important part of the investment market the nature of the property market may change in favour of owner occupation. With a higher level of property company ownership and owner occupation the quality of building design will change. A building's aesthetics and it's cost-in-use will become more important.

An example of this trend is a recently completed building on the edge of Kingston upon Thames known as Catherine House (4 & 5). It was funded by two development companies and made available to be let or sold during 1983. It very promptly let at about £12 per sq.ft. which was considerably higher than recent lettings in the Kingston-Surbiton area. The following schedule, which appears in the agents brochure, lists the principal features of the building:

*Full air-conditioning and central heating system, designed to allow complete flexibility of planning and to minimise running and maintenance costs.
*Wall carpeting throughout.
*Spacious marble entrance hall.
*Suspended acoustic tiled ceilings.
*Recessed fluorescent 83/84 "Power Saver" lighting tubes with automatic light reactor 'switch off' system. Lights are co-ordinated with window mullions to allow maximum flexibility for partitioning.
*For maximum energy conservation and minimum solar gain the building has double glazing, treated with gold and argon gas infill.
*Generous private parking plus 150 space public car park immediately adjoining.
*Two 12 person automatic lifts.
*Male and Female luxury finished washrooms on every floor.
*Sophisticated power/v.d.u./Telecom floor trunking system.
*Telephone systems incorporating 80 lines.
*6 Telex machines.

Figure 3

The Business Centre

SAXON HOUSE
HERITAGE GATE
DERBY DE1 1DD
(0332) 367268

Heritage Gate, Derby

FLEXIBLE ACCOMMODATION AND
SECRETARIAL SERVICE FOR THE
SMALL COMPANY

SERVICED OFFICE SUITES

SERVICED OFFICE costs. require from a single room
All-in-one office you require.
The size. upwards.
of 280 sq.ft. back-up.
● secretarial office
● Secured reception and office
● Shared copying facilities.
● Photocopying facilities.
● equipment facilities.

A survey undertaken by mechanical engineers, James Briggs & Partners revealed that, compared with similar suburban buildings, the
heating and cooling cost of Catherine House were very modest.

	Cost per sq.ft.pa.
Catherine House	66p
Similar Central Heated	68p
Similar Fully Air Conditioned	97p

An air conditioned building with windows that open and which can
also be used with just central heating, is fairly unusual.

Figure 2 clearly shows the sensible module of a building with a
lettable area which is easily divisible and which has a spacious
entrance hall.

The final area I wish to look at is that of serviced office suites,
a method of tenure commonly found in the USA. As I have previously
mentioned the UK property investment market has become dominated by
the system of 25 year leases where the tenant is responsible for fully
repairing, maintaining and insuring the building he occupies. The
serviced suite concept is the very opposite; the landlord is fully
responsible for the management, maintenance and insurance of the building. In addition he often provides furniture, telephones, a telex,
secretarial assistance and an accountancy service. The tenant rents
accommodation on a simple licence for an hour, a day, a week or
several months. The tenant or licencee can rent a single room or
several thousand square feet but the rent will necessarily be on a
gross basis.

In London two particular companies have specialized in offering
this kind of accommodation; Business Space Limited and Winter & Co
incorporating Spacebank 6 & 7. Outside London similar schemes have
been started up in towns as far apart as Norwich and Derby (8) (see
Figure 3). The concept has also been extended to industrial buildings
particularly where local authorities are trying to attract infant
industries. Some companies have also been successful in generating
income from otherwise worthless factories and textile mills using this
method of tenure. One such example is in the London Docklands.

So what has this got to do with design? Earlier in this paper I
explained how investors have become, due to the 25 year leasing
system, locked into an increasingly depreciating building and how
investors are relatively insensitive to the concept of costs-in-use.
The service suite concept is almost the reverse; the investor is not
locked into a long lease. Every year, or less if he wishes, he can
undertake a reappraisal of his property to see how he could reduce his
running costs, yet by more attractive interior and exterior design,
improve his net rental income. As an investor he is far more sensitive, on a year by year basis, to the concept of costs-in-use.

6. CONCLUSION

The question is whether the concept of serviced office suites becomes
more widely used as in the USA and is applied to larger buildings in
the UK. I believe that investors will increasingly reconsider the
concept of letting all their property with tenants being fully

responsible for repairing and insuring but not improving the accommodation they occupy. There are also signs that the UK market's penchant for 25 year leases is on the wane. Tenants may demand much shorter leases which, if only landlords realized, in many cases would enable improvements to be undertaken more frequently.

Even if much property continues to be let on an FRI basis, energy conscious designs, such as the example of Catherine House, will become more wide spread.

ACKNOWLEDGEMENT

I would like to acknowledge the helpful comments made by Tim Gauld of messrs Bonsor Penningtons, Chartered Surveyors, Kingston upon Thames.

REFERENCES

1. McIntosh A.P.J. and Sykes S.G. (1984) "A Guide to Institutional Property Investment" Macmillan.
2. Messrs Healey & Baker and Donaldson (1983) letting brochure for Millenium House, Kingston upon Thames.
3. Galbraith Prof. J.K. (1983) "From Economics to Architecture and the Arts" Paper given to RIBA 21 Nov. 1983.
4. Anon (1983) "Economising Energetically" Chartered Surveyor weekly 17 Feb. 1983.
5. Messrs. Bonsor Penningtons and Debenham Tewson & Chinnocks (1983) letting brochure for Catherine House, Surbiton.
6. Winter M.S. (1984) "Presentation of an appraisal of short term leasehold multi-occupancy offices furnished and serviced offices". Winter & Co. incorporating Spacebank, London.
7. Nissen R.P. (1984) "Short term office rental" and "Business Space Services" Business Space Ltd., London.
8. Messrs. Frank Innes and Weatherall Green & Smith, (1982) letting brochure for Heritage Gate, Derby.

PROFITABLE DEVELOPMENT AND PROPERTY WORTH

JOHN RATCLIFFE

Unashamedly the driving force of the property development industry
is profit. Both the nature and degree of development profit, however,
are subject to constant change according to circumstance and condition. Trading companies, for example, will seek to secure rapid
turnover and maximise short-term returns, whereas investing institutions will look for long term income growth and balanced portfolio
performance. Certain other development agencies active in the market
can take a more modest view of required levels of pure development
profit. Construction companies acting as developers may be more concerned with ensuring continuity of work and be content with an income
mainly based on the process of building. Major retail chains might
undertake direct development as a more effective means of creating
selling space and commanding shopping markets. Other commercial
organisations can often see distinct advantages from developing their
own accommodation for such diverse reasons as corporate image, special occupational needs or tax benefits. Profit itself, therefore,
is a variable within the development equation, alongside rent, yield,
cost, finance and time.

It should also be stated at the outset that this paper recognises
and accepts the concept of 'reasonable profit'. Whether it is expre--
ssed as a percentage of gross development value or total development
cost, or even as a proportion of initial development yield, profit
is taken to be a proper cost of production reflecting degrees of
investment risk and levels of management skill. Nevertheless, it
should further be appreciated that profit margins are notoriously
susceptible to extraneous influences often beyond the control of the
developer. Even minor adjustments in the yield perceptions of
investing institutions relied upon to buy-out the completed scheme
can have dramatic effects upon profitability. Similarly, changes in
the prevailing letting climate during the construction of a speculative development where rental guarantees have been given to the
funding agency can swiftly erode profit margins. Even where development is undertaken as a long-term investment, profitability is
dependent upon the vagaries of national and local economies.

Drawing upon the themes of the conference - quality, design, cost and profit - this paper aims to make a few general observations albeit vicariously, from the viewpoint of the developer as to what might be considered as an all-embracing objective - property worth. To do so the following questions are posed.

1.   WHERE WILL DEVELOPMENT FINANCE COME FROM?

Funding must, of course, be seen as a critical factor. While, however, we have witnessed a fundamental change in the structure of development finance over the last 10 years,it is possible to forecast that there will be just as significant a change again during the decade ahead. This process might be summarised as the rise and fall of the funds.

Throughout the latter part of the 1970's and the first part of the 1980's the growing dominance of the major financial institut-ions, especially the leading insurance companies and larger pension funds, has tended to dictate the shape and pattern of the prime commercial property development market. Many, if not most, of these funds are fast reaching a point of maturity where further massive switches into property are unlikely. Only a normal annual proportion of income will henceforth be directed towards the real estate market, and even that can go down as well as up in any particular year. Moreover, they may well find that substantial monies have to be invested in existing property holdings in order to maintain or improve rental income over time.
Delegates will have received a more profound paper on property institutions and their view of quality from Angus Mackintosh, but it is impossible to resist stating in the context of this text that many of their attitudes are seen by some as excessively conformist, blindingly unimaginative and dangerously imitative. Those investment based criteria currently employed by the financial institutions are increasingly being shown as insufficiently sensitive to demand in providing the sort of buildings that users actually want to occupy. With prospective tenants demand out-stripping acceptable institutional supply, it is, therefore, possible to predict a movement away from the institutionally led market and towards some of the following:

* Funding agreements with contracting firms
* A return to mortgage finance, with the accent upon more
  sophisticated variable rate loans
* Debenture stock funding
* End user funding
* Merchant bank syndication

With this reorientation in the development finance market will also come a re-emergence of the true role of the developer - as entrepreneur and not simply glorified project manager.

2. WHERE WILL DEVELOPMENT TAKE PLACE?

The increasing fragmentation of the property market means that skill is once more required in the early discovery and exploitation of 'hot spots'. Hunch and intuition alone are no longer sufficient attributes in the assessment of markets in order to determine new development opportunities. The element of 'fashion', however, must not be forgotten, for certain towns and cities can suddenly be in vogue. Nevertheless, with commercial and industrial activity no longer so constrained in terms of location, and with major corporations more and more taking a comprehensive and systematic appraisal of their total property needs, great care is called for in the analysis of demand.

Accessibility and environment are the twin determinants of prime location. The 1970's witnessed towns around London and to the West such as Reading, Bristol, Swindon, Watford and Luton grow enormously in popularity. The early 1980's saw the likes of Aylesbury, Chelmsford and Newbury come to the fore.
It is currently just possible to identify a further shift in demand towards even smaller growth centres such as Maidenhead, Ascot and Windsor, as environmental factors begin to outweigh, or at least balance those of immediate accessibility.

Good communications and proximity to markets remain essential; although improvements to the former render certain traditional concepts towards the latter redundant. Motorways continue to excercise an important influence, however, as the recent and exceptional level of interest around the intersects of the M25 with the M3 and M4 demonstrates. Heathrow also retains its own vital magnetism, as evidenced by the pressure for development in hitherto little known communities such as Cranford, Harmondsworth and Langford. A different kind of accessibility helps to explain the growth of centres like Brighton, Norwich and Cheltenham, whose regional role is probably paramount. And, in similar vein, the complementarity of specialist commercial activities in towns and cities such as Oxford, Cambridge, Leatherhead and Redditch provide the stimulus for growth and development.

As prospective tenants as well as intending owner occupiers become increasingly selective in terms of location, it is clear that the local environment in which a development is to be placed is at least as important as the precise level of accessibility of one site over another. Thus, there is a growing popularity for rural, semi-rural and natural settings.

A particularly high premium is placed upon such locations by overseas companies, especially American, who collectively contribute
a surprisingly large share of the demand for new commercial
premises. Unfortunately, many local planning authorities retain
an outdated attitude towards the location of business property
within and around smaller urban areas.

3. AT WHAT SCALE WILL DEVELOPMENT OCCUR?

Naturally there will be a continuing demand for new commercial
developments of all sizes and across all sectors: all that is
possible is to point to several trends which are likely to be
most apparent in the market . With the long lead times of large
scale development schemes, their contentious planning considerations,
complex engineering and design solutions and need for massive
injections of finance, it is not difficult to predict that in very
general terms small will be beautiful for the developer in established urban locations; whereas larger scale commercial development
will be increasingly attractive in more rural situations.

In the office sector, while the total amount of new floorspace
requirements from various sources is likely to rise, the rate of
growth in demand will probably be slower than has been experienced
over the past 25 years. What will change more radically is the
average size and type of office accommodation demanded, as well as
the chosen location. The requirement by major commercial
organisations, for instance, is not usually for more space but
for better space. In the same way, the rush of national and multi-
national takeovers frequently leads to a fall in the overall demand
for office accommodation, but can produce a fresh requirement for
new space more appropriately designed and positioned. Nevertheless,
in such circumstances, it is common to find that the headquarters
function is significantly reduced and the requirement for
decentralised office space enhanced. What could well happen is
that office functions and structures will become polarised between
very high order activities occupying top quality expensive office
premises in central area positions, and all other operations being
conducted in low-rise, low-density and relatively inexpensive
locations out-of-town.

Two other features likely to typify the office sector in respect
of the size of office units and scale of development are worthy of
mention. First, it seems almost certain that smaller individual
units of office accommodation will be in demand, with the accent
being on the provision of buildings in the 10,000 to 20,000 square
feet range for single occupancy. Even smaller redevelopment schemes
and refurbishments of 3,000 to 8,000 square feet for individual
occupiers will also be extremely popular, satisfying the 'own front

door' syndrome where self-contained buildings offering a sense of identity attract a premium price. Second, there will be a further and significant shift towards multiple occupation among business users, and a consequent demand for the development of purpose built multi-tenanted buildings which have the facility to be divided conveniently into discrete units of anything from 1000 square feet upwards.

It is equally difficult to generalise about the likely future shape of office buildings, except again to identify the principal direction in which the market is moving. The probable split between central city area and suburban or green field site location has already been mentioned. Suffice it to say that the former will be inclined to retain a concentration upon multi-storey development, while the latter will focus more upon two or three storey construction. The most significant feature in respect of building shape, however, is the optimum width of offices. The era of the shallow rectangular block is over, and so too, for that matter, is the very deep plan office layout. Few offices in future will be construct.ed to a width of much less than 40 to 45 feet, and much more than 50 to 55 feet. On large sites and for bigger buildings, therefore, a popular form of development shape will be a roughly square building surrounding a central atrium or courtyard. This trend will be reinforced by the increasing desire on the part of occupiers to have cellular offices with outside views, which, on average take up to about 60% of useable space. A different attitude towards the use of space, the relationship between private and communal space and the concept of efficiency ratios will thus be demanded.

In the retail development sector it is also possible to distinguish a few major trends which will influence the design, quality, cost and profitability of buildings over the next few years. These include:

*In the high street there will be more exploitation of adjacent deep standard unit shops and surplus back land plots in order to create small malls or precincts branching off the high street. Furthermore, an incremental upgrading of side streets and refurbishment of shop premises to establish a kind of 'Latin Quarter' atmosphere with a grouping of complementary service trades will take place in selected locations.

*The renovation and conversion of historic buildings, or buildings of special architectural merit, to produce a type of speciality centre, probably incorporating some form of catering and leisure facility will prove a popular and profitable exercise.

*Speciality shopping centres, carefully located and well managed, will generally be more widely developed. Some braver developers will create multi-level trading by converting existing high street stores - normally in absolutely prime positions.

*Planned central area schemes will tend to be smaller in scale, of a higher quality and better integrated into the towns they serve. Existing shopping centres will need extensive refurbishment to maintain their trading performance.

*District shopping centres of up to 200,000 square feet of retail floorspace comprising 20 to 40 standard units built around a superstore and often providing a range of civic as well as service facilities will be a popular form of larger scale shopping development.

*The construction of freestanding major superstores in edge-of-town or suburban locations will continue apace, as will the development of discount retail warehouses, with all the emotional planning and design reactions that are evoked by such development proposals.

The industrial development sector is just as segmented, and in the space available defies again all but the crudest of generalisations. Overall, however, there is an obvious and powerful trend away from pure manufacturing to a mixture of light manufacturing industry, assembly, laboratory, research, software development, packaging, storage, distribution, showroom, training, clerical, administrative and even recreational use. More specifically a few characteristics affecting demand and profitability can be identified. For example, some areas of the letting market are comparatively active and still attract the interest of the institutions. The electronics and computing industries are familiar market leaders. Retail warehouses in food, furniture, electrical goods and DIY represent a popular field of development on industrial estates. Foreign manufacturing companies continue to seek large storage and distribution centres. Transport based industry is relocating and service related industries are readjusting to changing needs and improved facilities. Some science park type projects and other special 'one-off' developments which concentrate upon a particular activity or theme are beginning to show that a challenging opportunity exists in that direction. The Industrial Building Allowance boom in small unit development is just about spent, but a residual demand for small workshops of various kinds in different locations remains. It should also be stated that at the other end of the market there exists a special demand for very large units, but there are not many developers prepared to undertake such schemes on a purely speculative basis.

4. WHAT WILL DEVELOPMENTS LOOK LIKE?

The quality of commercial buildings is increasingly the key to
present and future profitability. Unfortunately imaginative design
is still viewed with suspicion in this country - being somehow
regarded as costly and unmarketable. This is in sharp contrast to
the United States, for example, where American clients by comparison
are infinitely more adventurous in their approach to design and
demonstrably more aware that special character and style in a
building can have a public relations and promotional benefit to
potential occupiers and their business. It should also be
recognised that the architectural profession in the US seems much
closer to the actual processes of building production and design
cost analysis. Moreover, they appear to display a finer commercial
attitude towards development - perhaps due to the less restrictive
climate of professional practice that has prevailed there. On
the other hand, financial institutions have long been insufficiently
sensitive to the design and environmental quality of buildings,
and arguably have a great deal collectively to answer for in respect
of the slavish mediocrity that has characterised our development
industry until relatively recently.

For almost thirty years following the end of the second world
war the design of office buildings was uniformally dull and unimaginative. Architects were all too frequently nervous about the
mystical design and performance standards laid down by the major
financial institutions and their professional advisers, and as a
result tended to allow the control of design to pass out of their
hands. In matters of new office development it almost appeared
as if the accountant had gained total ascendancy over the architect,
and the approach to commercial development seemed to be reduced to
the provision of functionally clad and serviced floorspace. The
last decade can be said to have witnessed a constant improvement
in the standard of design of office developments. Nevertheless, the
trend back to brick as a facing material, prompted by demands from
property and planning interests alike, has given rise to a certain
amount of fresh critisism. This results from the rush of three and
four storey brick built developments, topped with lead or copper
mansard roofs and adorned by dinky dormer windows, some of which
are reasonably well done, some decidedly are not.

It is now possible to suggest, however, that we stand at something of a threshold in respect of office design where all concerned
with development are being asked to match space utilization to
occupiers requirements and building layout to a far greater degree
than before.

In exploring new dimensions in design it can therefore be stated that four themes increasingly will dominate the creation of good office space:

*Attractive external appearance
*Low energy cost
*Acceptance of new office technology
*Adaptability for organisation and acceptability to people

In similar vein, the design of industrial buildings historically has also been poor, often being more the preserve of the engineer than the architect. The adherence to strict zoning regulations tended until recently to constrain any serious consideration of the need for good design. It is only over the last few years from the late 1970's, that employers and management have become concerned about the working environment of their labour force, realising that pleasant and efficient buildings are synomomous with high productivity and satisfactory returns. Moreover, with the relaxation of rigid industrial zoning policies and a movement away from the separation of industrial processes into large estates, it became obvious that many small industrial developments would have to be sympathetically moulded into existing urban areas, and other large industrial developments would have to be sited in exposed and sensitive rural locations, often close to motorways where they would be clearly visible. The general quality of design for industrial development, therefore, has improved enormously, with such leaders in the architectural field as Terry Farrell, Nick Grimshaw and Norman Foster bringing a kind of respectability to what was previously a largely drab utilitarian approach. This qualitative improvement has been reinforced by the growing interest in the development of business parks, where, in terms of profitability, rental growth is likely to overtake that of traditional industrial property and the deficiencies in buildings erected over the past two decades will become ever more apparent. This is an extremely important market, and one which is all too often misunderstood by many sections of the built environment professions.

Even in the retail development sector, which has always shown a relative degree of flair and adventure in architectural terms, considerable uncertainty regarding some aspects of commercial design persists . Are glass roofs good or bad? Should there be full environmental control or not? Closed or enclosed centres? Will people shop on more than one level? Are mixed use schemes viable?

Architecturally, however, across all sectors of the commercial property market the auguries are favourable. Demand for high quality space combined with a competitive letting market has encouraged developers to adopt design solutions that would not have been countenanced a few years ago.

There is greater use of natural materials and less brutalistic treatments. More emphasis is placed upon planting, lighting and insulation; and the provision of attractive and controllable internal micro-climates for the staff of business premises is more common. External appearance and extensive landscaping attract the attention of developer and occupier alike. Increasingly tenants desire to become more and more involved in the finishing and fitting-out of buildings and generally, bespoke rather than off the peg buildings are becoming the order of the day.

5. HOW IS CONSTRUCTION ORGANISED?

A paper principally devoted to development profit cannot fail to make some mention, however contentious, of the way in which commercial buildings are procured. With notable exceptions the building industry in the United Yingdom remains badly organised, highly fragmented, poorly motivated, disputatious and of low productivity. The time worn adage that it costs twice as much and takes twice as long to produce a building of worse quality in this country than it does in the United States still rings too true to permit complacency. From admittedly limited experience it would seem that virtually everything in the United States is capable of being finalised before tender. Drawings are complete, comprehensive and accepted; specific and detailed descriptions of the work involved and the materials to be used are produced and agreed. It helps of course that sub-contractors and suppliers are more committed and better informed than over here - being able to provide shop drawings and details for which they will accept full responsibility at an early stage in the pro-curement process. These drawings can then be incorporated into the design and specification without tedious reproduction by the arch-itect. Further, the availability of finalised design information at the tender stage encourages tight competitive bids and leads to fewer claims. Even where variations take place, they are usually the result of a clients' changing requirements and can be priced and agreed before the work is carried out. In general there would appear to be much closer teamwork between those involved together with a realistic set of incentives aimed at encouraging the contractor to make proper savings and not fabricated claims. How unlike the picture here.

In a year which opened with the publication of the British Property Federation's Manual for a System For Building Design and Construction, some comment upon the objectives of their proposals and the reactions of the principal professional institutions to them is pertinent. The BPF unrepentently sets out to obtain a better deal for clients, with the prime aim being to produce good buildings, more quickly and at low cost, thereby fostering a fast and purpose-ful building process.

It is their intention that the attitudes of all those involved in the processes of design and construction should profoundly be changed so that the overriding ethos of the professional team is collaborative rather than conflicting. More certainty is meant to be injected into the process by requiring the client developer to establish more precisely the objectives to be attained by a particular design prior to tender, and the interests of the client are consolidated by placing them in the hands of a Clients Representative .

The new system also aims to remove much of the duplication of work that frequently takes place between designers, quantity surveyors and contractors under the present method of building procurement. To this end, one consultant, the Design Leader, is to be responsible for all pre-tender design work, and it is expected that a significant proportion of the design should, at the discretion of the client, be handed over to the contractor. The contractor, in turn, is encouraged by the system to identify means of reducing costs, and thereby share in any resultant profits. In order to prevent over-detailing and the generation of unnecessary work the system replaces the traditional Bill of Quantities in the building contract with a priced Schedule of Activities. Disputes and disagreements between any of the parties are to be settled in an expedite manner by an independent expert known as an Adjudicator, who has no other duties in administrating the building contract. Finally, a fixed fee structure with incentives is introduced as opposed to a scale of fees which provides little motivation to design and build within a target price.

The initial reactions of the main professions affected by the proposed new system run true to type. Architects accept the recommendations relating to cost control and the abandonment of the Bill of Quantities but defend their project management and full design responsibilities. Quantity Surveyors, conversely, welcome the changing role of the designer but resist any challenge to the conventional preparation of an independent Bill of Quantities. In separating the role of the Clients Representative from that of the architect and the quantity surveyor, however, the BPF Manual helps clarify certain fundamental responsibilities within the construction process, and reflects the dissatisfaction felt in many development quarters about the conflicts of interest which arise with the architect as designer, project manager and administrator of the building contract. Similarly, even within the more progressive ranks of the quantity surveying profession there is a growing recognition that the traditional Bill of Quantities is not necessarily the most effective or reliable vehicle for cost analysis and control, and an emerging movement towards alternative forms of assembling and documenting cost information and advice.

Many quantity surveying practices will need to reassess the basis upon which professional services are provided, their nature, and the point at which they are given. Nevertheless, the clear identification of the role of the Clients Representative presents many opportunities for those with the necessary management skills as well as professional expertise, and the likelihood must be that the new system will foster the growth of multi-disciplinary practices.

Looking for a moment at the developer, however, it should be stated that a keystone of the proposed system must be the added weight of responsibility placed upon the client . It is, therefore, to be hoped that the BPF will themselves be able to promote a heightened level of awareness on the part of individual developers of their extended obligations to perform efficiently, decisively and within the given timescale prescribed by the Manual. Nevertheless, until a fair degree of familiarity and confidence in the proposed BPF System has been developed, it is inevitable that those involved - clients, consultants and contractors - will experience some difficulty in apportioning and agreeing the various responsibilities and relationships throughout the design and construction process.

6. WHAT KIND OF MANAGEMENT WILL BE NEEDED?

To both generate initial profitability and maintain income flow through time the management of commercial property generally needs to become very much more aggressive. Developers and investors alike will have to learn to 'live' with their buildings in the same way that the American property industry does. The major funds, in particular, have always been far too anxious to disengage from their property management role and responsibilities, and will need to get more involved in the physical condition and performance of their property assets and the level and quality of support services they provide, especially as the demand for multi-occupied space effects the prime investment and development market.

In terms of tenure and occupation, more flexible leases as well as more flexible buildings are already being demanded by tenants, particularly by those occupiers in the vanguard of the new technology industries and specialised professional services. Tenants are increasingly looking for shorter leases with one-sided break clauses, or a series of renewal options, and happy then to rely on their rights under landlord and tenant legislation. On the other hand, the pattern of rent reviews in multi-occupied buildings can usually be arranged to benefit investors, and such properties could soon out-perform the single letting development. Alert landlords, therefore should find rising rents at more regular reviews more than sufficient compensation for any perceived loss of security, and quickly come to welcome the improved cash flow.

The area where a high level of professional property management
can progressively be seen to be closely related to development
success and investment performance, rather than as a simple
administrative procedure, is that of planned shopping centres.
It can thus be argued that much of the experience and expertise
gained in retail development and management could equally be applied
to the office, industrial and leisure fields. In this way, some of
the following matters are worthy of consideration across the sectors
of the commercial property market:

*Image - not merely for marketing a property at the outset of
development, but also for subsequent letting and rent review
negotiations, it pays to establish and sustain an appealing corporate
identity. The naming of a building or estate, the creation of a
special logo, the erection of well designed sign and nameboards at
entry or access points and around buildings and estates, the construction and maintenance of attractive 'gateways' or entrance halls,
and the publication of professionally presented marketing material
all make for a sense of community and a successful letting record.

*Tenant mix - in the same way that there has been an appreciation of
the importance of selecting the right mix of tenants in the management of industrial estates, which will spread to the contrived
complementarity of use in office premises.

*Landscaping - what used to be regarded as an unnecessary and
expensive luxury, often imposed as a condition of planning approval,
is now usually seen as an essential attribute of any successful
commercial development by investors, developers and occupiers alike.
In fact, it is becoming quite fashionable to retain large, striking
natur al features within estate plans, or to create artificial
environments or artifacts as a focus to a building or group of
buildings. In this context, it has been found that the imposition
of positive covenants in the lease upon a tenant is rarely the best
way of looking after common landscaped areas. It is normally much
better for a developer to arrange for the overall management of all
facilities and services, and recover the cost by way of a properly
accounted service charge.

*Security - even for the smallest buildings or estates security is a
matter of considerable concern to tenants, although how much security
is provided, of what kind and who pays, is obviously a matter of
negotiation between landlord and tenant.

*Utilities - some developers undertake the bulk purchase of oil and
gas which they make available on a metered supply. Special water
supplies and steam power are also provided on certain industrial
estates, and central sprinkler installations are another facility
commonly provided.

136

Refuse disposal with common central compactors are also sometimes
supplied where the demand for such a service is thought likely to
be high. Otherwise the supervision of cleaning services and the
monitoring and enforcement of conditions relating to the health
and safety of buildings are some of the day to day problems which
confront management and affect profitability.

*Business and personal services - it is becoming more and more
common for a developer to provide a range of support services
to tenants in multi-occupied property. These include reception,
mail-handling, telephone exchange, typing, computing and even
accounting. Increasingly developers are also conscious of the
need to consider the possible provision of such personal services
as shopping, catering, health care, banking, transport, leisure
and recreational facilities in larger schemes. Moreover, both
landlords and tenants are slowly coming to recognise the mutual
benefits that can be gained from the formation of a tenants
association in multi-occupied premises.

CONCLUSION

In summary then how will profitability in the field of commercial
property development be ensured and property worth maintained?

*Funding institutions will have to adopt a less cautious and
conformist approach towards the provision of development finance,
otherwise developers will have to look elsewhere for support.

*Locational analysis will have to be more carefully conducted, for
while communications and environment will increasingly dictate the
level and pattern of demand for development, other factors such
as the age and occupational profiles of the surrounding population
as well as the current state and future prospects for the local
economy, will all have to be more closely attended. Throughout
the property world when it comes to sites and siting there will be
a greater 'need to know', both at the outset of development and
on a continuing basis.

*Attitudes towards the size, shape and layout of buildings will
have to change. Traditional conventions regarding cost and rent
equations, and the valuation and levy of rents and charges for
private and communal space and services will have to be challenged
and reordered.

*Greater variety in design and better quality of building will be
demanded. Definitions and interpretations of property 'use' will
need to be re-thought, and attitudes towards the density and mix
of development schemes re-appraised.

*The parochial and defensive stances struck by so many consultants engaged in the process of the design and construction of buildings will have to be tested and formed. A more harmonious approach to the task of bringing buildings in on time and to cost needs to be induced.

*In an increasingly competitive climate good management and good marketing will play an ever more important part in assuring the original and continued success of development projects.

Section V
Techniques

THE MEASUREMENT OF QUALITY AND ITS COST

DEREK BEESTON, Department of the Environment (Property Services Agency)

The views are the author's own and not necessarily those of the Department.

1. QUALITY VERSUS QUANTITY

In dictionary terms quality is the antithesis of quantity although in building design the distinction is sometimes blurred. For example a design could be described as being of better quality than another because it had higher ceilings. Generally though, most would agree that the difference is clear: quantity is readily measurable whereas quality is not.

It may be that the nearest we could get to a definition would be that quality comprises the unmeasurable features of a design. If so, the development of a method of measuring an aspect of quality converts it from quality to quantity!

Luckily the definition of quality does not matter unless progress is impeded by any unclear thinking resulting from want of it. If we are inclined to say that because something is classified as quality it cannot be measured we may at certain crucial times give it too little weight in decision making. Whether we like it or not decision making increasingly tends to be based on a sort of balance sheet of measurable and costed considerations. We may intend to take other factors into account, and cost benefit analysis is a formal expression of this, but many people are worried by the late stage at which such consideration is usually given and the tendency to omit it if time and profit are at risk.

Thus those who believe that by resisting the measurement of quality they are keeping the Philistines at bay may be producing the opposite effect to what they intend. Refusal to measure quality may prevent account being taken at a malleable stage of a project of the very things that resistance is intended to protect. Major decisions are made by clients more used to balance sheets than aesthetics.

Most architects and many other designers would say that measurement is impossible and in any case wrong. Nevertheless the development and acceptance of methods of quality measurement would be useful. It does not follow that it is worth developing methods of measuring all aspects of quality but simply that nothing is sacred and almost everything is potentially measurable if the effort is justified.

141

2. MEASUREMENT OF QUALITY

What are we measuring when we try to measure quality? We are back to the problem of definition.

A useful beginning is to set down some aspects of quality, such as the following, and decide whether they can be measured objectively with respect to design, materials and workmanship.

The following aspects of quality will be dealt with in this paper.

Reliability (including durability) - objective.

Fitness for purpose - is objective where the requirement is accurately specified - elsewhere subjective.

Appearance - subjective.

If we can find ways of measuring them we shall also need to specify a base line for each aspect. This could be the minimum requirement; then quality would be defined as the amount by which the design and its execution exceeded the minimum requirement.

3. RELIABILITY

At present the formal evaluation of reliability in an existing building or a design is confined to mechanical and electrical plant. Even here the usual method is to rely on accepted good practice, such as the use of rules to guide the amount of redundancy in the provision of base load diesel generators or water chillers.

4. RELIABILITY DATA

To extend the evaluation of reliability to the rest of the building would require the establishment of a bank of data on the history of use of materials and techniques. This would be a healthy development for its own sake, and if it is to be done the use of the data for reliability evaluation would ensure that the data collection was structured. The PSA method of reporting defects is a good start but sample studies of histories of buildings will also be needed. In the private sector it may be necessary to buy data.

For each component the time to failure should be recorded and supplemented by the number which did not fail estimated from the sample studies of building histories. Together they would give rates of failure per year.

The descriptions of components would need to be systematic and the meaning of "failure" decided in each case. Typically "failure" would be the need for replacement, but another category may also be needed for repair costing more than a specified proportion of replacement cost.

The most difficult problem in collecting failure data is to classify the circumstances in which the component or material was used and its relationship to the rest of the building. It is here that the planning of a data collection system must be nicely judged to get significant information without overloading the collection system. Circumstances of use must be held in the data bank and summarised for the user when he withdraws failure data.

None of this would be easy or cheap but many of the problems have already been solved for mechanical and electrical equipment, a large data bank for which exists as part of the Sytems Reliability Service set up by the UKAEA. This bank has been used by the author, among others, for appraisal of the reliability of mechanical and electrical plant.

5. EVALUATION OF RELIABILITY

Whether the evaluation is to be of an existing system or a design there are two distinct ways of using the same reliability data. Both require the system to be interpreted in the form of dependencies of components on each other. For each component a list must be made of other components which directly depend on it for successful operation and which would be rendered ineffective if it were not working. The reason for it not working could be either because it had itself failed or because it was ineffective due to the failure of a component on which it depended.

These dependencies can, if required, be represented by a dependency diagram. One version of this is called a fault tree.

There are sometimes conditions in which perhaps only 3 out of 5 components are required to be servicable for a sub-system to be effective. For anything but a simple system a computer evaluation is necessary so complexities of this sort are easily dealt with by one or two lines of program.

Sometimes the recording of dependencies is simplified by the insertion of a dummy component whose failure rate is zero and which cannot be identified with any particular component. A common example is power supply.

When the dependencies have been worked out the two methods diverge.

The first method is used by most analysts including the Systems Reliability Service. In essence the average number of failures per year for each component is multiplied by the average proportion of a year it would be out of action having failed. In simple cases this is its repair or replacement time. This product is called its "unavailability". It is the proportion of a year for which the component can be expected to be out of action due to its own failure. The rules for combining probabilities are then used to combine these unavailabilities, in accordance with the dependencies, to give unavailabilities of sub-systems. Finally unavailabilities of sub-systems are combined to give a single figure for the unavailability of the whole system. It is the proportion of a year for which the whole system can be expected to be out of action and is a measure of its reliability.

143

The method has three major limitations. The rules for the combination of probabilities assume independence of failures whereas the underlying cause of failure of one component can often have some effect on the probability of failure of other components in the same system. A series of bad winters affects the probability of failure of many components so that failures cannot be regarded as independent. This difficulty can be overcome but the complexity introduced makes it impossible for those not closely involved in the calculation to understand it. This leads to uneasiness and unwillingness to use the results.

The second shortcoming of the method is the reduction of the result of the reliability appraisal to a single figure. This means that, when one design is compared with another the one with the lower overall unavailability must be accepted because there is no further information upon which to base any other decision. It may be that an unavailability of 0.10 for one design would be manifested by one system failure lasting a tenth of a year whereas an unavailability of 0.20 for an alternative design would be produced by 20 system failures lasting an average of a hundredth of a year and ranging in duration from 1 day to 10 days. Some clients may be happier with the latter.

Thirdly, the method cannot satisfactorily handle time-related events such as the buffering effect of a storage tank or the time taken by repair staff to arrive depending on the time of day and day of week.

These shortcomings are especially important when extremely high reliability is required, as in some defence systems. To overcome them the author developed a method which goes back to first principles and represents the reliabilty data in the form of failure rates and out of action times without combining them into unavailability figures. The method simulates the operation of the plant hour by hour. Each component is represented by its failure rate, its average outage time and the way in which other components depend on it for their contribution to the plant's operation. Running the simulation for 8766 simulated hours produces a diary of outages and their durations in a sample year. This sample year can be run hundreds of times with the same input data but, in Monte Carlo fashion, with the results differing in a random way.

The results can be summarised in any way desired, such as a frequency table of system outages of various durations, separately for each component whose failure was responsible for the system outages. Highly flexible investigations can be made using the method and users learn about the problems of maintaining the system, such as the likelihood of overloading maintainance resources. If it were required and the data were made available it would be easy to add cost of repair to the simulation.

The computer looks at each component in turn and compares a random number with its probability of failure in one cycle to decide whether the component is to fail in that cycle. Another random number governs the outage time to be drawn from that component's distribution. The shape of the distribution of component outage times must be assumed. A lognormal distribution has been found to be typical, but there is room here for further research. In each cycle components whose outage time has elapsed are brought back into service. The computer next examines the state of the system and records it for the diary of events. Especially it decides whether applying the rules it has been given result in the whole system or a sub-system being out of action for that cycle. When a cycle is complete another cycle begins.

The length of a cycle can be an hour, for refined analysis, or as long as a day for rough evaluations. Whatever it is the computer adjusts the failure rates to represent the probability of failure in that period and translates the outage times to numbers of cycles.

There are other details to be considered but problems are few and the realism which can be obtained is striking. Whenever an existing plant is simulated in this way users are impressed by the familiarity of the diary as it unfolds. They recognise patterns which they had ascribed to causes but which the simulation produces by chance. They are made to realise that certain design practices run risks which will result in system failure eventually and can watch calamity strike or be only narrowly averted after years of satisfactory running.

Any amount of realism can be injected by introducing new rules for dependencies and the calculation of outage times. Failure rates can be made conditional on the time of year and on other failures. They can be gradually degraded to simulate aging of components. The plant's performance can be calculated quantitively in terms of its task, such as the amount of power produced or capable of being produced. This can be matched against the requirement or load which can be varied according to rules with perhaps a random element.

Where failure data are not available a high value must be assumed. Sensitivity analysis can be carried out by systematically varying the failure data and repeating runs.

For large systems with hundreds of components, working out the dependencies is laborious, but it must be done if reliability is to be evaluated whatever method is used, and the exercise is usually instructive. The second method requires a fairly fast computer if overnight running is to be avoided.

Simply following rules of good practice can result in more expensive plant than is necessary or in insufficient reliability. Such methods have been shown to be too crude for cases were high reliability is required and some clients who used to be satisfied with professional assurance are beginning to insist on quantification of reliability.

To return to building reliability other than mechanical and electrical, its evaluation requires the ability to express dependencies of components on one another. This way of thinking is not used explicitly at the moment but the effect of doing so would be to improve understanding of design weaknessess. There is a parallel here with the advantages of considering cost in realistic terms rather than in artificial quantities of work.

6. FITNESS FOR PURPOSE

This is too big a subject to be treated here in detail but a
generalisation is possible. It is that the quality content of
fitness for purpose is measured by comparing with a standard
requirement for such buildings and measuring the discrepancies
from it.
There are some respects in which this is quite easy because
the requirement is specific and measurable and the requirement
for the design or building being evaluated can be directly compared with the standard. Ceiling height and thermal insulation
are examples. For other matters there are two sorts of problem.
The user requirement may be specifiable only in qualitative
terms. Such requirements may be stated clearly enough to be given
a rating but are more likely to be omitted from the brief and
left to the architect to decide upon. Others may not be covered
by the brief because they are of interest to the user but not to
the client. However, if they could be measured the client may
become interested because they may affect the acceptability of
the building. In both these categories, there are some requirements, such as flexibility of use, which are not measurable and
cannot even be given a rating until they have been subjectively
assessed.
Independent assessment is the obvious way of obtaining a
rating of fitness for purpose but suitable methods could be
devised for self-assessment by the design team provided that the
incentive to conceal deficiencies could be removed. It may help
to ask how much more (or less) would have to be spent to achieve
a defined standard. Subjective assessments can be quantified
using the well tried skills of market research. They are entirely applicable to the exploitation of professional experience and
can go a long way towards the elimination of the effects of
prejudice.
There are some aspects of fitness for purpose which ought not
to be left only to professional assessment. These are matters
relating to pleasing users of the building. Techniques for obtaining quality preferences from professionals, clients and users
will differ but market research methods can be used for all.

7. MEASUREMENT OF OPINIONS OF QUALITY

It is a part of the normal design process for the designer to
make judgements about subjective aspects of quality and appearance. He also needs to be able to communicate quality ideas
because, although his client will have some general ideas about
the level of quality he wants, the designer will have to do a
great deal of interpretation and develop an understanding of the
client's mode of expressing his ideas of quality.
Because of the nature of their training and task designers
tend to have a paternalistic attitude towards their client's and
users' needs. They often say or imply that they know what the
client needs better than he does himself. This is often justified

but the necessary understanding can only be developed by good communication between client and designer. The designer must have sufficient information about the way in which the building is to be used to make good judgements about the client's requirements, and this is as true of quality needs as of anything else.

Communication about appearance has the same problems as about the other subjective aspects of quality so it need not be separately dealt with.

Communication with the client is seldom as good as both would like and communication with the users of the building is often impossible. Without such communication the designer is not in a good position to make judgements about many features of the design, and especially about quality. Quality must be one of the most difficult things to describe so it is likely to be a cause of dissatisfaction when the client sees the finished building. The designer would be helped by being able to present to the client costed examples of what could be provided at the various quality levels. If the same could be presented to the users so much the better. Both the designer and the client would often benefit from better communication with the user.

Designers use check lists to help them communicate by providing a structure for eliciting information. This is well on the way towards an opinion survey in which actual questions, not just topics, are set down in the form of a questionnaire. Such a degree of formality is hardly necessary when the client is represented by only one person, but if the client is many-headed a questionnaire may be advisable. This would bring out disagreements or conflicting requirements more clearly than reading a brief.

Questionnaire methods are applicable to all a client's requirements but are especially helpful when there is the extra problem of communicating ideas of quality. It is because ideas about some aspects of quality are difficult to convey that a more structured method than a checklist is helpful.

To use a questionnaire method it is necessary to be very clear about exactly what is to be asked. This is healthy and is not necessarily a feature of checklists.

A great advantage of questionnaire methods is that, if they are designed suitably, costs can be attached to quality options. This was exploited in a method devised by the author to assist architects who were writing design guides for domestic accommodation for the armed services.

The method was intended to overcome two difficulties. First, users and, to a lesser extent, clients can often only say what they want when they are given examples of what is possible. Second, it is necessary for those choosing design options, especially when quality is involved, to have to consider the cost implications of their choices.

It was a questionnaire method but the questionnaire was very different from the usual sort. The users were asked to select from sets of design options. For each option a cost was quoted and the respondent had a restricted number of cost points to spend.

Careful choice of design options ensured that the results
would have great significance for designers while at the same
time being within the competence of respondents who had no more
than the written instructions to guide them. Before costs could
be attached to the options some design assumptions were necessary
but they were found to be entirely reasonable by those called
upon to use the results.

The items for inclusion in the questionnaire were obtained
from a series of interviews with users. The interviews were
structured only in so far as the general topics which were raised
were compared with a checklist to ensure that discussion ranged
over all areas. The users' informal comments were analysed to
provide designers with subjects. From these architects produced
design ideas based on the analysis of the discussions. They were
a mixture of various sorts of drawings and descriptions.

The designers' ideas were costed by quantity surveyors and the
costs translated into points. They were put into the form of a
questionnaire in which users were asked to select from menus
within a specified total points cost. Some pages from this
questionnaire are at the end of this paper.

The evaluation of quality was only one of the objectives but
several quality preferences emerged. For example most male officers preferred a traditional appearance for the dining room of
their mess but women liked a light, modern appearance. An unexpected finding was that cavalry officers preferred size to
appearance because they were able to pay for wood panelling from
their own funds after the mess was built.

8. COST DATA BANKS

When quality ratings of existing buildings have been made they
will be a valuable addition to a cost data bank. They should
account for some of the unexplained variability of costs.
Quantity surveyors may consider this to be a sufficient reason on
its own for measuring quality.

The quality measurements should be classified into major
groups, perhaps reliability, fitness for purpose and appearance,
and a rating recorded for each. When the rating system has been
in use for long enough data analysis will establish the connect-
ions between quality and cost, then the quality ratings can be
treated like any other descriptors. The analysis of historic
data will provide factors to be used to adjust cost data to allow
for differences between the buildings from which the data came
and that for which they are being used.

It is important to avoid overlap between adjustments for
quality and the more usual adjustments. For example whereas it
may be useful to adjust cost per unit area for ceiling height it
would be wrong to adjust a quantity based price index because
storey height, and therefore ceiling height, affects quantity.

Until the quality ratings have attained a sufficient degree of
acceptance to be used like any other adjustment factors, they
could simply be recorded in the cost data bank and used

informally for a final subjective adjustment by the estimator. Though informal this adjustment should be guided by statistical analysis of all data in the bank to provide as much understanding as possible of the way they vary with respect to quality.

SUMMARY

Quality is difficult to define but this should not be allowed to impede the development of methods of measurement, because nowadays anything about a building which cannot be measured and costed is likely to be given scant importance by the client.
    Until quality is routinely measured there will be serious omissions from the adjustments which can be made to costs in data banks and hence an unnecessary inaccuracy in estimating and a gap in the cost advice which can be given to designers.
    It would be premature to claim that quality measurement could be immediately instituted but there are some areas, reliability and opinion measurement, where techniques are already available and in which a start has been made on data gathering.

Background

1. Earlier this year a team from the Property Services Agency of the Departmen of the Environment interviewed more than 200 Mess members in the UK and Germany to learn their views on the types of Messes they will want in the future. This information is needed to update accommodation scales and to guide the design of new Messes and help in the alteration of existing ones.

2. Some needs are already clear from the interviews alone. For example, there is generally a wish for a separate entrance leading to the living quarters and for a means by which residents can get a drink without going into the public area. There is no need to obtain more information on these.

3. Other questions could not be satisfactorily dealt with in the interviews, usually because of the difficulty of taking cost into account. For example, everyone would like a dining room large enough for all members and their guests, but if this means giving up something else the requirement is less clear This questionnaire will complete the information by calling for one requirement to be weighed against another.

The Questionnaire
Idea

1. The following yellow sheets set out a menu of possibilities from which you are invited to make a selection so that we can see the sort of Mess you want. At some stage before Messes are built or adapted cost limitation has to be applied and we think you should have a chance to say how.

2. This questionnaire makes planning your Mess like choosing a meal from a menu within a price budget. The prices on the menu are represented by points, except that in each section the cheapest item has been made zero to ease the arithmetic. Other prices in the section and the total budget have been reduced equally.

The Questionnaire
How to complete it

1. Record your selections by noting the serial number and points rating of each on the record sheet which follows the "menu".

2. First time through do not feel limited by cost, but select the items you would want if cost were no object. (This will show us whether there are some you would not have at any price.) Next time through try to reduce the total points below the target figure.

3. To make the method of recording clear the following is an example of how the first two attempts might look.

4. It would not reduce the validity of the survey if, before beginning to make your selections, you were to discuss the items in the lists with a few other Mess members to ensure that you have seen all the implications. When it comes to choosing, however, please work alone because it is only the views of the random sample of Officers which are required.

5. When you have finished please use the enclosed envelope to send the questionnaire to PSA.

Dining Room

	SIZE of dining room Choose one of the following four	POINTS
A	Large enough for all members and their guests	90
B	Large enough for all members and opening on to a hard area outside which could be covered by a tent to give room for all members and their guests for functions between May and October	0
C	Large enough for all members and capable of extension by opening partitions (see Fig 9 which gives one possible layout) to give room for all members and their guests	10
D	Large enough for all members and capable of extension by both methods B and C above to give room for all members and their guests	5
	APPEARANCE of dining room Choose one of the following two	
A	Wood panelled, traditional appearance	20
B	Simple, modern appearance	0

Garages

Choose one of the following four

		POINTS
A	Lock-up garages on the scale of one per resident	20
B	Space in a large lockable garage for all residents' cars	10
C	Lock-up garages for half the residents' cars and covered parking for the rest	10
D	Unreserved space for parking	0

SYSTEMATIC QUALITY APPRAISAL

KENNETH DRAPER, RIBA. Building Design Partnership

1.   INTRODUCTION

1.01 At the BDP annual conference in 1975 we took as our theme
'The Pursuit of Quality' and explored for the first time the use of
a structured method of quality appraisal. Our principal guest on
that occasion was Paul Kennon of Candell Rowlett and Scott (CRS)
the Houston based multi-discipline practice and it was his
colleague, Bill Candill, who had first directed our attention to a
method of appraisal, in his book 'Architecture by Team'.

1.02 We had already discerned many parallels in the development of
CRS and our own practice, and one quote in particular from
Candill's book struck home at that time.   In commenting on CRS,
Candill had written '...the quality level of our products had
always been comparatively high, although not always reaching the
highest level. Frankly, our reputation had been more for process
than for product. The input excelled the output', and he went on
to describe project appraisal as used by CRS at all stages of a
project and for a very wide variety of projects.

1.03 While the BDP design method has always placed great emphasis
on the 'design session', at that time there was a feeling that
something more was needed.   Something that was sharper in its
critical content, and more specific in terms of comment and
direction. If we could adopt a common language in our debates,
criticism could be used positively as a positive step towards
achieving a more consistent quality of product. The design session
was still important but it needed to be complimented.

1.04 The design session as used in BDP is primarily a discussion
bringing together those directly 'on line' for the project and
others of different disciplines or from different locations with
specialised experience or accumulated wisdom. The purpose is to
provoke and test the solution emerging from the teams efforts. We
see the design process as a sense of considerations of alternative

proposals, some being accepted and some rejected through value judgements of one kind or another by an individual or group of individuals. Appraisals in this sense form a natural continuation of this procedure and if we could become more aware of this, the greater the likelihood of improving the quality of our product. Quality appraisal, would therefore be a further tool to be used in the design process, it would not be an end in itself.

1.05 The 1975 conference was the first occasion on which we used formal appraisal and as a result of that experience the following brief was established for further development and implementation.

'It was agreed that assessment of jobs should be carried out at regular intervals and the method adopted should be directed towards securing contributions from the whole firm's experience, wherever located, to improve quality. The method adopted for BDP need not be precisely the same as the CRS method but it was established that this was a good basis to work on. Consideration should be given to those in the firm who had the ability and acceptability to form an appraisal panel. Key qualities for the appraisal panel were seen as being authority, consistency, credibility and continuity. Design sessions should continue in the normal way and the appraisal sessions were seen as complementary'.

Since that time quality appraisal has been absorbed into the working pattern of the firm and with the benefit of experience, both those being judged and those making the judgements have been able to contribute to a refinement of the method as first considered in 1975.

2   SYSTEMATIC APPRAISAL - A METHOD

In describing the method certain points of reference as used by BDP must be established.

2.01 What are the objectives? We see these as:

- maintain and improve quality in all BDP products
- ensure that all projects satisfy a minimum BDP acceptable standard and compare well with achievements of other design practices.
- optimise opportunities inherent in any project
- demonstrate the essential part which criticism plays in creativity
- influence approach to jobs and working methods
- underline unity of the firm by encouraging cross-pollination of experience, ideas and standards between offices
- promote clear and crisp presentation
- develop individuals during the design process
- educate by participation in or observation of the process

155

2.02 What is appraisal? We feel it may be defined as:

- an authoritative valuation or estimate of comparative worth;
- a summary judgement on examination against a known set of questions;
- an evaluation of a project at any stage of development from concept to completion.

2.03 What are the basic criteria against which we will make these judgements? From the very nature of our product we discern these to be:

FUNCTION - how the project works for people and things.
FORM   - what the project looks and feels like.
ECONOMY   - maximum effect with minimum means in relating function and form.

2.04 How do we identify quality?:

Quality will be the result of realising the interdependence of these three criteria, considering them simultaneously, bringing them into balance in complete integration and raising each to the highest magnitude.

2.05 BDP philosophy has always been to seek unity and balance in its design, appropriate to the type and scale of project - whether it be a monument or a utilitarian shelter. Generally, therefore, equal weight should be given to all three characteristics.
    It is interesting to read the following extracts from 'A Survey of Quality and Value in Building' - a BRE report published in 1978:
    'In this paper quality is defined as the totality of the attributes of a building which enable it to satisfy needs, including the way in which individual attributes are related, balanced and integrated in the whole building and its surroundings'
    'Although the preparation of the brief and the layout design are crucial, all subsequent parts of the design, construction and use of the building have to be done well to ensure quality and value. The achievement of quality and value is therefore intrinsic to the whole process...'
    'In this paper value is defined as quality in relation to cost. Maximum value is then in theory obtained from a required level of quality at least cost, the highest level of quality for a given cost, or from an optimum compromise between the two...'

2.06 Against this background how do we undertake the process of quantifying quality? In essence by marking or grading against a number of key characteristics for each of three basic criteria of Form, Function and Economy; a numerical interpretation of a value

156

judgement, as most of us accept at some time in our educational and working experience.

We use a scale of 1 to 10 to evaluate the success or otherwise of a particular characteristic on the following basis:

Characteristic Value:

Perfect	10
Excellent	9
Very Good	8
Good	7
Fair	6
Acceptable	5
Poor	4
Fair below acceptable	3
Critically bad	2
Complete Failure	1

To focus consistently the application of these evaluations we have developed a series of 'question sets' that amplify the key characteristics of the three basic criteria, and which will serve to direct the jury and the team in a similar way. As currently developed these are as follows:

FUNCTION

A   Is there a clear concept arising from the main functional determinants?

B   Are the interior and exterior spaces grouped, sized and shaped to satisfy the detailed functional relationships, and to reinforce the basic concept?

C   Does the circulation pattern allow people and things to flow efficiently?...

D   Is space provision adequate and properly disposed both for function and effect?...

E   Have ideas for future growth and change informed the concept in a controlled way?

F   Will the physical environment envisaged continue to give comfort in thermal, aural, and visual terms?

FORM

G   Is the relationship to site and its surroundings well exploited in the positioning, approach, massing, and expression of the building in its landscaping?

157

H   Is the handling of <u>form and space</u> an imaginative and meaningful expression of functions and of the spirit of the times?

I   Is there a creative <u>integration of structure and services</u>?

J   Does the design ensure a <u>psychological effect</u> derived from function and expressive of the right image?

ECONOMY

K   Is there a simplicity and restraint in planning and technology suggesting an effective <u>economy of means</u>?

L   Has <u>balance and value</u> been achieved in the handling of spaces, and in the choice forms, systems, and materials for present functions and future changes?

M   Have <u>operation and maintenance</u> considerations had an effect on the concept and its details?

N   Have <u>construction</u> programme and methods influenced, or been influenced by, the design?

2.07 The results of considering these questions, making a judgement against the predetermined set of values may then be quantified and the average mark established for the project for each of the three principal criteria. As an example Function might achieve 6.33 within a range of 5 to 7 as recorded against each question and an overall total of 38. Similarly the results for Form and Economy may have produced 5.5 and 6.25.

2.08 While these represent the component marks the appraisal must go further and identify the strengths and weaknesses with regard to the balance achieved across the project as a whole, and this we demonstrate graphically by representing the three marks as forces set at 120 degrees from one another, and then completing the triangle described by joining the three separate values.

2.09 Differing projects will thus have triangles of different shapes and areas. In the example described previously with marks of 6.33, 5.5 and 6.25 the graphic representation would be as shown, and the area of the triangle would be 47.08 using the formula ab + bc + ca x 0.433.

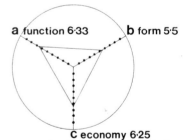

a function 6·33   b form 5·5

C economy 6·25

It is this resultant calculation of the area of the triangle that we define as the 'quality quotient' for the project in that it reflects both the components of the evaluation and the way in which they have interacted.

158

With a maximum evaluation of ten for each of the three criteria, a project meriting this level of judgement would describe a perfectly balanced triangle of maximum area as shown. Such a project would achieve a 'quality quotient' of 129.9 - 'a triangle of perfection'.

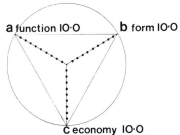

2.10 To make a more realistic comparison we could similarly consider two projects with consolidated marks for Form, Function and Economy of 5, 5 & 5 and 8, 3 & 2 respectively and generating 'quality quotients' of 32.5 and 19.9. These would be represented graphically thus:

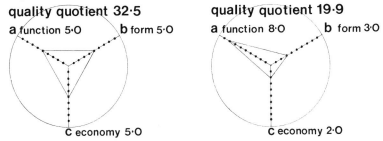

2.11 The higher quotient of 32.5 acknowledges a medium level of achievement in component marks and a balance in the disposition of these marks, whereas in the second example a relatively high mark for one component has in overall terms been offset by lack of balance in the solution as a whole. Function has dominated at the expense of the other two criteria in that instance. A difference in shape of the triangle represents comparative approaches to balance : a difference in area represents comparative degrees of success of the whole design.

2.12 In terms of describing the range of quality quotients we have adopted the following table:

Perfect	130
Excellent	105
Very Good	83
Good	64
Fair	47
Acceptable	32
Poor	21
Far below acceptable	12
Critically bad	5
Complete failure	2

3   SYSTEMATIC APPRAISAL - AS PRACTICED

3.01 BDP experience to date has shown that irrespective of the merits or otherwise of the actual process, the objectives will only be achieved if all those involved are equally committed to the idea.

3.02 The team must respect the jurors: the jurors should be selected primarily for their qualities as individuals, for their particular experience and for their capacity to appreciate the wholeness of a project. Wherever possible, the make up of the jury should reflect the various disciplines appropriate to the scope of the project. The use of jurors drawn from a regular panel will ensure a true understanding of the process and some consistency in how it is applied.

3.03 Equally, the jurors must respect the team, and maintain an open and receptive approach to understanding the problem and why and how solutions have been developed.

3.04 The team must be disciplined in their use of systematic appraisal. In terms of timing we would expect appraisals to take place at concept, outline proposals, scheme design, detail design and completion.

3.05 The process must be both manageable and affordable. The team must be able to communicate succinctly with a minimum of prior preparation: an hour should suffice for presentation and for the jury to question the team, with a further 1½ hours for the jury to mark and comment and to report and discuss the results with the team.

3.06 It would be wrong to ignore the doubts expressed by some when we first embarked on this more formal and systematic method of appraisal, but acceptance has grown with experience. Considered in abstract, it may appear to be too cumbersome a tool to be effective : in reality it is simply 'taking the pulse' of the job. We believe that appraisal has been a positive influence on the development of many of our projects. Possibly some have suffered a degree of confusion, but we would certainly claim a better understanding as to their 'state of health'. We would also claim a greater awareness of the talent we have, and how best to apply this in pursuing the highest level possible in the quality of our product.

160

REFERENCES

1.   Candell, B (1973) 'Architecture by Team'.

2.   White, B, AA Dipl (Hons) FRIBA. (1976) 'Project
       Appraisal' - One Firms Approach, BDP.

3.   BRE (1978) 'A survey of Quality and Value in Building'.

4.   White, B, Chairman Quality Appraisal Group, (1980) 'Quality
       Appraisal', BDP QP publication.

VALUE AND ITS ASSESSMENT IN THE CONTEXT OF HOUSING ADAPTATIONS
FOR PHYSICALLY-HANDICAPPED PEOPLE

ALAN MORRIS, Polytechnic of the South Bank
CAROL UNDERWOOD, London Borough of Croydon

1. INTRODUCTION

We who are employed to put resources to effective use in building
development give our close attention to the client's view of value.
We must learn to avail ourselves of means he has developed of
measuring that value. The client usually will have covered a
considerable distance in reaching his present understanding while
accepting that there is yet more to be explored.

This paper outlines the position reached by the client in the case
of housing adaptations for physically-handicapped people. Here the
client* has developed a method of assessing handicap. The method,
developed initially for monitoring improvement in patients undergoing
treatment in hospitals, may also assist towards arriving at value in
planning a programme of housing adaptations. Handicap, a significant
factor in the client's make-up of value, may usefully be considered
separately, on its own.

We may learn here something of the framework that is called for
whenever value or measures of value are illusive in the development
context.

2. CONTEXT

The benefit the physically-handicapped person will derive from
adaptations to the home to meet his condition is often unquestionably
greater than the additional cost incurred. Yet finance is short and
the number of physically-handicapped people in our society is growing
appreciably : so the needs of all are not met. Choices between one
person's need and another's are necessary, calling on cost. So, too,
are choices between ideal and less ideal forms of adaptation. We
give details below (Figure 1) of the proportion of physically-
handicapped found in the population and also Government
recommendations for their modes of housing provision.

---

* The term client is used here to signify Occupational Therapists
employed by Local Authorities responsible for making provision
for physically-handicapped people.

Figure 1   The population and recommended modes of housing provision
            for physically-handicapped people

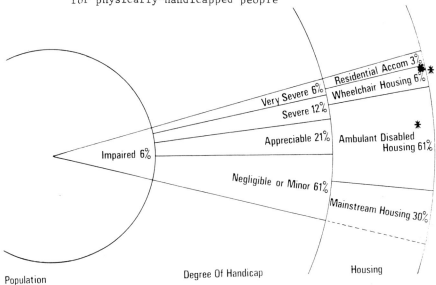

Population   Degree Of Handicap   Housing

Adaptations to handicapped persons' existing homes constitute a considerable proportion of the housing provision in Figure 1: a large
part of <u>mainstream housing</u> and also of <u>ambulant disabled housing</u>.

3. NATURE OF THE CLIENT'S PROBLEM

The following are examples of the evaluation problem with which the
client is faced in the context of housing adaptations for
physically-handicapped people :

> • A young man, aged 19 years, paralysed from the waist down
>   following a road traffic accident, lives with his parents
>   in split-level accommodation. Is it better to carry out
>   adaptations to this person's unsuitable accommodation,
>   which will be difficult to adapt satisfactorily, or to
>   suggest the family move to a purpose-built house that will
>   not be completed for one year?

---

*   Ambulant disabled housing (or Mobility Housing) is ordinary
    housing built to prevailing public authority costs and space
    standards but designed so that it is convenient for disabled
    people to live in.
** Wheelchair housing is housing for chairbound people who can
    manage independently or with family support. Goldsmith 1976.

- A mother, unable to climb stairs, needs to maintain care and discipline within the family. The growing family could benefit from some extra room in the house, although this aspect is not essential. Is it better to build a ground-floor extension to the house or to instal a through-floor lift giving the mother access to the first floor?

- A handicapped boy, unable to go upstairs to his own room during the day, has need of access to a bathroom and toilet. He also has need of privacy and his own social contacts. He lives with his family in a single-access terrace house. There is space in the garden for a ground-floor extension but all the materials for the construction would have to be carried through the house. Is it better in this case, again, to build a ground-floor extension or to instal a through-floor lift?

We have, here, problems each of which calls for handicap to be overcome in a cost-effective way recognising there is a social overlay to take into account. For this discussion, we will put this initially, that the value which motivates adaptations to housing for physically-handicapped people broadly identifies with overcoming handicap.

4. HANDICAP

The nature of handicap first needs ex plaining. Handicap is disadvantage or restriction of activity caused by disability. It is larger than disability - that is, the handicapped person's physical impairment. A disabled person's handicap may be a function of such factors as his personality, and his weight, as well as his disability. It may be also a function of his housing.

Handicap must be permanent and substantial in order to warrant adapting housing with the use of public funds.

5. HANDICAP ASSESSMENT

Handicap, being different from disease or disability, calls for a form of assessment of its own. Assessing handicap is the particular province of the Occupational Therapist. For the purpose of handicap assessment, Occupational Therapists have developed tests of function to give a person's handicap profile in relation to activities of daily living.* There are various forms of such assessment that have been developed. The design of the assessment varies according to the context in which it is being used.

The following (Table 1) is a form of assessment used for general screening of the population. Such assessments are necessary for determining the nature and extent of handicap in a particular administrative area, in order to carry out the provisions of the Chronically-Sick and Disabled Persons Act **Open University 1977.**

---

* These are known as ADL assessments.

Table 1   A measure of incapacity

Activities	Capacity to perform activity		
	Without difficulty 0	With difficulty but with-out help 1	Not at all or only with help 2
I Personal tasks			
1 Go out of doors on own			
2 Go up and down stairs			
3 Get about house on own			
4 Wash down or bath			
5 Dress and put on shoes			
6 Cut own toenails			
7 Get in and out of bed			
8 Brush and comb hair			
9 Feed self			
10 Go to toilet on own			
II Household tasks			
1 Clean floors			
2 Make a cup of tea			
3 Cook a hot meal			
4 Do the shopping			
III Physical and mental faculties			
1 See (even with spectacles)			
2 Hear (even with hearing-aid)			
3 Speak			
4 Organize thoughts in lucid speech			
5 Sit or move without falls or giddiness			
6 Control passing of urine			
7 Control passing of faeces			
8 Manage other special dis-abilities without help			
9 Co-ordinate mental facul-ties in performing personal services			

Source: Sainsbury 1970
'*Note:* One point should be made clear about the number of activities in the list. For purposes of broad approximation in general surveys of a population the items in the first two sections would be enough. But in classifying a minority of individuals, as between 'severe' and 'moderate' incapacity for example, it was found that attention had to be paid to general physical and mental faculties' **Open University 1977**.

A maximum score of 46 is obtainable in the assessment. An individual's total score is added up and graded as follows:

slight incapacity	0-7
moderate incapacity	8-14
severe incapacity	15-21
very severe incapacity	22 and over.

'Prevalence surveys of this kind have been developed as a crude
index of capacity to manage personal and household activities.
This particular approach, using a short list of selected activities,
is designed to make it possible to screen a large group of people and
draw conclusions about incapacity in different age groups; compare
persons with multiple disabilities; uncover the magnitude of need;
and, subsequently, plan better services.

For the purpose of assessing the incapacity of individuals a
different form of assessment is necessary, which goes into greater
detail. Returning to Table 1, many factors are included in each item
so that it is difficult to identify precise difficulties. For
example, cooking a hot meal includes preparation, moving materials
round the kitchen and using oven and hot plate. Functional
incapacit y and building hazards are not distinguished so that there
is not any way of knowing whether the person has a functional or a
housing problem. Going up and down stairs may be impossible because
of arthritic knees or because there are 54 steps down to the street'
Open University (1977).

An example of the more detailed form of assessment for assessing
people's individual capacity is shewn overleaf. Precise difficulties are now identified and the individual's capacity is separated
from the suitability or otherwise of his housing.

Table 2 A more detailed form of assessment – activities of daily living **Cornell and Broom (1977)**

COMMUNICATION	9
Receive Info: verbal	
non-verbal	
written	
Convey Info: verbal	
non verbal	
written	
Telephone	
Task comprehension with instructions	
Independent	
Totals	2

MOBILITY	17
Sit	
Stand	
Walk	
Climb - stairs up	
down	
- slope up	
down	
Aids w/chair	
Transfer - bed on	
off	
chair on	
off	
floor on	
off	
In own home	
Access own home	
Independent	
Totals	1

FOOD/LIQUID	8
Utensil hold	
Food on	
Cutting	
To mouth	
Chew	
Swallow	
Drink	
Independent	
Totals	6

TOILET	7
Bowel Control	
Bladder Control	
Clothes	
Transfer	
Paper	
Menstrual - Hygiene	N/A
Independent	
Totals	2

WASHING	15
Face	
Rt. side top half front	
back	
Rt. side bottom half front	
back	
Lt. side top half front	
back	
Lt. side bottom half front	
back	
	5

WASHING CONTINUED	
Transfer Bath	
Shower	
Drying	
Taps	
Soap	
Independent	
Totals	1

UNDRESSING	16
Off Head	
Off arms	
Off trunk	
Over bottom	
Down legs	
Off feet	
Fastenings off	
Appliances off	N/A
Calipers off	N/A
Stockings/socks off	
Corsetting off	N/A
Rt. side top half	
Rt. side bottom half	
Lt. side top half	
Lt. side bottom half	
Independent	
Totals	8

DRESSING	17
Over head	
+ in position	
Over feet	
+ pulled up	
Over bottom	
Fastenings	
Foot - wear on	
Foot - wear fastened	
Appliances	N/A
Stockings/socks	
Corseting	N/A
Rt. side top half	
Rt. side bottom half	
Lt. side top half	
Lt. side bottom half	
Independent	
Totals	6

GROOMING	13
Hair - do	
wash	
Teeth	
Nails - finger nails	
toe nails	
Make - up	N/A
Shaving	
Handkerchief	
Rt. side top half	
Rt. side bottom half	
Lt. side top half	
Lt. side bottom half	
Independent	
Totals	4
% ability	37%
Section Score	0

The form of assessment in Table 2 is to determine a person's overall ability score. This is obtained in the following way :

There is a box alongside each activity. In completing the form, all activities in which the patient is competent are left blank. Those activities in which the patient is not competent are shaded in.* The unshaded boxes are totalled (see bottom of each section) and the score thus obtained compared with the total score possible (stated at the top of each section).** Then the overall percentage is computed.

The information on the assessment form is transferred to a summary sheet (Table 3).

Table 3 Assessment Summary **Cornell and Broom (1977)**

ADL — Personal		
Name *Mr X*		Date —
Age *26*		
Diagnosis/Disability *Cerebral Abcess & Anoxia*		
Dominance Ⓡ/L		
Total Nos. of units possible	=	*102*
N/A (Not applicable to this Pt.)	=	*7*
Applicable = 100% = y	=	*95*
Raw score = x	=	*35*
% ability score = x/y x 100/1	=	$\frac{35}{95} \times \frac{100}{1} = 36.8 \ (or\ 37)$
Section score	=	*0*

Score for summary chart (circle appropriate grade)

%	Grade
0 =	0
1 - 24 =	1
25 - 49 =	②
50 - 74 =	3
75 - 99 =	4
100 =	5

37 = Grade 2

---

*   Assessment forms have been developed elsewhere to incorporate quality of performance. How far it is helpful to attempt such detail is a matter of debate.

**   If the patient is independent in the whole of a section then this is recorded, by leaving unshaded the box located alongside the section heading. If the patient is not wholly independent in the section then that box is shaded in.

The assessment form above, providing a structured approach to handicap assessment, makes a significant contribution towards breaking
down such problems as those outlined in Section 3 into constituent
parts. The form shewn was developed in a hospital context, however
- not at all with priorities for housing adaptations in mind. In
order for the form to be used in determining housing-adaptation
priorities some expansion on the housing handicap side is desirable.*

5. HANDICAP AND COST

If value equates with overcoming handicap then in assessing activities of daily living we have the beginning of a means of assessing
value: the greater the handicap overcome, the greater the value
achieved.

This value must be weighed against cost: the cost of the particular means that are necessary of adapting the housing. The cost of
adaptations varies considerably with different types of provision and
house layouts (see Table 4 for an example). In general, the greater
the handicap to be overcome the greater is the cost.

Table 4   Costs in overcoming handicap, relating to stairs.

	Approximate price
1. Grabrails and mopstick	£100
2. Paramount rail	£500
3. Stair-climbers	£1500 - £9000
4. Through-floor lift	£3500 - £5000 plus builder's work

6. VALUE : BEYOND HANDICAP

Handicap, meanwhile, is not the whole issue. The social overlay to
handicap that was referred to earlier in the paper may be a powerful
factor in assessing what represents best value. The age of the handicapped person is an important factor: young people often have a
bigger claim than the elderly. So, too, are family considerations.
A mother's need for access to children's bedrooms is very persuasive;
and, again, avoiding extra trauma to a family caring for a handicapped
son (examples from Section 3) **

---

*   Existing terrace houses are found to fall into a small number of
     classifications for the purpose of adaptations Morris 1983.[3] This
     could provide a suitable basis from which to expand housing
     handicap on the assessment form.
**   The mother retained the care and discipline of the family through
     the installation of a through-floor lift. The handicapped boy was
     also provided a through-floor lift: to endure the construction
     of a ground-floor extension, everything being brought through the
     house, was too much to ask of the family already coping with
     handicap.

Social factors have not been the subject of any published attempt at structured assessment. They are very diverse, so that a basis for a structured assessment would present considerable difficulty. The social factors may, even so, be separated from handicap and assessed as a separate constituent of value initially, in the appraisal.

7. CONCLUSION : VALUE AND ASSESSING VALUE

We have outlined here the nature of the value sought in carrying out adaptations to existing housing for physically-handicapped people. Also outlined are the means that are used in assessing handicap – moving some way towards measuring value. A process is followed drawing out value from its constituents – handicap, cost, social factors.

Assessing handicap is considered at some length in the paper. There are two aspects to this which need noting.

In the first place the forms of handicap assessment that are in use stem from a hospital environment. They tend, in consequence of this, to focus on <u>intrinsic</u> handicap: disadvantage arising from an individual's own characteristics; and to place less emphasis than is warranted in the housing context on <u>extrinsic</u> handicap: disadvantage arising from the individual's environment. In order to bring about development in measuring handicap, joint work between those who understand intrinsic and extrinsic handicap is required. Short studies undertaken – for example exploring the suitability of different lifts for installing in houses of different construction materials* – would build up mutual understanding while advancing the subject.

In the second place, the illusive nature of what is being measured gives cause to be reticent in making overt judgements on matters of quality in the assessment. There are doubts whether introducing levels of assessment much beyond a general impression serves a useful purpose – less, still, when they will be interpreted by a third party.

These matters are for resolving with experience. For assessment to develop – and with it our appreciation of what constitutes value – individuals must be free to practise and experiment – and then discuss. Developing an open, questioning, frame of mind towards assessment is important: where value is illusive, assessment provides a basis for reflection and illumination. The wish is always for decisions to be on the best basis available**.

Assessing value in the context of housing adaptations for physically-handicapped people is an imprecise art and will long remain so. By separating out the parts, however, reflecting upon them and where possible introducing structured forms of assessment, an improved focus on value will be obtained. In its turn this will stimulate further insight.

---

*   Such studies provide excellent material for project work in Surveying degree courses.
** The points here are applicable to assessment in any development context where measures of value are illusive.

References

Cornell, A. and Broom, S. (1977). 'Useful assessment form'.
Occupational Therapy, Vol. 40, No. 8.

Goldsmith, S. (1976) 'Designing for the Disabled'. Architectural
Press.

Morris, A.S. (1983) 'Advance special housing units for
physically-handicapped people in Rehabilitating existing
buildings'. Proceeding, CIB 83, Vol. 2.

Open University (1977) 'The Handicapped Person in the Community'.
Course notes.

Sainsbury, S. (1973). 'Measuring Disability'. Social
Administration Research Trust.

RISK ANALYSIS IN RETAIL STORE DESIGN: A CASE STUDY

ALAN WILSON, Liverpool Polytechnic

Introduction

The cost of a commercial building must be considered in the context of the benefits gained from the building. However, at the time at which feasibility studies are performed, the quantitative assessment of both benefits and costs is fraught with uncertainty. Thus, commercial building decisions should realistically be viewed as calculated gambles. There are emerging techniques for analysing such gambles.

Since it is too easy for researchers to describe these techniques from behind a safety screen of vague generalisations, this paper presents a detailed analysis of the building gamble facing a commercial client. The case study has been compiled from the recent experiences of the author in analysing such decisions.

The Problem

The directors of a large, multi-storey, department store in a provincial town wish to remodel a large area of sales floor space. they envisage a significant investment in a new sales "concept" which, by the integration of interior design, mechandising and advertising, will establish a strong image in the customer's mind, and in effect create a new store within a store, hopefully expanding their customer base.

Fortunately, the director charged with investigating the feasibility took the sensible step of employing a Risk Analyst with experience of this type of problem. It might prove useful to develop the case by considering the subsequent dialogue.

ANALYST: I wonder if we could explore a little more fully your motivations for this proposed change and the extent of the remodelling you have in mind.

DIRECTOR: Our business is fairly successful, but we are conscious that we may have too much floor space, or at least too much similar floor space. Of course, every business like ours must make regular changes in presentation of stock in order to stay fresh, and we are happy that we are satisfying our regular customer base. We hope that we may extend this

customer base by attracting the growing number of more affluent professional and business people in the area which our regular market research tells us is not being satisfied. We intend to draw together the more expensive items from several of our departments, which would include both demand and impulse goods, and to supplement these with 'top of the range' brand names across certain produce lines, ranging from clothing to furniture.

This policy proved very successful in our other store, which is in a town about one hundred miles away, but that was on a much smaller scale than we have in mind here.

I have to admit here that we do have a strong suspicion that if don't act to extend our customer base, then rising levels of unemployment in the town will mean that our turnover will decrease in the coming years.

As to the scale of the remodelling envisaged; our floor plan lends itself to the proposed change. We would like to adapt a 20m by 30m wing on the first floor and put in a new staircase to the same space on the second floor, making a total gross floor area of 1200 m^2. We would like new false ceilings and, at least in part, false elevated floors. In addition, there will be extensive and stylish lighting fixtures, carpeting, and use of colour, with the best sales fitments we can find. Overall, the use of superior surfaces and textures such as leather, chrome and glass, with a sophisticated colour scheme will contribute to the effect we are hoping to achieve.

As you may have guessed, I have already had preliminary discussions with an architect and interior designer.

ANALYST: O.K., I think I now understand enough of the problem to begin to rationalise the options. It will prove useful to begin to structure the problem as a decision tree even at this stage. There are two alternate courses of action which we need to consider. Obviously, you are particularly concerned with the remodelling option, but never forget the zero option of not doing anything; at the very least it allows us a datum for comparison.

Each of the alternatives could lead to various results taking place, each leading to a different outcome. In fact, there are a good many possible events and outcomes, as we shall see later, but for the time being, in order to get us into the analysis, I will assume just two events are associated with each option. These will result in six possible outcome permutations (Fig. 1).

We are now going to derive some numbers to help us to choose the best course of action. The numbers at this stage will be very approximate compared to those which we develop later. But, it has been my experience that most people respond more favourably to a gradual build up of complexity rather than going for total accuracy with its consequent complexity initially. Let's see how it goes.

If we were to leave this 1200 m^2 as it is, what do you anticipate would happen to trade?

DIRECTOR: Well, it may continue at the present level, or it may decline. Currently, annual average sales are £600 per square metre per year and our net profit before tax is about 8% of sales turnover.

ANALYST: If we take one of the events as trade continuing at the present level, can you give me a figure for decline in sales which you think is just about as likely as the status quo?

DIRECTOR: Well, it is difficult, but we have been thinking about this issue recently within the firm and I think a decline in takings of 30% in twelve months time is an equally likely possibility.

ANALYST: Good. Now could you give me two equally likely outcomes in terms of sales if we go ahead with the remodelling?

174

Let's assume that we have got over the initial disturbances and we have settled down to a steady state condition, say, 12 months ahead. Don't worry too much about the figures at this stage, there will be plenty of time to refine them later.

DIRECTOR: Well, I'm convinced that sales will increase. At our other store they have more than doubled. I think the two figures I would consider equally likely are 100% and 200% increase on current sales. In addition, the net profit on this type of goods is slightly higher , say 10%.

ANALYST: Now we come to the cost of the remodelling. As I have a quantity surveying background, I have some expertise here, but at this stage, it is very difficult to estimate likely costs. Not only are we not in a position to seek tenders for the work, but we do not even have a design at this stage. Of course, the position of uncertainty matches that which we have just gone through regarding the likely benefits of the proposal. You might then be a little surprised to learn that quantity surveyors and cost adivsors in the building industry are not in the habit of expressing this uncertainty in terms of probability, but rather they tend to ignore it, or try to fool themselves and their clients by their "judicious" choice of suitable averages and by using empirical "contingency allowances".

DIRECTOR: Nothing would surprise me about the building industry. Our experiences in the past have been very bad. I take all figures for costs and time given by building professionals with more than a pinch of salt.

ANALYST: Well, I hope that you will feel a little happier with this approach. Later on, we are going to consider the cost uncertainties in much more detail, but for now to retain consistency, I am going to take a range of costs and display two values which I think are equally likely within the cost range.

To explain briefly, even though we have no design as yet, we do know the dimensions of the space and so the major remaining uncertainty is the quality and hence cost of the final components to be used. I have separated the one-off items such as the stairs and, using approximate quantities, I have attempted to include ceiling, floor and services costs, etc. My main worry is the cost of the special shop fittings. You must realise that I feel quite uncomfortable expressing all this uncertainty, even within a two figure band. It is only because I know we are to treat the problem more fully later, and in the interests of getting an early picture of the problem that I have succumbed to this

DIRECTOR: simplistic process. The two cost figures I have derived are £140,000 and £230,000. We will allow 10% for all fees, etc. By the way, we intend to finance the building work from loan capital and not equity capital. We will take out a loan for the full amount of the building work for 12 months.

ANALYST: We are nearly in a position to put figures to the outcomes in our decision tree. There are two more factors to be considered, each of them prone to uncertainty.

During the construction works, the sales space is wasted, so we must include for the loss of earnings. The construction time can, to some extent, be determined by you as the client and can be reinforced contractually by liquidated and ascertained damages. We will allow four months in this first calculation, but will later treat it as an uncertain variable.

Another factor which I need your help with is the future time period over which to discount benefits. How many years do you think is appropriate?

DIRECTOR: Well, I realise this is important. Obviously, the longer the period we take, the less significant will be the construction costs compared with the benefit to income. But then in the retail sector, fashions change fairly quickly so we would be fooling ourselves to take too long a period. It will take some time for this new concept to become known in the public mind and so we need to give it a fair chance. I think a "pay back period" of three years might be a sensible first test. Forgive me for using this term, I realise it is frowned upon by modern investment analysts, but it does usefully describe the principle.

ANALYST: O.K. Now we can calculate the outcomes in our decision tree. Figure 1 shows the results. (The calculations are shown in full in Appendix 1)

Before we consider the results in detail, I would like to mention the issue of taxation. This can have a large effect uypon the numbers used in the analysis since, as you know, all expenditure on plant which serves a function in contributing to your trade is allowable against your future tax bill. Recent cases have broadened the definition of plant to such an extent that I anticipate a large proportion of the cost of remodelling will be allowable.

176

FIGURE 1

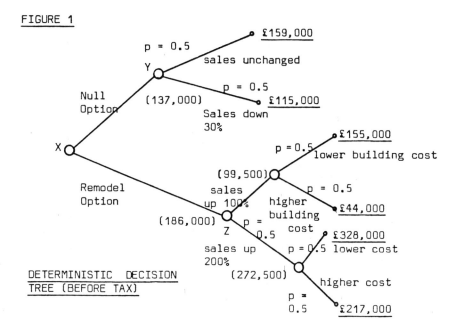

DETERMINISTIC DECISION
TREE (BEFORE TAX)

DIRECTOR: I am glad you mentioned this because we have found that professionals in the building industry often do not understand the significance of the tax allowances to us. If they did, they might formulate their documentation differently.

ANALYST: Well, in this case, I have carried out a complete post tax analysis. Whilst the value of each outcome is very much reduced, their relative rankings are not greatly altered, since assuming you continue to be profitable, taxation effects reduce both expenditure and income. The post tax analysis is subject to even more uncertainties than pre tax analysis, especially those arising from internal accounting practices. In this case, I am confident that our decision making can be based upon the pre tax position without loss of accuracy.

All the figures shown in Figure 1 have been discounted or compounded to a common base datum, namely the completion of the remodelling. We have derived a deterministic decision tree showing six possible outcomes from your choice of actions and with each outcome qualified. How does it seem to you?

DIRECTOR: Well, it seems fairly straightforward and it gives me a clearer view of the problem than I had been able to achieve so far. It seems to me any decision is a gamble in which in the most favourable outcome I could stand to make a profit of £328,000 and in the most unfavourable, the profit would reduce to £44,000. Since this last figure is much less than I am currently making from the space, this could be seen as a loss. The gamble is becoming clearer. What do the figures in brackets mean?

ANALYST: Well, in the case of a deterministic decision tree like this, we can express the result of several possible outcomes of an event by a single number. This is called the Expected Value, and it is a first approximation of the worth of the possibilities. For instance, if you chose the null option, two outcomes of the event at Y are possible. We can summarise these by calculating a weighted average by multiplying each outcome by its probability of occurrence and adding all the results for each event. In this case, the probability of each even it equal so the expected value is equal to the arithmetic mean which is £137,000. At event Z by the same reasoning, the expected value is £186,000.

DIRECTOR: This then seems to suggest that I should take the remodelling option as it has a much higher expected value. But I am a little worried that whilst this course of action has a high possible profit, it also has a fairly high possible loss. I did not realise the decision was going to be so cirtical. I appear to have a one in four chance of making a large loss.

ANALYST: Well, having outlined the problem, we are now going to deal with it in a little more detail and, hopefully, examine the possibility outcomes in a more rigorous way, but I hope you will agree that the above analysis was useful in setting scene.

It is often useful at this stage to transform the monetary values used so far in the analysis into units of utility. These units would express more accurately the worth of the actual gains or losses to the business at this point in time. It is misleadingly simple to assume that a company would react to a spectrum of possible outcomes from, say, -£100,000 to +£100,000 in a linear way. It is a basic tenet of business economics that companies attempt to maximise utility. A utility scale can be developed for a client by structured questioning to see how he values certain gambles and, in particular, asking what certain value of money he would sell the gamble for. Space does not permit a utility analysis in the present case.

178

So far we have made a major assumption which is at odds with reality. We have assumed many values and we have assumed that there are only two distinct possible outcomes at each of our event nodes. In reality, there are many many more which need to be considered.

DIRECTOR: Is this process known as Sensitivity Analysis?

ANALYST: No, we are going a little beyond that. Certainly, we can use Sensitivity Analysis to examine the effects of certain assumptions. For example, we could examine the influence of different discount periods on our results. And there are other factors which lend themselves to this approach. But there are several factors in our analysis that can take on such a range of possible values that if we allowed them all to vary the total number of possible permutations of results would be enormous, in fact, infinite. We are going to use a technique which will make use of your expertise in refining the possible benefit assessments and my expertise in refining the building cost assessments and combine the results to produce a representative sample of this inifinite set of possible outcomes. The technique derived originally from research in nuclear physics and is called Monte Carlo Simulation. One of the problems of this approach is that because of the complexity it is no longer possible to respresent it by our simple deterministic decision tree. From now on, at our event nodes, it will be necessary to envisage an 'event fan', i.e. a large number of possible outcomes.

Let us proceed to consider the likelihood of each possible outcome within the event fan.

So far, when we had an uncertain event we summarised the picture by choosing two equally likely values. But now we need to know what other values are possible and what is the spread of likelihood amongst all the possibilities. The two values chosen could, in fact, come from any probability distribution as in Fig. 2.

Now we must look at the events in our decision tree and see if we can derive for each of them a more realistic distribution of the possible values. Let's look first at estimated benefit to turnover of the new sales "concept".

179

DIRECTOR: Well, I am quite relieved about this, I have felt a little unsure of those figures all along. Before I proceed, I will have to think in a little more detail about the product range, number of customers, etc. Can we meet again next week?

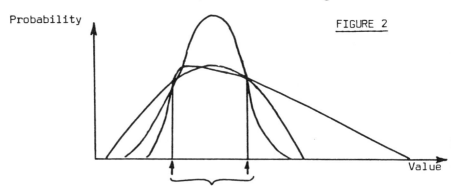

Two equally likely values

ANALYST: O.K., now let us proceed. Is it still appropriate to think in terms of percentage effect on turnover?

DIRECTOR: Yes, I think that measure is as good as any.

One of the problems here is that since we don't have a detailed design, we cannot know how the customers will react. How customers react to aspects of design is something which interests me, but I don't have any scientific data. I guess for the analysis I shall have to assume that the design is the "best" which we can come up with.

ANALYST: We will try to build up a probability picture from first principles. There are many ways of doing it. We could, for example, concentrate on equally likely value bandwidths or we could just consider significant single figures. We will try the latter. Can you give me a percentage increase in turnover which you would regard as the maximum probable value. Notice that I don't want the maximum possible value, since this would be enormous and extremely unlikely. Just give me a figure which you think might not be exceeded in more than, say, one in a hundred guesses.

DIRECTOR: Well, I would say 220%.

ANALYST: Now let's try for a figure describing the bottom of the range.

DIRECTOR: I think 50% is suitable.

ANALYST: Now please give me your single best guess of the likely
              effect on turnover.

DIRECTOR: Well, I have taken the advice of several people on this and
              the most popular figure seemed to be 125% increase. .

ANALYST: That's fine.   I would now like to point out that this picture
              is slightly different from your earlier two figure estimate.
              From this simple triangular distribution of probability, Fig.
              3, it appears that 100% and 200% are not equally likely.   It
              looks more like 100% and 135% have equal likelihood.   Do
              you want me to change things?

FIGURE 3

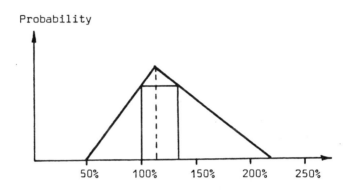

Increase in Sales

DIRECTOR: No, if I am honest, I was aware I have been optimistic in my
              earlier assessment.

ANALYST: There are many other probability profiles which we could
              have generated, but I believe in keepings things as simple as
              possible and this triangular form is easy for us to handle and
              most people can relate to it.   Now, if you could give me a
              similar profile for the null option event fan, I will go on to
              consider the probabilities surrounding the cost of the
              building.

              Let us consider what gives rise to the uncertainty around
              the cost of the building.   It would seem that the cost of the
              building should have less inherent uncertainty than the
              benefit assessment we have just gone through, since much of

the building cost uncertainty is circumstantial and may be controllable. If the work is finally let on one or more fixed price contracts, then at that stage there will be little uncertainty around the cost, that which remains arising from unforeseen circumstances from which, with care, the client can be contractually screened. At the stage at which we are doing this feasibility study, though, we are far from that level of detail.

ANALYST: A point worthy of mention at this stage which emerges from our analysis is that we can afford to spend quite some money to gain better information on the gamble facing you. Such money would be well spent in going ahead and getting a detailed design produced. It may even be worthwhile going further and paying contractors and sub-contractors to actually produce tenders for the work. Such is heresy amongst the building professions.

But back here at our current feasibility study, there are many factors contributing to the uncertainty of cost estimation: the lack of design, contractual form, availability of labour, market conditions for building work, availability of special materials and components, to name but a few.

DIRECTOR: Well, yes, but surely professional advisors are aware that that the cost estimates they make early on are subject to uncertainties. I have even seen the more enlightened ones give a figure in the form of $£X \pm y$.

ANALYST: Well, yes, and this is perhaps better than no attempt to allow for uncertainty, as in the single figure point estimate, but in fact to express the uncertainty in this way is not making full use of the skill of the advisor. An expression of the above kind means that the cost has an equal chance of occurring between $£(X-y)$ and $£(X+y)$. In fact, they rarely mean this, they actually mean that although the range is between these limits, the probability of occurrence is not equal throughout the range.

Another approach may be to use bland statistical descriptors such as Variance and Standard Deviation. Now these are a movement in the right direction, but they are not entirely adequate, because in order to be processed into meaningful cost estimations, they require assumptions about the distribution of probability such as symmetry and normality which don't seem to be supported by experimental evidence.

182

I much prefer to build up the uncertainties in a particular project from first principles. In fact, with the help of a micro computer it is almost as easy to do as traditional estimating, but expresses much more meaningful information. As usual with renovation work, the quantities of the completed work are fairly accurately known, the major uncertainty is the quality specification. My major problem lies in deciding at what level to break the cost down. It is possible for me to consider every item of work to be carried out in very great detail; all of the direct resources to be consumed, each and every material, every labour type, and indirect resources of plant and management, etc. I do not think it appropriate to go down to this level for two reasons. Firstly, I expect much of the work to be carried out by specialist sub-contractors and so it appears appropriate to reflect this in our cost breakdown. Secondly, the greater the disaggregation we employ, the more probabilistic data we will need and the greater will be the troublesome influence of covariant effects between the items. By choosing suitable aggregated cost subsystems we can procedurally reflect both the realities of the contract and improve the accuracy of our analysis.

DIRECTOR: Makes sense to me. Please note that I think I can get you a firm single price quote on the fitments.

ANALYST: The results of my deliberations are shown in Table 1. As you will see, I have made use of triangular probability distributions again.

ANALYST: Having now all the raw data we require, we can now proceed to the analysis. Figure 4 represents our new probabilistic decision tree. As before, there are two alternate courses of action possible, but now rather than have two misleading simple consequences of these options, we have event fans representing the spread of possible consequences. At this stage, we can simply feed the data into our desk top micro computer with Monte Carlo simulator and see what happens.

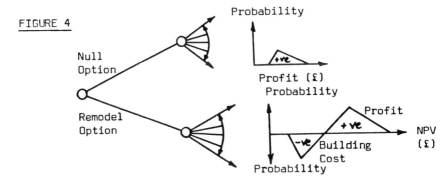

FIGURE 4

183

TABLE 1

SUB SYSTEM		COST ESTIMATE (£)		
		Minimum	Most Likely	Maximum
1.	Stairway	12,000	15,000	19,000
2.	False Ceiling	17,400	17,400	19,500
3.	False Floor	2,400	3,600	4,800
4.	Carpeting	13,920	20,880	34,800
5.	Services	15,080	23,200	40,600
6.	Decorations	7,000	18,000	20,000
7.	Fitments	45,000	45,000	45,000
8.	Screening	3,800	3,800	5,000
9.	General Building Overhead $(x\Sigma_1^8 C_i)$	0.14	0.18	0.35
10.	Design Fees $(x\Sigma_1^9 C_i)$	0.08	0.01	0.15

DIRECTOR: I have read of the dangers of using "black box" computer aids. How much of what happens in the simulation process do I need to understand?

ANALYST: A good point. If we consider the remodelling option, for example, we can no longer subtract the cost of building from the increased profit because we have admitted that each of these is prone to uncertainty and instead of each being a single figure, each is now represented by a range of figures, each of which has a certain likelihood of ocurrence. There are analytical, that is to say mathematical, ways of dealing with the subtraction of probability distributions. But such methods are complex and are not always appropriate. In many cases, to many people the mathematical analysis itself would be a "black box" as witnessed by the commonly encountered lack of statistical understanding. Simulation is a pragmatist's method. Rather than try to derive general analytic solutions, it simply allows us to pick a value from each of the distributions and doing

this a large number of times, typically 1000, gives some
confidence that the consequent pattern of results will
represent the true picture. It is a sampling technique.

DIRECTOR: So it is as if the computer has done the modelling and
checked out the effect on income and hence profit, one
thousand times. That is certainly something I couldn't do
manually. This is exciting, I am anxious to see the results.

ANALYST: Well, Fig. 5 tells the story. It shows the results of a
number of computer runs. It is the cumulative probability
distribution. The vertical scale shows the probability that
the Net Present Value of that course of action will be equal
to or less than the amount identified on the horizontal
scale. The two courses of action are represented by the
two curves.

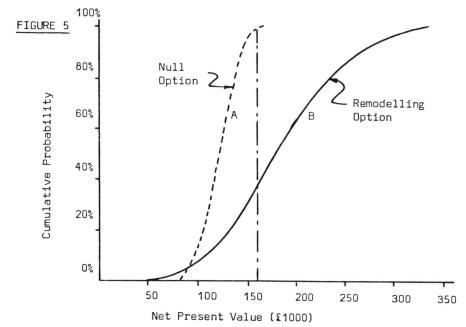

DIRECTOR: This looks impressive, but I am not sure I know how to
interpret the information.

ANALYST: Don't worry, most people react this way initially, but it is
well worth just spending a little time getting familiar with
this form of presentation since it will enable much improved
decision making. We have added a new dimension to our
earlier deterministic analysis, so there is much more

185

information to be assimilated, but I think you will agree that the decision is important enough to spend a little time over.

DIRECTOR: I think I probably know the answer to this in advance, but is there a single simple result arising from the analysis?

ANALYST: No. This was a business gamble and still is. What we have done though is to use the best information available to put figures to the risks involved. This will enable you to make your decision with much more confidence. In fact, in this case, I think the best course of action is fairly clear, but let us consider the curves in a little more detail first.

It is immediately apparent that the remodelling option has a much shallower curve than the null option. The N.P.V. of the remodelling option goes from £30,000 up to £340,000, whilst that for the null option goes from £81,000 to £167,000. Thus, the curves cross and this occurs at a probability of around 6%. Threfore, there is a better than 90% chance that the profit will be greater in the case of remodelling.

A single value which is quite useful and easily understood when describing a probability distribution is the median. The median value is that value which divides the population into two equal parts, i.e. it has a 50/50 chance of being exceeded. In the case of the null option, the median value is £124,000, in the case of remodelling it is £192,000. A large margin in favour of remodelling.

Finally, comparing these scenarios with the status quo. It can be seen that the remodelling option has a better than 60% chance of exceeding the current profitability of the floor space in question (£159,000).

DIRECTOR: Thank you, it is now clear to me that we should go ahead with the remodelling, and thanks to the improved quality of information which this analysis has given me, I anticipate no problems in persuading my fellow directors.

186

APPENDIX 1

DETERMINISTIC ANALYSIS - BEFORE TAX

A. Profit at 8% on £600/m^2/year discounted
    over 3 years at 8%   =   £148,435

    Opportunity value of capital not used in construction
    (loan at 12% for one year, internal rate of return
    at 18%)   =   £11,100

                    Total, (say)   £159,000

B. Profit calculation as above but reduced by
    30%   =   £103,904
    Opportunity value of capital
    as above   =   £11,100
                    Total, (say)   =   £115,000

C. Profit at 10% on £1200/m^2/year discounted
    over 3 years at 8%   =   £346,348
    Cost of disturbance (loss of sales) for
    4 months   =   -£19,180
    Building cost £140,000 + fees +
    interest   =   -£172,480
                    Total, (say)   =   £155,000

D. Profit as for C   =   £346,348
    Cost of disturbance as for C
                                =   -£19,180
    Building Cost £230,000 + fees
    + interest   =   -£283,360
    Total, (say)   =   £44,000

E. Profit at 10% on £1800/m^2.year discounted
    over 3 years at 8%   =   £519,523
    Cost of disturbance   =   -£19,180
    Cost of building as
            in C   =   -£172,480
    Total, (say)   =   £328,000

F. Profit as in E   =   £519,523
    Disturbance Cost
    as before   =   -£19,180
    Cost of Building as
    in D   =   -£283,360
    Total, (say)   =   £217,000

SPECULATIVE HOUSING: A CRITICAL APPRAISAL OF THE DESIGN QUALITY OF
SOME STARTER HOMES

SEBASTIAN LERA, Royal College of Art

ABSTRACT

We were asked to appraise a number of plans of speculative houses -
mostly starter homes - produced by some major building firms. We
describe how we went about the appraisal, and the resulting comments
we made about the plans.

INTRODUCTION

As an ever increasing number of people become owner occupiers they
will be living in houses built speculatively for the private market.
We recently assessed a number of plans of speculative houses, with
particular regard to the amount of space they provided and the
efficiency with which this space was planned. Twenty-two homes were
studied, ranging from a bed-sitting room flat of 24.3 square metres
to a four/five bedroom house of 153.3 square metres; most of the
plans, however, were of one and two bedroom starter homes.

In our appraisal we have tried to take some sort of balanced view
of the circumstances surrounding the construction of these houses.
On the one hand, developers often pay high prices for land and in
order to maximise profits they need to fit in as many houses as
possible with as little as possible of the site taken up with roads
and paths; narrow frontage houses are generally more economical to
construct in these circumstances. Also these starter homes are
characterised by being designed as a vital first rung on the ladder
of home ownership. Their primary requirements are to be
comparatively cheap to buy, to be mortgageable, and to hold their
resale value. Standards of planning and design are secondary,
especially where improving standards would increase costs. And,
after all, developers have little difficulty selling starter homes.
On the other hand, purchasers' willingness to buy these homes is not
necessarily an indicator of design quality, since large builders can
offer a variety of incentives from cheap mortgages to free fittings,
which may influence strongly purchasers' attitudes.

On balance we have tried to recognise some of the forces that
cause these houses to be planned as they are, while also pointing out
that some of those we looked at were not well planned.

We have purposely avoided illustrating the particular houses and identifying the developers, since the aim has been to make some general points about current speculative housing, and not to criticise any particular developer or design.

METHOD

We worked from floor plans and elevations at a scale of 1:100. From them we were able to locate possible furniture positions and areas required for activities.

Our initial approach was to refer to the Parker Morris report (1), Space in the Home (2), and the mandatory minimum standards for local authority housing established in the 1960s (3). We intended to compare the extent to which the plans satisfied the space standards established therein. However, we quickly found that the starter homes did not comply with those space standards of (1), nor with the plan arrangement and furniture listed in (3).

We therefore had to establish our own basis for the appraisal. For a two bedroom starter home we decided upon the following list, parts of which are taken from Parker Morris:

(1) A porch. Enclosed porches act as draught lobbies and provide useful space for coats and outdoor shoes. A roof and one side wall provide shelter while searching for keys; just a roof is better than nothing.

(2) External dustbin cupboards are essential in terraced houses with no rear access.

(3) A hall or lobby internally acts as a draught lobby and is useful for coats and hats.

(4) The living room. The living room should have enough room for at least three single chairs, or a two seater settee plus chair, plus television set and small tables. Where the living room is also used for dining, there should be room for a dining table (folding) and two chairs. These are absolute minimum requirements; space for an extra chair, for example, would be a bonus. The minimum width should be 3300 mm.

(5) Kitchen. Kitchens generate steam and smells; an enclosed kitchen prevents them wafting through the house, and reduces problems of condensation. The layout should consist of worktop-cooker-worktop -sink/drainer. There should be space for a washing machine and a refrigerator. Worktops with cupboards under are 500 mm or 600 mm deep; a disadvantage of the shallower ones is that most cookers stick out at least 600 mm, so with 500 mm deep units the cooker sticks forward in an awkward way both visually and practically. It is desirable for sinks to be underneath windows, to provide a view when washing up. It is a little awkward to have the sink right next to a corner. There should be the potential for eating in the kitchen.

(6) Staircase. A staircase in a living room allows the warm air of the room to rise and escape, and cold air to descent into the room; an enclosed staircase, or one rising from a hall, prevents this, and is to be preferred. Straight flights or stairs with quarter landings are considered safer than those with winders (non-parallel nosings). If there are winders, it is better for them to be

at the bottom of the stairs, then there is less far to fall.

(7) Main bedroom. The bedroom should be able to accommodate a double bed metric size 2000 mm x 1500 mm/6'6" x 5'0" with a bedside cabinet each side, plus a chest of drawers and a wardrobe, unless there is one built in.

(8) The second bedroom should accommodate a bed 2000 mm x 900 mm/ 6'6" x 3'0", a chest of drawers which may be next to the bed, assuming there is not room for a bedside cabinet, and a wardrobe unless there is one built in.

(9) Bathroom. This may be the minimum practical size provided the layout of the fittings is not constricting.

(10) Airing cupboard. For preference this should open off circulation space (eg. the landing) rather than off a bedroom or bathroom. The closer the airing cupboard is to the hot water taps the less water will be wasted waiting for it to run hot, and the less hot water will be left to cool in the pipe afterwards. Airing cupboards should also be internal, to conserve heat, rather than against an outside wall.

(11) Additional built-in storage. It is convenient to have some space where items used at intervals can be kept hidden from view (eg. ironing board, vacuum cleaner, sports equipment, tools).

Using this list we tabulated the floor areas and the features of the houses. We rated each plan for its overall success, taking into account all the individual features but weighted towards the main rooms of the house.

RESULTS

Studio flats
There were three studio flats among the dwellings appraised. They all comprised a bed-sitting room, kitchen and bathroom. One had a separate hall, another a separate dressing area. Their planning owes as much to boat and caravan design as it does to house-building; we suspect that they demand a well-organised lifestyle, free of hobbies that need much equipment, and free of extensive entertaining.

One bedroom dwellings
There were six dwellings in this group - one flat and the rest two-storey houses. The houses were based on a minimum size for a double bedroom plus bathroom, airing cupboard and landing; these rooms dictated the area of the first floor, and the ground floor was given the same area. This resulted in very small living rooms and kitchens. One house rather cleverly had low eaves so that the ground floor could be large and the bedroom could fit into first floor roof-space. Generally the narrow frontage houses were less well designed with deep, narrow sitting rooms; the flat had the best layout as the living room was not used to gain access to another room, nor did it contain the staircase. Our appraisal suggested that the narrow frontage houses were more difficult to design well; they tend to have narrow living rooms which also have to serve as circulation space, making furnishing awkward. The most successful

plan was that of the flat, where the living room was not used to gain access to another room, neither did it contain the staircase. In all the houses the staircase was in the living room. We believe this is likely to cause draughts in winter, and is particularly undesirable where the front door also opens directly into the living room, as it did in four of the five houses.

Two bedroom dwellings

There were ten of these, all two-storey houses. The two we most preferred were up to the Parker Morris space standard for a three person dwelling at 61 square metres. Two gave a choice of eating in either the kitchen or living room. The least pleasing feature was that five of the houses had the staircase in the living room, and of these, three had the front door also opening into the living room. We believe that the occupants must suffer from uncomfortable draughty conditions in winter. Our other main criticism was the small size of some second bedrooms; the narrowest of which was just 1.7 metres wide.

Three and four bedroom houses

There was just one three bedroom house, one four bedroom house, and one very large four/five bedroom house in our sample. The three bedroom house was the bare minimum in size we considered desirable at 72.7 square metres (compared with a Parker Morris standard of 76.5 square metres for a four person dwelling, including storage). Each room would accommodate a reasonable amount of furniture. The plan was weak, however, in having a ground floor arrangement where the front door entered a small lobby containing the staircase. The lobby gave access to the living room, but it was necessary to pass through the living room to get to the dining room and through the dining room to get to the kitchen.

The four bedroom house also had an open plan ground floor. This might not seem disadvantageous, except that the two smallest bedrooms were little more than box rooms and we wondered about the desirability of an open plan living/dining room in a house where two of the bedrooms were too small for any activity apart from sleeping.

The largest house was a generous 153.3 square metres. It had many attractive planning features not only generous in area but well laid out such as the utility room acting as a draught lobby for the back door. The only omission we noted was that of an enclosed front porch.

DISCUSSION

This was a quick appraisal, prepared for a publishing deadline and done at a cost of about £50 per dwelling. It would have been interesting to compare additionally:

(1) the construction of each house and thus capital costs comparisons;
(2) in depth appraisal of the space utilisation of the dwellings with more detailed reference to activities and to potential furniture layouts;

191

(3) computer simulations of heating, energy consumption, running costs and air movement perhaps along the lines described in reference (4).

We hope that our brief appraisal has captured most of the relevant factors, even if we have not been able to be precise about, for example, the likelihood of draughts.

It would be interesting to know what, if any, of these factors are studied in depth by those who prepare the plans. From the results of our brief evaluation, we suspect almost none.

This is a particularly disappointing conclusion, and one where we should be pleased to be proved wrong. It appears that despite all the research undertaken for public sector housing in the 1960s, and widely disseminated since, houses are currently being built in large numbers at space standards which were considered inadequate a quarter of a century ago. And, despite the sophisticated computer simulation and modelling techniques currently available, and widely disseminated in the architectural press and elsewhere, some of the houses we have appraised have such poor performance characteristics that they can be detected at a glance.

If the past is a guide to the future, these houses will probably be occupied for at least a century. Designers, therefore, have a duty and a responsibility to give real consideration to questions of space utilisation. People are comparatively adaptable, and it might be unnecessarily restricting to apply space requirements rigidly. At the same time there is a well-established body of knowledge of domestic activities and their spatial requirements which, it appears, some designers are failing to incorporate.

When designers fail to be self-regulating in their adoption of standards, regulations are introduced to enforce standards mandatorily. As a result of this study we began to think hard about the desirability of imposing standards. We are unwilling to suggest that it is possible or desirable to regulate in order to improve design quality, and are loathe to suggest extending the Building Regulations. Nevertheless, the Building Regulations help to protect the nation's housing stock and ensure reasonable standards of safety, longevity and energy consumption. In fact, they also cover subjects such as open space, ventilation and the height of habitable rooms. If dwellings continue to get smaller and smaller, so the likelihood of regulations covering space standards and planning arrangements seems to increase. From the results of this study, with the greatest reluctance, we conclude this might be no bad thing.

REFERENCES

1. Ministry of Housing and Local Government (1961) Homes for today and tomorrow, HMSO.
2. Ministry of Housing and Local Government (1968) Design Bulletin 6: Space in the home, HMSO.
3. Ministry of Housing and Local Government (1967) Circular 36/67: Housing standards, costs and subsidies, HMSO.
4. BRE: Options on Energy, Architects Journal (23 May, 1984) pp.47-63

Section VI
Life Cycle

LIFE CYCLE COSTING - A MEANS FOR EVALUATING QUALITY

ROGER FLANAGAN, University of Reading

INTRODUCTION

Life cycle costing is a method of appraising quality and total cost
of buildings whereby the economic consequences of available alternatives are analysed to arrive at the optimum cost solution. The
technique is equally applicable to existing buildings or for consideration of an element of a building.

At the present time the professionals in the building industry
like to discuss and pay lip service to life cycle costing. However,
the reality is that whilst the theoretical concepts are well
developed and well understood, there is very little practical
application on building projects. A major constraint of the life
cycle approach is that capital expenditure tends to be viewed
differently to revenue or operating expenditure. The design team
is concerned with meeting the client's brief which will have an
imposed budget capital cost limit. This budget constraint will
often hinder efforts to reduce future costs, which might well be
justified on the basis of a life cycle cost approach.

The relationship between the initial capital costs of construction and the running costs of buildings over their life cycle
has changed dramatically over the past decade. For example, the
capital cost of constructing an office block at 1984 prices will
be in the region of £600/m^2 of the gross floor area. The running
costs for 1984, excluding interest charges, will be in the region
of £200/m^2 per annum: inflate this cost at 5% per annum over, say,
a 25 year period and the importance of running costs becomes a very
real issue. The implications of these long term costs should have
a much stronger influence on design decisions with respect to
buildings and building elements than is currently the case. The
shift from understanding the theory of life cycle costing to using
it in practice will not be easy. The technique is heavily dependent
upon using running cost and performance data gathered from previous
projects and forecasts made about future events. Despite the progress in forecasting techniques, the fact remains that there is no
fallible way to predict the future; forecasting is not an exact
science. Nevertheless, it is an area of decision making which cannot
be left to go by default. A number of assumptions about future

195

events will be made when forecasting. A constant criticism of life cycle costing is that it uses assumptions and forecasts which are no more than best guesses. There is nothing wrong with the 'best guess', as long as it enables the design team to make better decisions than would otherwise have been made. Note the use of the phrase 'better decisions'. Forecasting will not guarantee correct decisions, but it will improve the basis upon which decisions are made. As Lord Keynes said, 'It is better to be almost correct, rather than precisely wrong'.

This paper puts forward the proposition that when active minds are applied to the best available data in a structured and systematic way, there will be a clearer vision of the future than would have been achieved by intuition alone. The effort will be justified even if it merely leads to the rejection of a few demonstrably wrong decisions.

Life cycle costing comprises four main components: the input data, a series of techniques, a system for applying techniques to data, and the output. Figure 1 shows a diagram of the life cycle system. (See diagram on next page).

UNDERSTANDING QUALITY

One of the fundamental aspects of life cycle costing is that it attempts to measure quality in a building. Different users of a building will have varying views on the meaning of quality. The concept that the lowest capital cost always means poor quality is not correct. Careful design can provide a good quality building at a low initial capital cost.

The user will measure quality in subjective terms by considering the comfort levels of lighting, heating, cooling, minimum waiting time for lifts, and so on. Most importantly the user is concerned with the way the building performs the function for which it was required.

The designer will view the quality of the building as an entity with its impact on the visual environment, while the energy manager will consider a quality building as one that performs adequately with minimum fuel consumption. The maintenance manager's assessment of a quality building means it must have high reliability with little or no maintenance problems, while the cleaning manager assesses quality to be associated with the types of floor, wall and ceiling finishings.

Hence there are varying perceptions of the meaning of quality. There is a need to take account in some quantitative manner of these varying perceptions at the design stage. Life cycle costing provides such a tool to consider the relationship between initial capital and long term costs.

THE INPUT DATA

Life cycle techniques are heavily dependent upon data which are collected from buildings in use. The data will be for complete buildings and building systems. Types of data will be:

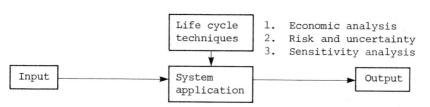

<table>
<tr><td></td><td>Life cycle
techniques</td><td>1. Economic analysis
2. Risk and uncertainty
3. Sensitivity analysis</td></tr>
</table>

Input ───────→ System application ───────→ Output

1. Data collected on occupied buildings in use.

2. Assumptions about how the building will be used (occupancy pattern, frequency of cleaning, etc).

3. Forecasts about future events (life cycles, maintenance, patterns, inflation, energy use, etc).

4. Client's input (discount rate, time period of analysis).

5. Client's tax position.

6. Measurement information on the proposed building (floor area, wall area, etc).

7. Alternative options to be considered.

1. Life cycle cost plan.

2. Life cycle cost analysis.

3. Life cycle cost management.

4. Full year effect cost.

Figure 1. The Life Cycle Costing System.

- cost (fuel, annual and cyclical maintenance, cleaning, insurance, etc)

- performance (the units of electricity used, therms of gas used, what items of maintenance have been undertaken, etc)

- occupancy (periods of occupancy, number of occupants)

- general descriptive (superficial floor areas, areas of glazing, functional use of spaces, age of building, general condition of building)

Collection of these data is complicated by five factors:

- There is no standard method for collecting and recording running cost or performance data. Whilst the owner may collect data for accounting purposes, it is unlikely to be in a format suitable for life cycle costing purposes.

- There is a long time lag between the design stage of a building and the time when running cost data becomes available. Whereas with the capital cost the tender price is a measure of the accuracy of the budget forecast, the check on the forecasts of running costs may not be available for several years.

- Building use is dynamic. Alterations will be made to the fabric and the pattern of occupancy at frequent intervals. Buildings must be seen as complex systems and it has to be recognized that there is an imperfect understanding of the full set of interrelationships and interdependencies between the components of that system. For example, if window design is considered; the energy usage is a function of the window area, the type and thickness of glazing, internal and external temperature differences, exposure, aspect, building usage, and location.

- No two similar buildings will have identical running costs. The data will relate to different buildings, in different locations, with different occupancy, at a fixed period of time. The data will be highly variable and require careful analysis.

- Data are not costless to collect, there must be a perceived value to the owner in capturing the information.

The method of recording the data has not been adequately researched up to the present time. Cost data are easier to collect than performance data, but cost data on their own are of little use. For example, some forecast of the useful life spans of building components used within the building needs to be made. Typically, if the life of a boiler is considered. The traditional 'bath tub' curve of Figure 2 is an idealized characteristic of failure rate with respect to time for a boiler during its useful life.

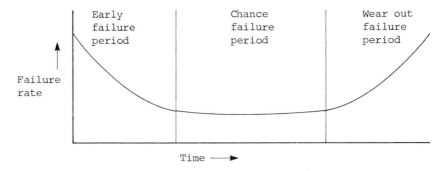

Figure 2. Failures Against Time.

Immediately after the installation, the failure rate is high, due to manufacturing faults, faulty installation and commissioning. After the initial period, the failure rate settles down to a steady low figure where accidental or random faulures occur. The final stage is the wear out stage where the failures are frequent. Unfortunately no common distribution, like normal or exponential, fits this curve. The distribution which fits the curve is called a WeiBull distribution.

In a boiler the unit can be considered as: a) boiler shell, b) burner, c) pumps, d) controls. The burner is the most likely component susceptible to failure and this will be affected by the maintenance pattern. Correct maintenance decisions and hence an indication of maintenance costs, can only be taken if there is an idea of the useful life periods for the various components in a building. Very little published data exist on useful life periods. However, some data for mechanical plant are available as shown in Table 1.

Table 1. Useful Life Periods

Equipment	Years
Boilers, cast iron sectional	20
Burners, (gas and oil)	20
Pumps	15
Pump bearings	7
Calorifiers and cylinders	22
Oil tanks	25
Galvanized cold water tanks	20
Standby generator set	25

(Source: M J Patel, 1984)

There is a need to develop a data base and data base management system that will provide structuring facilities which are capable of expressing the complex relationships that will exist between the data items used for life cycle cost calculations.

LIFE CYCLE TECHNIQUES

A large body of literature exists on life cycle techniques involving the principle of compound interest. The operation of compound interest is fundamental to all formula used in life cycle costing and reflects the time value of money concept - namely that money has the capacity to multiply with time.

The discount rate is selected to reflect the investor's time value of money. It converts costs occurring at different times to equivalent costs at a common time. There is no single correct discount rate for all investors. Selection of the discount rate should reflect the level of return on alternative investment opportunities or on the cost of borrowing money. The earning rate available on alternative investment opportunities should take precedence over the borrowing rate as an indicator of the appropriate discount, if that earning rate exceeds the borrowing rate.

The selection of the time study period over which the life cycle cost is calculated will vary with the client type. A client may wish to consider the study period as being the holding period that is expected to maximize speculative profits. This may well be shorter than the physical or functional life of the building.

All future cash flows should be estimated to reflect the rate of general price inflation by using a nominal discount rate which includes inflation. Where certain items, such as energy are expected to change at rates significantly different from the general rate of price increase, they should be estimated on the basis of the specific rate of price increase expected; be it faster or slower than the general rate of price increase.

The real rate of interest then becomes:

$$R = \frac{1 + i}{1 + e} - 1 \qquad\qquad (1)$$

R = real rate of interest
i = interest rate
e = inflation rate

Figure 3 shows interest rates and the general rate of price inflation plotted against time for the UK. Interest rates on the money market behave erratically, whilst the inflation rate follows a steady upward trend.

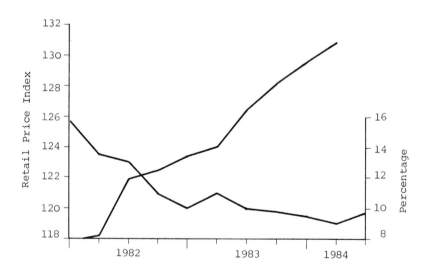

Figure 3. Interest Rates and Inflation.

200

The impact of taxation for the private sector investment should not be ignored. Many of the running costs of a building represent an expense which is tax deductible. The impact of tax can have the effect of turning a loss making project into a viable one.

There are problems that may arise in dealing with risk and uncertainty when assessing life cycle costs for projects. Uncertainty and risk need greater definition. Uncertainty can be taken to be anything that is not known about the outcome of a venture at the time when the decision is made. Risk is the measurement of a loss, identified as a possible outcome of the decision.

A decision is said to be risky and uncertain if it has several possible outcomes. The majority of life cycle cost calculations can be expected to involve uncertainty. Assumptions and forecasts are made about the cost of energy, cleaning, maintenance, and so on. These financial flows are not known with certainty. Information should be considered in the context of what could happen, what should happen, and what will happen. Furthermore options that are 'more risky' than others should be identified.

The preferred method of incorporating risk and uncertainty into life cycle costing involves sensitivity analysis. By using that knowledge which is available regarding the probable variability in cost and revenue streams it is possible to generate a picture of the probably outcomes for the project.

A simple example shown in Table 2 considers the worst possible outcome and the expected outcome of two alternative options. By putting an objective value of the probability of the outcome as a 'weighting', the variance and standard deviation can be computed as the measure of dispersion. Option II is more risky than Option I since it has greater variance.

Table 2

	Possible Outcome (O_i) (1)	Expected Outcome (E) (2)	Deviation from E $(O_i - E)$ (1) - (2) =(3)	Squared Deviation $(O_i - E)^2$ (3) x (3) = (4)	Probability of Outcome (p_i) (5)	Weighted Squared Deviations $p_i \times (O_i - E)^2$ (5) x (4) =(6)
OPTION I	30	37.5	-7.5	56.25	0.2	11.25
	35	37.5	-2.5	6.25	0.3	1.875
	40	37.5	2.5	6.25	0.3	1.875
	45	37.5	7.5	56.25	0.2	11.25
				Variance		26.25
				Standard Deviation		5.12
OPTION II	10	37.5	-27.5	756.25	0.2	151.25
	20	37.5	-17.5	306.25	0.3	91.875
	45	37.5	7.5	56.25	0.3	16.875
	80	37.5	42.5	1806.25	0.2	361.25
				Variance		621.25
				Standard Deviation		24.92

The use of weightings is either explicitly or implicitly a form of probability analysis.

Another approach to handling risk would be to treat all estimates as if they were totally certain, and then allow risk intuitively somewhere within the decision making process. This approach may be justified in cases where no knowledge is available about the size and likely direction of deviations from the estimates. It is a view based on the same reasoning as that which suggests it may be best to assume that a wobbly cyclist will continue straight on if there is no way of predicting the direction of his next wobble.

The 'no knowledge condition' is unlikely to occur in practice. Usually some information is available about the probable direction of deviations that are likely to occur, and perhaps of their magnitudes. Dealing with these in a purely intuitive way is undesirable, since such an approach is unlikely to identify those elements to which the final decision is particularly sensitive.

It would be feasible to discount 'more risky' projects more heavily to incorporate a risk premium in the time value of money. For example, by assigning different projects to different risk classes, and applying different discount rates to each class of risky investment. This approach has the major disadvantage that since the discount rate is compounded in the discounting process, the risk premium is also discounted.

SYSTEM APPLICATION

The system application involves using life cycle techniques on the components of life cycle cost. The list shown in Table 3 shows a list of typical items that should be considered for the effective implementation of a life cycle costing system. The list is not complete, but it does show the hierarchy of levels of information.

Any system of cost categories is merely a filing system, a series of pigeonholes designed as reminders of possible costs to be included. These costs are initial costs or costs incurred on an annual, cyclical, or intermittent basis during the life of the building.

While the system by which data are classified is crucially important, any system can provide results that are only as good as the original data will allow. If the basic data are inaccurate, then no amount of modelling or sophisticated analysis will give results that are anything other than inaccurate.

For many of the cost components there is an imperfect understanding about the interaction between the building and the building user. There is only a 'sketchy idea' of how buildings are used.

For example, if cleaning is considered. The cost of cleaning is related to the types of covering; the frequency of cleaning; the accessibility of coverings; the use pattern of the building; the number of large areas, corridors, and entrances; the nature of soiling in the building; the superficial areas to be cleaned; time constraints on cleaning times. There is an understanding of the soiling patterns in carpets, see Figure 4 which shows the accumulation of soil in a corridor carpet after 20,000 footsteps.

Table 3

Level 1	Level 2	Level 3
Operation costs	A Fuel (where possible apportion fuel bill to appropriate categories) i) gas ii) oil iii) coal iv) electricity v) other	A1 heating A2 cooling A3 hot water A4 ventilation A5 lifts, escalators & conveyors A6 lighting A7 building equipment & appliances A8 special user plant & equipment A9 other
	B Cleaning	B1 internal surfaces B1 i) user B1 ii) circulation B2 external surfaces B2 i) windows B2 ii) external fabric B3 lighting B4 laundry & towel cabinets B5 external works B6 refuse disposal B7 chimneys & flues B8 other
	C Rates	C1 general rates C2 water rates C3 effluents & drainage charge C4 empty rates C5 other
	D Insurances	D1 property insurance D2 mechanical & electrical services/combined eng. D3 boilers D4 electric motors & pumps D5 fixtures & fittings D6 public liability D7 employer's liability D8 loss of profits or rent receivable D9 special perils D10 lifts, sprinklers & boilers statutory insp. D11 other

	E Security & Health	E1 security services E2 pest control E3 dust control E4 other
	F Staff	F1 porterage F2 caretaker F3 commissionaire F4 lift attendant F5 gardening F6 uniforms F7 other
	G Management and Administration of the Building	G1 builder manager/ occupancy manager G2 plan manager/engineer G3 building management consultancy fees G4 telephone charges G5 stationery and postage G6 other
	H Land Charges	H1 ground rent H2 chief rent H3 easements H4 other
Maintenance costs Main structure	A Main structure	A1 substructure A2 frame A3 upper floors A4 roof structure/roof covering & rainwater drainage A5 stair structure/stair finish/stair balustrade A6 external walls A7 windows, external doors & ironmongery A8 internal walls & partitions A9 internal doors & ironmongery A10 other
Decorations	B External decorations	
	C Internal decorations	C1 wall decorations C2 ceiling decorations C3 fittings C4 joinery C5 other

Finishes/ Fixtures/ Fittings	D Finishes/Fixtures/ Fittings	D1 internal wall finishes D2 internal floor finishes D3 internal ceiling " D4 internal suspended ceilings D5 fixtures D6 fittings D7 curtains & furnishings D8 other

Plumbing,
Mechanical &
Electrical
Services
	E Plumbing &
	Sanitary Services
		E1 sanitary appliances
		E2 services equipment
		E3 disposal installation/
		   internal drainage
		E4 cold & hot water
		   services/water mains
		   supply
		E5 other

	F Heat Source	F1 boilers, controls, plant & equipment F2 fuel storage & supply F3 other

	G Space Heating &
	Air Treatment
		G1 water and/or steam
		   (heating only)
		G2 ducted warm air
		   (heating only)
		G3 electricity (heating
		            only)
		G4 local heating
		G5 other heating systems
		G6 heating with
		   ventilation (air
		   heated locally)
		G7 heating with
		   ventilation (air
		   heated centrally)
		G8 heating with cooling

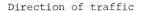
Direction of traffic

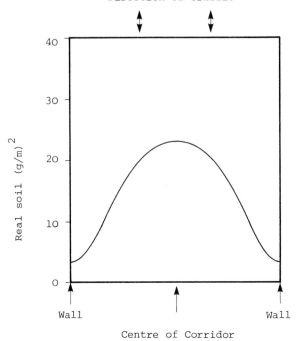

Figure 4.   Soiling Patterns in a Carpet.

(Source: E M Brown (1982))

What is lacking is the bridge between the theory and the way that data can be used by the design team to ensure the best quality at the optimum cost.

OUTPUT FROM LIFE CYCLE COSTING

The output from life cycle costing will be either:

- A life cycle cost plan which is appropriate for proposed buildings or building systems. It takes explicit account of initial capital costs and subsequent running costs, and expresses these various costs in a consistent, comparable manner by applying discounting techniques. In essence its main use is to facilitate the choice between various methods of achieving a given objective.

206

- Life cycle cost analysis involves the collection of information on the running costs and performance of occupied buildings. The main use of life cycle cost analysis is to provide a historical data set on the operating costs of buildings.

- Life cycle cost management is a derivative of life cycle cost analysis as it focusses attention on those areas where running costs might be reduced.

- Full year effect costs are an estimate of the running costs of a proposed building or system. The costs are not discounted, but an allowance is made for the effect of inflation.

Inevitably the output from life cycle costing poses some unresolved issues:

1) How reliable will the results be?

2) What will the life cycle costing service cost to the client?

3) What professional indemnity can be offered, bearing in mind the numerous assumptions that must be made?

4) Which professional discipline is best qualified to undertake the life cycle costing service?

CONCLUSIONS

Life cycle costing is very largely a numerical and analytical concept that will quantify only those things that can be quantified. By itself life cycle costing does not make decisions. Decisions are made by people. In so doing they must be expected to use a set of values of which the numerical results of an analysis may look very precise and convincing, but there is no substitute for professional skill and judgement.

Future research effort should be concentrated on the data requirements of life cycle costing. It is an unfortunate fact that current sources of historical data do not have sufficiently wide coverage for effective use for life cycle cost calculations. Consideration must be given to the design of data capture systems that relate running costs much more closely to individual building functions and components.

Life cycle costing does have an important role to play as a means for evaluating quality - the step from theory into practice is not a major step.

BIBLIOGRAPHY

1) Flanagan, R. and Norman, G. (1983) 'Life Cycle Costing for Construction', Surveyors Publications Ltd, London.
2) Patel, M.J. (1984) 'Replace or Repair', Building Services, Vol. 6, No. 2, pp 43-44.
3) Byrne, P. and Cadman, D. (1984) 'Risk, Uncertainty and Decision-making in Property Development', E. & F.N. Spon, Ltd, London.

4) Butler, H. and Petts, C. (1980) 'Reliability of Boiler Plant',
   Technical Note TN1/80, Building Services Research and Information
   Association, Bracknell.
5) Brown, E.M. (1982) 'Fundamentals of Carpet Maintenance',
   P.A. Brown and Associates, Leeds.

AN APPRAISAL OF WARRANTIES AS A PRODUCT SUPPORT POLICY
FOR THE CONSTRUCTION FIRM

RICHARD J GROVER, Portsmouth Polytechnic
CHRISTINE S GROVER

Introduction

Improving the quality of the finished product is a strategy often
employed by a firm to differentiate its product from those of its
competitors. The objective is to persuade the consumer that the
competitors' products are not perfect substitutes for its own so that
they are willing to pay a premium price in excess of that offered for
a generic product. Alternatively, the product may be offered at a
competitive price in order to increase the firm's market share as
consumers come to regard it as offering better value for money than
competing ones. The ultimate objective in either case is to increase
profitability and, thus, the rate of return on the firm's assets.
 If a product is to be marketed on the basis of its quality, a
product support policy is required to secure and maintain product
differentiation. This is a means by which consumers may be persuaded
that they are buying more than just a generic physical item, but
rather a product whose intrinsic properties offer services that are
not available from competitors.[1] Product support can range from a
guarantee to replace faulty items; through customer support after the
sale by such means as after-sales servicing, extended warranties, and
guaranteed availability of parts; to services designed to meet the
customers' needs throughout the product's useful life, for example,
by maintenance training or post-sale upgrading of the product as
subsequent improvements are made. A product support strategy changes
the emphasis in marketing away from the initial fixed cost of the
product to the consumer towards a consideration of his life cycle
costs. The strategy in offering a product of improved quality involves persuading the consumer that his costs-in-use of it, including
those associated with reduced uncertainty, will be lower than for
competing products. Thus the consumer may be willing to pay a premium initial price in order to secure perceived lower operating costs,
or to buy the product in preference to a similarly priced alternative
whose lifetime costs are perceived to be higher. The strategy may
also be a defensive one designed to protect the market share of a
product whose quality has been subjected to criticism so that the
producer offers to bear a higher proportion of its lifetime costs
(as perceived by the consumer), including those arising from

uncertainty.[2]

This article explores the problems a construction firm must face if it is to use quality improvements to gain a competitive advantage for its products. These fall into two main categories. Firstly, as the marketing of quality involves a product support strategy, the firm must calculate the costs of such a policy. Although there are a variety of product support policies that may be employed, we focus upon one, the offering of an extended expressed warranty. The contingent nature of the claims made under warranties produce greater complexity in costing the policy than a product support policy whose costs are inevitable. The firm must balance the costs of maintaining what may prove to be excessive reserves against the consequencies of being unable to meet claims of an unknown quantity, size, and timing. A variety of strategies are open to a firm seeking to minimise the costs of its warranty policy. It is also an area where construction firms may, in the future, be compelled to meet additional costs as a result of pressure from governmental agencies to extend warranties or judicial decisions on product liability.[3] The cost implications are similar to those from the use of warranties as a means of product support but without the potential benefits from product differentiation since the firm's competitors must make similar provision. The second group of problems faced by the firm are concerned with how to make quality an effective marketing tool in order to enjoy the benefits of a premium price or an increased market share. We examine the difficulties involved in shaping product support services, such as extended warranties, so that the product is perceived by the customer as meeting his requirements and fulfilling his expectations of the complementary services offered in addition to the generic product. As a consequence of the way in which the industry is organised, there are specific problems in this respect faced by construction firms which serve to make quality a more difficult attribute to market than in some other industries.

Modelling Warranty Costs

Expressed warranties are just one method of product support available to a producer. They are contractual obligations entered into by the producer by which he promises the customer certain services, usually for a specific period of time. In particular, they serve to shift some of the costs of uncertainty from the customer to the producer for a specified part of the product's life. The services they offer form an integral part of the satisfaction the customer expects to derive from the product. They, thus, provide a means of product differentiation by enabling a producer to offer different services from those of his competitors. As well as this promotional function, warranties may also be used as protective devices in an attempt to limit the producer's liability. The promotional use of a warranty requires expenditure by the producer who must meet not only the costs of operating it but also of its promotion if it is to fulfill its role in product differentiation. A protective warranty does not involve services to the customer and, so, does not result in direct expenditure. Many warranties, such as that offered on new homes under the National House-Building Council scheme, exhibit both promotional and protective features. It is used by builders for promotional purposes, though its widespread use constrains its effectiveness as a means of product differentiation. It serves to reduce uncertainty for the purchaser, for example, by bearing the risk that a builder of defective premises may become insolvent before remedial work has been carried out. The protective elements are its exclusions such as those for defects arising as a result of wear and tear, gradual deterioration, or from normal dampness.[4]

In order to estimate the potential costs to the producer of a promotional warranty and the size of the reserves he must retain in order to meet potential claims, it is necessary to compute the failure rate function of the product. From this may be derived the probable number of failures at each stage of the product's life and their likely costs. Study of the failure rate function enables the producer to determine the point at which the costs of the warranty outstrip its potential benefits as a means of product differentiation in order that he may decide what proportion of the product's life the warranty should cover. It aids the drawing up of protective clauses to limit the producer's liability through the identification of when the absence of exclusions is likely to prove excessively costly.

Failures of building components can be classified as burn-in, random, or wear-out. The classic bath-tub curve shows a failure rate function comprising these three types of failure. Burn-in failures

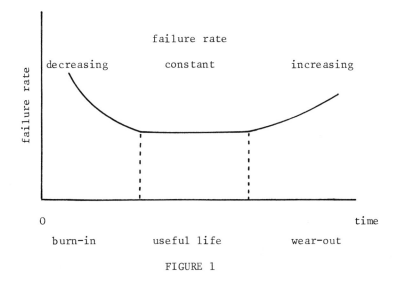

FIGURE 1

decrease rapidly with time. These are usually due to design faults or bad workmanship in construction or installation. It would be unwise for a producer to use warranties as a means of product differentiation unless he was confident that poor workmanship could be controlled and design faults rectified before the product went into commercial production. The useful life part of the curve represents random or stochastic failure. Such failures may occur due to environmental hazards such as misuse or poor maintenance and may be due to factors over which the producer may be unable to exercise direct control. The wear-out phase is represented by an increasing failure rate as the product reaches the end of its useful life. The protective features of the NHBC warranty serve to ensure that the claims made under it are for burn-in failures rather than those associated with the useful life or wear-out stages. If the objective is to use the warranty in order to differentiate a product from the generic item by means of its quality, then a warranty that covers the useful life stage can be used to demonstrate that the producer believes his product to be capable of withstanding normal abuse and that he does not anticipate wear-out failures during the period of the warranty. He is prepared to indemnify the customer against its premature failure. Warranties that cover just the burn-in stage can play only a small part in product differentiation in view of the limited complementary services additional to the generic product that they offer.

A useful representation of the types of failure is the two-parameter Weibull probability distribution function

$$f(x) = \frac{\beta}{\alpha} \left(\frac{x}{\alpha}\right)^{\beta - 1} \exp - \left(\frac{x}{\alpha}\right)^{\beta} \tag{1}$$

and the Weibull cumulative probability distribution function

$$F(x) = 1 - \exp - \left(\frac{x}{\alpha}\right)^{\beta} \tag{2}$$

where α and β are the scale and shape parameters respectively. A value of $\beta < 1$ gives a decreasing failure rate, $\beta = 1$ gives a constant failure rate or exponential distribution, and $\beta > 1$ gives an increasing failure rate. The bath-tub curve can be represented by using a mixed Weibull function. The distribution is widely used to model failure rates and has been successfully applied to building problems.[5]

The costs of a warranty are determined by the number of claims made during the warranty period plus those of administering the scheme, including validation of claims and processing the rebates, and the costs of promoting it as an effective means of product differentiation. The actual cost of each claim depends upon the nature of the warranty. For example, suppose that the benefits to the customer from a product decline as a function of the duration of ownership. If the warranty provides for equivalent reinstatement in the event of failure, then the cost per claim will also decline during the warranty period. If, however, the warranty covers the repair of the item, the costs may increase. Replacement of failed units by new ones may serve to produce a constant use.[6] The objective in offering a warranty is either to secure a premium price or an increased market share. Part of the additional revenues secured as a result of the warranty should be assigned to a sinking fund to meet future claims. As the additional revenues and costs from the warranty occur at different times in the product's life, they must be placed on a time equivalent basis in order that an appraisal may be made of the effectiveness of the warranty. The future costs should be discounted to a present value and the investment income from the sinking fund taken into account.

Traditional approaches to the calculation of warranty costs assume stochastic or random failure. The burn-in and wear-out failures are not considered to be covered by the warranty. Design faults are usually assumed to have been eliminated prior to commercial production and installation problems are ignored. The approach is consistent with the assumption that the producer is liable to meet burn-in failures irrespective of whether an expressed warranty is given so that the additional costs of a warranty must arise from useful life failures. Wear-out failures may be excluded by limiting the period covered by the warranty.

W W Menke has estimated the cost of warranty claims by using the mean time to failure (MTTF). His treatment assumes a decreasing warranty rebate scheme and calculates the total warranty reserve fund to be allocated to cover anticipated claims for a given production size during a specific production period.[7] His model has been extended by H N Amato and E E Anderson to allow for discounting so that future warranty costs may be placed on a time equivalent basis and to allow for the effects of inflation on the product price that must be charged to cover warranty costs.[8] They argue that Menke's model, by ignoring these influences, overstates the required warranty reserve and the product price that must be charged to cover the warranty's costs.

In a further study Amato, Anderson and D W Harvey have established a more robust model which recognises that failures need not necessarily be random.[9] They calculate the total warranty costs for production over T years where the warranty lasts for w years as

$$C = \int_{o}^{T} \int_{o}^{w} Q(u) (R(t) f(t) dt) du \qquad (3)$$

where $Q(u)$ is the demand (number of units sold) at time u, $R(t)$ is the failure cost at time t, and $f(t)$ is the probability density function of failure at time t. The equation calculates the probability of failure at time t during the warranty period w and multiplies this by the resultant cost of this failure. This is repeated over the production period T. Since warranties may be given for products consisting of different components, each with a different failure function and repair and replacement costs, equation 3 may be expanded to allow for n different components:

213

$$C = \sum_{i=1}^{n} C_i = \sum_{i=1}^{n} \int_{0}^{T} \int_{0}^{w} Q(u)(R_i(t)f_i(t)dt)du \qquad (4)$$

The warranty costs must be attributed to the appropriate accounting period so that revenues are matched to the costs of earning them. Rebates are paid for warranty claims some time after sales. Equation 3 has been modified by Amato et al so that the rebate costs stemming from sales in the current period, which may be reported in any accounting period in which failure could occur, can be computed. Claims evaluation costs are likely to vary with sales and may similarly be computed for each accounting period. Warranty administration costs, though, are likely to be fixed rather than to vary with sales and this raises the issue of how they should be apportioned. They could be apportioned between accounting periods according to the proportion of total product failures each periods' sales may be deemed to have contributed, though this involves treating a fixed cost as if it were a variable one. Alternatively they could be treated as selling expenses attributable to the period in which they were incurred. Whilst such an approach recognises their contribution to marketing, it overlooks the claims against future earnings that the warranty produces. The preferred approach of Amato et al is to regard a warranty as giving rise to a liability to continue its administration until the end of the warranty period so that the present value of such future obligations are charged against the sales revenue of the period in which the obligation was incurred. The way in which warranty costs are attributed is liable to influence pricing decisions with consequential effects on the competitiveness and profits of the firm. The method chosen for apportioning warranty fixed costs is critical if a full cost rather than a marginal cost pricing policy is adopted.

The failure rate function plays an important part in deriving the expected cost of a warranty but knowledge of it may be limited until the warranty has been operational for some time. Proxies for the true failure rate function include failure rates from similar products and those derived from accelerated life tests undertaken on samples of the product during its development. An added complication in modelling the failure times is that available data does not necessarily reflect the exact failure times but exhibits a considerable delay before failure is detected and claims made under the warranty. Customers may not claim under a warranty if the transactions costs of doing so exceed their expected benefits so that the extent of failure and of customer disatisfaction with the product may not be apparent from the claims data. The way in which claims are evaluated may also influence the data. Failures which are falsely rejected serve to reduce the costs of the warranty though have adverse implications for the strategy of product differentiation by quality, whilst unwarrantable claims which are accepted give the appearance of a higher failure rate.

An alternative approach is adopted by S C Saunders.[10] Where there is an increasing failure rate, samples provide lower tolerance bounds of confidence or service life with specific assurance of no failures. N R Mann and S C Saunders assume a Weibull failure rate and calculate a warranty period derived as a function of ordered observations and determined by production size, sample size, and assurance level.[11] An estimation can also be made of the number of failures before the end of the warranty period. A better estimate is available given three rather than two failure observations from the sample. In addition, knowledge of the Weibull shape parameter will give a better tolerance bound for failure. These calculations can be used to determine the warranty period during which a producer might expect a given number of claims so that it may be set so as to place a limit on a maximum number of claims and likely maximum cost.

Problems may also occur in deriving the cost function from a failure rate function. The cost of meeting a claim can change over time as a result of inflation. If the warranty provides for a rebate of the original purchase price in the event of failure, then the purchaser will refund less in real terms than he received. If the warranty provides for repair or replacement, the consequences depend upon the relationship between their inflation rates and the general price index. If their costs rise at a faster rate than the average, and this has not been anticipated by the producer, then the warranty reserve fund may prove insufficient to meet the rising real costs of claims. A similar result may arise if claims evaluation or overhead administrative costs rise faster than anticipated. The level of costs may be different for different claims. In particular, delay in making a claim or in processing them after failure has occurred may result in higher costs. Figure 2 illustrates two alternative cost

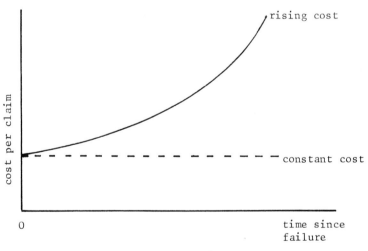

FIGURE 2

functions after failure has occurred. In the case of the constant cost function the cost of the claim does not vary according to the time since failure. For some products, though, failure causes other parts of the system to fail so that costs rise as a result of any delay in remedial work. In the case of construction products, such as foundations or in damp proofing, failures are liable to result in rising warranty costs if there are delays between failure and the claim being met. A rising warranty cost function presents the producer with particular problems in its administration. Whilst he may control the time taken in claim evaluation and the speed in paying a rebate or carrying out repairs, he may find he has little control over delays between failure occurring and it being discovered, or between discovery and a claim being lodged. Delays between failure and the lodging and processing of claims may result in the necessity to make provision for the payment of rebates, claims evaluation, and overhead administrative costs after the period covered by the warranty.

A variety of strategies are open to a producer seeking to minimise the costs of a warranty but the precise means by which this may be done will depend upon the nature of the failure rate, the cost functions, and the sales functions. However there is scope to reduce costs by the judicious manipulation of warranty periods, exclusions, and claims procedures. For example, for products with an initially high failure rate a burn-in period, which would increase the producer's costs, may be balanced against warranty costs and the repercussions of warranty claims on the product's reputation.[12] If as seems likely that there are diminishing returns to a warranty in terms of its effectiveness in product differentiation, it will be necessary to control costs by such means. The producer need not bear the costs of a warranty himself. Since a warranty is essentially a form of insurance for the customer against failure, the warranty costs may be passed on into the insurance market upon the payment of a premium. The consumer may find it advantageous to purchase insurance with the product rather than to make his own insurance arrangements due to the possibility of lower transactions costs and the benefits of bulk purchase being passed on to him in the form of lower premiums. Insurance markets exist for particular elements of buildings such as central heating systems and have recently developed for complete commercial and residential properties.[13]

An important factor in determining the optimal strategy for minimising warranty costs is how the cost function behaves after failure has occurred. If the cost of claims are constant with respect to the time since failure, the cost to the producer can be minimised by delaying rebates. If costs rise with delay, then the producer must seek to minimise the period between failure and the claim being settled. This could involve ensuring that the delay between failure and a claim being made was minimised. The producer could inspect the product in order to ascertain its state. The cost of inspections must be balanced against the savings resulting from the early detection of failure. The type of inspection policy chosen will depend upon the failure rate function. Periodic inspection can be technically efficient where failure is random though it may not always prove cost effective. For an increasing failure rate, where the cost function rises steeply, a policy of sequential inspection reduces the time interval between inspections which can ensure that the risk of failure between each one remains constant. Where random failure

gives way to a rising failure rate, it may prove prudent to delay
the commencement of inspections until the failure rate increases. A
delayed start for inspections may also prove optimal where inspection
costs are high. The periodic, sequential, and delayed periodic
policies are illustrated in Figure 3. The choice of an optimal
inspection policy is complex and we have discussed the issues in
more detail elsewhere.[14]

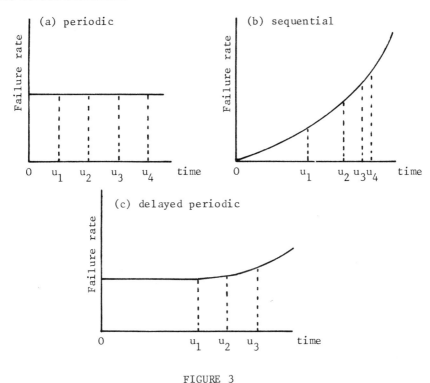

FIGURE 3

Warranties and Marketing

Well-designed products of high quality can fail financially if they
do not reach those potential customers seeking their attributes or if
they are perceived as possessing different qualities from those the
purchasers desire. It is customers' perception rather than the
intrinsic qualities of the product that are important as durable
products are bought in the expectation of a flow of future satisfaction. Product support policies can help to influence customers'
expectations as well as differentiating an item from the generic
product by means of additional services. M M Lele and U S Karmarkar
suggest however that product support policies often do not meet

customer expectations and so fail to differentiate a product from its competitors.[15] They argue that this is often due to policies such as warranties not being viewed by producers as part of their marketing strategy. The responsibility for them may therefore be diffused within the firm and the policies may not be considered until late in the product's development. The policies may reflect the interests of the producer rather than the customer so that, for example, warranty costs are measured in terms of engineering reliability or parts availability rather than the downtime per failure, and warranty conditions designed to be protective rather than promotional.

The devising of an appropriate product support policy for a construction product involves complex trade-offs between its effectiveness and the costs of the strategy. As we have already shown, the appraisal of any individual element within a product support policy, such as a warranty, is likely to be complex in its own right. Moreover, there may be an interaction between one product support policy and another. Two or more policies may be complementary and reinforce the benefits each offers to the customer, as for example with a warranty and after-sales servicing. Policies may also be wholly or partially substitutable in terms of the benefits produced for customers.

It is necessary to measure customer expectations and use these to segment a market in order to provide an effective package of product support policies for each part. Failure to appreciate different customer expectations in different parts of a market are likely to result in some areas of support being over serviced whilst others are neglected and in over- and under-pricing in different segments.[16] The measurement of customer expectations presents problems. Although it is possible to measure how different types of warranty alter the customers' lifetime costs, purchasers do not react to costs as such but to the disutility they are expected to produce. Since utility functions are normally presumed to be non-linear, the change in satisfaction is likely to be at a different rate from a linear change in lifetime costs. Thus, for example, a doubling of the downtime resulting from a product's failure may result in a more than doubling of customer satisfaction. J P Brown[17] has derived an expression for the expected total discounted cost to the consumer, measured in utility (K), of having to replace an asset at time t after having bought a guarantee of level R and length G:

$$E[K(R,G,t)] = L[P(R,G)] + L(A - R) \int_{O}^{G} e^{-pt} f(t)dt$$

$$+ L(A) \int_{O}^{\bar{G}} e^{-pt} f(t)dt \qquad (5)$$

where L stands for the utility loss function, P = premium payable, A = cost of the asset, G = the greatest possible asset life, p = the discount rate, and f(t) = failure rate function. The expression comprises three terms: the loss due to paying the premium, the expected discounted future loss due to failure of the asset during the guarantee period, and the expected loss due to failure after the guarantee period. The use of such an expression enables the consequences of different methods of sharing the risks of failure amongst buyers, sellers, and insurers to be analysed, and the optimal amount and duration of the guarantee for the purchaser to be derived.

The utility that the consumer obtains from a product may be derived by means of conjoint analysis so that a price may be set that reflects the consumers' value-in-use.[18] The repercussions of changes in product support policies on the value-in-use may also be examined. Conjoint analysis involves the identification of the benefit components of a product and the quantification of their relative values in price equivalent terms. Consumers' perceptions of the product's attributes relative to its competitors and the value placed upon them are used to determine price differentials. There are particular problems in applying such an approach to certain construction products. The seller often does not deal with the final purchaser but with his professional advisers or even his appointed main contractors. Their utility functions may differ from those of their clients. When selling to large corporate bodies, it may not be the utility function of the organisation that is of relevance but that of a sub-section or of an individual employee who is responsible for decision making in this particular area.

The main utility from warranties derives from their ability to alter the balance of risk bearing between the producer and the customer. A survey of consumer goods warranties in the USA by J G Udell and E E Anderson pointed to their marketing value in situations in which the costs of uncertainty to the consumer could be significant.[19] They found that their use was associated with products of greater complexity, with lengthy repurchase cycles, and those produced by small and relatively unknown producers, situations which are consistent with those found for many construction products. Their research failed to find an inverse relationship between the use of warranties and the degree of expertise of the consumer, indicating that the possession of technical knowledge by the customer was not regarded as a substitute for shifting uncertainty on to the producer by means of a warranty.

Warranties provide a means by which the customer can exercise greater control over his lifetime costs than by using conventional procurement methods.[20] The latter lay down minimum acceptable standards of performance or of inputs which producers agree to meet. The producer has little incentive to do more than comply with these standards in situations in which improvements would involve him in additional costs and a consequential loss of profits. With a warranty procurement the supplier assumes responsibility for all normal maintenance and repairs and is paid a price determined prior to the contract. He has an incentive to minimise costs over the entire life cycle, and not merely the initial costs, as his profit is dependent upon the subsequent maintainability and reliability of his product. The supplier, and not the customer, is inevitably the one who finally determines the quality of the product. Under a warranty procurement system the contractor has an incentive to improve quality as this should increase his profitability. Under a conventional procurement system there may be an incentive to minimise quality, subject to the constraint of a minimum standard, if this results in a lower initial cost and, thus, the appearance of a more competitive price. The contractor may also be able to escape penalties where the realised performance differs from what was anticipated, particularly if poor workmanship does not become apparent until some time after the sale, or there is no clear definition of what constitutes failure. With a warranty procurement system he cannot escape penalties if the product has poor maintainability or reliability. Minimum standards are likely to present particular problems where these are set in terms of inputs rather than outputs or performance.

The system of competitive tendering employed in the construction industry for bespoke buildings has many features in common with the conventional procurement system. The client and his advisers draw up specifications for the various components. Usually these are based on British Standards and so specify minimum levels of inputs and performance. Contractors tender for the work and, other things being equal, the lowest bid is accepted. The procedure presumes that contractors offer identical products so that price variations would be due to differences in efficiency or profit margins rather than quality. The contractor is therefore under pressure to minimise the initial cost to the customer rather than to offer services in addition to the generic product which may reduce lifetime costs.

L W Hardy and E Davies, as a result of research into the marketing of fencing, suggest that it is difficult for a specialist sub-contractor to offer the range of services associated with the warranty method of procurement by, say, providing a total security service rather than supplying and erecting fencing systems.[21] They argue that

"this is because the traditional definition of roles in the industry stipulates that the client's advisers, that is, the professions, are in possession of all knowledge and skills required to draw up detailed specifications."

The specialist sub-contractor may however possess experience and expertise beyond that of the client's advisers but this is not drawn upon in formulating specifications. A further problem may be that the client's advisers at the initial construction stage are likely to be different persons, drawn from different professional backgrounds, than those employed to advise on the subsequent management of the premises. Hardy and Davies report that the public sector seems responsive to installation and maintenance proposals which would be a means of shifting uncertainties over future expenditure on to the supplier in a period of financial restraint. The strategy they propose is for the producer to develop a strong brand image so that specifications name the brand or "other similar approved" products embodying the same features. If the quality of the product was the result of a production process that might be patented, the producer would clearly be in a strong position to differentiate his product from those of competitors, who would find difficulty in matching its features.

Speculative building would appear to offer greater scope for product differentiation through quality since charging a premium price is feasible. Moreover, the developer may retain a long term interest in the building after its completion through a management contract or by adding an investment interest in it to his portfolio and so may have an incentive to minimise lifetime costs. Alternatively, as in housebuilding, he may produce a large number of buildings under a common brand name and would have an incentive not to prejudice future sales.

A policy of product differentiation on the basis of quality is often regarded as incompatible with being a volume producer. Quality tends to be associated with more costly production methods due, for example, to the use of more expensive components or to higher standards of workmanship. The firm in using it as a means of product differentiation must incur higher marketing, promotion, and product support costs. It is often presumed that higher costs will result in higher prices so that the strategy of producing products of enhanced quality is deemed incompatible with large scale market penetration. The strategy is thus considered more suitable for a small firm seeking to create a niche that is protected from competition from lower cost volume producers. The quality of its product may be further differentiated by its exclusiveness so that premium prices are charged which result in a higher return on capital than for a volume producer.

Whilst quality may result in higher total costs, it does not necessarily cause average costs to be higher or require a premium price to be charged. If the pursuit of quality leads to higher fixed or overhead costs rather than to increases in variable or direct production costs, then there is scope for reaping economies of scale by which the overheads may be spread over a greater volume of production so as to lower average costs. Costs such as advertising and those of establishing and administering a warranty have substantial fixed elements that are unrelated to the level of sales.[22] If product differentiation through quality enables a firm to increase its market share, then it can result in the firm being able to reap economies of scale, not only from the fixed costs produced by the strategy itself, but also on those that would be encountered by producers of the generic product. The fixed costs from higher quality and its promotion can also serve to protect the firm's market share from competitors who, unless they produced on a similar scale, would be unable to enjoy the same economies. Variable costs may also fall if increased sales enables the firm to secure reduced input prices as a result of bulk purchases. Thus the strategy could be compatible with lower average costs and prices, volume production, and a higher return on the firm's assets.

There is evidence from USA that lends support to the view that quality and a low relative cost position are not incompatible. L W Phillips et al have suggested that quality tends to influence the return on investment indirectly via market position rather than directly as would be implied by the niche approach. They found that of their six business groups only the capital goods group had a significant positive relationship between relative quality and relative direct costs, and only the consumer durable group one between relative quality and relative marketing expenditures. Overall, quality appeared to exert a beneficial effect on relative direct costs via the market share, but without compensating trade-offs in other expenditure areas. They suggested that the economies of scale from an increased market share may be reinforced by a learning curve effect not present in products on an inferior quality.[23]

It may prove difficult in the construction industry to secure the increase in scale of operations that the policy of product differentiation by quality appears to require to be successful. Hardy and Davies argue that public sector purchasers tend to spread contracts amongst a variety of bidders and to favour local contractors. Whilst this may be justified as providing evidence of the openness of tendering and in ensuring that there is not undue dependence on a single supplier, it limits the market share of each firm.[24] The pursuit of the strategy in the private housebuilding sector may be limited by consumers' mistrust of advertising and their scepticism about the quality of products, so that the marketing of product's differences from its competitors may prove difficult. In a recent survey of consumer attitudes towards marketing, 48 per cent of English consumers stated that they believed that differences among competing brands were insignificant and only 18 per cent thought that manufacturers' advertisements were a reliable source of information about quality and the performance of products. They tended to favour government testing of products, regulation of marketing, and setting of minimum standards of quality.[25]

Conclusions

We have argued that a product does not sell itself by its quality alone but that customers must perceive its attributes and that these must match those they desire. There is therefore an important marketing role for product support policies to draw attention to the product's attributes and to differentiate it from the generic item. We have focussed our attention on one such policy namely the use of an expressed warranty, though much of what we have said about the marketing use of warranties is applicable to other product support policies. We have shown how the costs of a warranty to the producer may be modelled once the failure rate function is known and suggested some strategies by which the producer may control his warranty costs. We have suggested that product differentiation by means of quality presents particular problems for construction products as a result of tendering practices and the roles played by professional advisers and main contractors. However, since product support policies such as warranties alter the customer's lifetime costs, the current interest in life cycle costing may produce a more favourable environment in the future.

A number of factors are likely to produce a growing interest in warranties for construction products and services. One major developer, Barratts, has employed an extended warranty on its residential properties as a part of its marketing strategy. An insurance market against structural defects has developed in recent years and there are proposals to extend the limited warranties available against defects to the majority of purchasers of residential properties and of home improvements.[26] Maintenance contracts, which include a warranty element, are available for certain types of construction products. As we have indicated, there are advantages in a warranty procurement system over the traditional competitive tendering for the supply of a product or services which are likely to appeal in particular to the public sector in a period of financial restraint. The appraisal of warranties can be complex, both in their costing and in assessing their marketing benefits, so that we would counsel caution in their use.

References

1. The approach adopted here is that goods comprise a number of intrinsic properties which is closer to the normal marketing view than that found in traditional economic theory – Lancaster, K J, (1966) 'A new approach to consumer theory', Journal of Political Economy, 174.
2. An example would be Barratt's extension of the warranty on its houses in 1983 to 20 years following adverse press comment on timber framed houses. The company is one that pays particular regard to its marketing offering consumers a range of services with its products – Spring, M (1983) 'Inside the Barratt machine', Building. One should also note the view of the National House-Building Council that an extended warranty for a specific type of construction, such as timber frame, could imply that it represented a greater risk even when there was no evidence that this was so – NHBC (1983) A Review of the Evidence about Timber Frame Dwellings, app B.
3. For the marketing implications of a change from fault-based to strict product liability see Noon, J (1981) 'Marketing Management and Products Liability', European Journal of Marketing, 15.
4. Tapping, A P DeB, & Rolfe, R (1981) Guarantees for New Homes: A Guide to the National House-Building Council Scheme, 2nd edn.
5. The properties of the Weibull function and its applications are discussed in Grover, C S & R J (1982) ' Explorations towards an optimal inspection policy for the failure of building components', Brandon, P S (ed), Building Cost Techniques: New Directions.
6. The cost implications of different warranty policies for different types of failure rate function are explored in Mamer, J W (1982) 'Cost Analysis of Pro Rata and Free-Replacement Warranties', Naval Research Logistics Quarterly, 29.
7. Menke, W W (1969) 'Determination of Warranty Reserves', Management Science, 15.
8. Amato, H N, & Anderson, E E (1976) 'Determination of Warranty Reserve: An Extension', Management Science, 22.
9. Amato, H N, Anderson, E E, & Harvey, D W (1976) 'A General Model of Future Period Warranty Costs', Accounting Review, 51.
10. Saunders, S C (1968) ' On the Determination of a Safe Life for Distributions Classified by Failure Rate', Technometrics, 10.

11. Mann, N R, & Saunders, S C (1969) 'On evaluation of warranty assurance when life has a Weibull distribution', Biometrika, 56.

12. Nguyen, D G, & Murthy, D N P (1982) 'Optimal Burn-in Time to Minimize Cost for Products Sold Under Warranty', IIE Transactions, 14.

13. Gaselee, J (1983) 'Insuring against structural defects', Chartered Surveyor Weekly, 5th May 1983; 'Building Defects Insurance', Estates Gazette, 5th March 1983.

14. Grover & Grover, loc cit.

15. Lele, M M, & Karmarkar, U S (1983) 'Good Product Support is Smart Marketing', Harvard Business Review, 61.

16. As an example, the aluminium window and door market may be segmented into a replacement market where the product is offred with design, installation, and warranty services and those for new premises where just the product is supplied – Price Commission (1978) Prices, Cost and Margins of Metal Doors and Windows for Domestic Purposes.

17. Brown, J P (1974) 'Product Liability: The Case of an Asset with Random Life', American Economic Review, 64.

18. Christopher, M (1982) ' Value-In-Use Pricing', European Journal of Marketing, 16. For an example of the application of conjoint analysis to residential properties see Findikaki-Tsamourtzi, I, & Dajani, J S (1982) 'Conjoint Analysis of Residential Preferences', Journal of the Urban Planning and Development Division, Proceedings of the American Society of Civil Engineers, 108.

19. Udell, J G, & Anderson, E E (1968) 'The Product Warranty as an Element of Competitive Strategy', Journal of Marketing, 32. Their survey was of 365 warranties of which 39 per cent were classified by companies as primarily protective, 26 per cent primarily promotional, and 34 per cent about equally protective and promotional.

20. Knight, C R (1974) 'Warranties as a Life-Cycle-Cost Management Tool', EASCON '74, Electronics and Aerospace Systems Convention, Washington DC, 7-9 October 1974.

21. Hardy, L W, & Davies, E (1983) 'The Marketing of Services in the UK Construction Industry', European Journal of Marketing, 17.

22. As an example, Barratts spent £2.2 million on direct advertising of residential property in 1981, Wimpey £0.35 million, and Leech £0.4 million. This amounted to approximately £190 per completed dwelling by Barratts, £47 by Wimpey, and £200 by Leech – Mintel Market Research (1982).

23. Phillips, L W, Chang, D R, & Buzzell, R D (1983) 'Product Quality, Cost Position and Business Performance: A Test of Some Key Hypotheses', Journal of Marketing, 47.

24. Hardy & Davies, loc cit, pp.14-15.

25. French, W A, Barksdale, H C, & Perreault, W D (1982) 'Consumer Attitudes towards Marketing in England and the United States', European Journal of Marketing, 16, pp.21,23.25.

26. 'Guarantee for five years proposed to end repeated house surveys' Times, 25th May 1984; 'Builders offer guarantees', Times, 6th April 1984.

OPERATIONAL COSTS OF OFFICES

ROB SMITH, Davis Belfield and Everest

Rising operational costs were often quoted as one of the contributory factors during the apparent rush a few years ago to relocate from London. At the time, The City seemed doomed as a business centre as more and more firms decided to move out. However, analysis of the figures from the now defunct Location of Offices Bureau, suggests that a slack property market, in terms of both letting and development, highlighted the shift away from the capital as one of the few positive trends.

Many companies quoted rapidly rising local authority rate levels as the most significant cost factor influencing their decision, which was an argument supported by the much publicised acts of profligacy from within many town halls. In actual fact, a number of other service charges combined to create the surge of interest and, despite the fact that relocations have dwindled away, operating costs in the UK still represent a considerable constraint on development.

This is illustrated in the World Rental Level: Offices Survey published in May 1984 by Richard Ellis. As can be seen from Table 1, the City of London once again topped the league table of total occupation costs per square foot in the world's major commercial centres.

The Survey indicated that although prime office rents in Tokyo and New York are higher than in the City of London, service charges combined with rates make London the most expensive place in which to operate, while outside of London and Paris the service charges in Manchester and Glasgow are higher than anywhere else in Europe.

It is often argued by agents that tenants' readiness to pay high costs in The City reflects London's dominant position in the international financial market; indeed, there are various predictions that City rents could go as high as £45 per square foot in 1985. To a degree the total occupation cost of offices reflects the absolute limit that the market will stand at any point in time, with the result that any growth in rental levels is to a large extent dependent upon movements in those elements of cost that are outside of the tenants' control, such as rates and insurances.

Prior to the onset of the recession, high construction costs
(tender prices paid by clients) were often quoted as one of the
reasons for feasibility appraisals floundering. At that time,
tender prices were still buoyed up by the boom in investment in the
early seventies.

As one of the most significant variables in a financial appraisal,
along with rent and investment yield, a major increase in tender
prices relative to the other two has a very damaging effect on
viability.

Graphs 1, 2 and 3 illustrate movements in rents, investment yield
and construction costs over the past 12 years and Graph 4 shows the
relative movements in rental growth and tender prices.

The sensitivity of these three variables was illustrated in
Valuation and Development Appraisal* which listed the downside risk
using a theoretical appraisal. This approach showed that a
previously viable development would just break even (ie no profit
would be made) if rents fell by 17.5 - 18%. Similarly, an increase
in investment yield of slightly over 21% or 27.5% in the case of
building costs would have a similar result.

Over the last three years clients have in fact benefitted from
falling, or at worst, stable, tender prices, so that the argument
as to high construction costs no longer applies. At a time when
rental growth seems to be improving and in the short term at least
there is little likelihood of a significant increase in tender
prices, it would seem that the prospects for office development are
vastly improved. However, the concern mentioned above about
operating costs now seems to be firmly established in the minds of
an increasing number of tenants and potential owner occupiers.

In the case of rates, the days of rapidly escalating rate increases
may be over, but the commercial property market is still having to
live with the consequences. The effects of the last few years
of high local authority spending are still being felt by occupiers
and, of even greater significance as far as future development is
concerned, this expenditure has created widespread and deep dist-
ortions in the local rates burden.

The relationship between office rent and rates has been studied
by David Furbur[+]. His research, which was carried out in Liverpool,
concluded that rates do not increase in proportion with rent. Graph
5 shows that the heat may have gone out of the situation, but
compound growth in rates over the last ten years has been running
at over twice that of both rents and tender prices.

A similar comparison by David Furbur of service charges with rent
indicated even less correlation than that between rates and rent.

One of the problems associated with this type of analysis is the
paucity of full and reliable operating cost data. Whereas capital
costs are collected and collated in great detail, the enormous
diversity in the building interests of investors, owner occupiers
and tenants leads to considerable variation in the priority given
to property management.

Inevitably, the tenants' data will be incomplete, for even under
an up to date commercial lease, his obligations are likely to be
limited to the interior and finishes. For a complete picture of

the cost of occupancy, it is necessary to gain access to the actual costs which go to make up the service charges paid by the tenant. These charges cover the landlord's costs of repairing and maintaining the structure, the exterior and the common parts, plus the cost of environmental services, security systems etc.

Even owner occupiers rarely have a centralised source of data. Moreover, despite the fact that in all other aspects of their business, they may well be cost conscious, setting targets and financial projections, the management of the building is often deemed to be of secondary importance. Where there is a lack of property management, the value of the asset is, of course, at risk and unnecessary expenditure is unavoidable.

This is well illustrated where preventative maintenance is ignored, with the consequence that longer term damage is caused, such as where the failure to maintain rainwater goods eventually results in fabric damage and unnecessary future expenditure.

The obvious need in respect of operating costs is to establish reliable bench marks for each outgoing against which items of expenditure can be evaluated.

The histogram in Table 2 gives an indication of typical service charges/operating costs for a medium size air conditioned office block in Central London.

The water and sewerage rates are non-negotiable; however, the benefit of the discounts and commissions from the various insurances to cover contents, fire, lifts, etc. is often not passed on to the tenant or fully exploited by the owner occupier.

The variation in energy consumption figures for office blocks can be enormous. Detailed investigation often reveals outdated environmental systems which have been poorly maintained and the absence of any energy management system.

Audits that show limited expenditure on building fabric, decoration and repairs often conceal a lack of maintenance which inevitably leads to very much increased expenditure in the future.

Because of the direct impact on usage of a failure or malfunction in the building services, it is not unusual to discover high expenditure figures, but this is often due to the deferral of routine maintenance of mechanical and electrical services and lifts. Regular servicing and maintenance is essential if excessive repair costs are to be avoided.

Because office cleaning is an essential day to day requirement the costs are often overlooked, despite the fact that they can be in excess of 10% of the annual occupancy cost. Where high cleaning bills occur, they can be as a result of the use of direct labour, which can cost half as much again as contract labour employed as a result of competitive tendering. The task itself can be minimised with careful detailing at every point of entry to reduce shoe-borne dirt. In addition, the period in the day when cleaning is carried out can have a knock on effect on energy consumption, such that evening cleaning in the winter can add as much as a third to the units of electricity used for lighting.

The decorations form one of the most noticeable office operating cost features albeit that re-decorating is generally one of the least significant annual cost items.

One cost that is of increasing importance to many tenants and owner occupiers alike is that of office adaptations. This is largely due to the effect of changes in computer and communication technology. Major refurbishment or fitting out may only recently have been completed and yet many firms are finding it necessary to re-organise furniture and partition layouts, adjust environmental services and service workstations to accommodate changing staff groupings and relationships.

Table 3 lists some of the operating costs for a number of owner occupied air conditioned offices that were analysed by DB&E a few years ago. The wide variation in operating costs under each heading illustrates how difficult it would be to set meaningful operating cost targets from such a small sample; however, with detailed audit information, useful cost parameters can be established.

Returning to the question of the effect of operating costs on rents and as a result, development, David Furbur could not find a direct link between rents, rates and service charges. However, his research was carried out in 1979, prior to the changeover to a buyers' market, and there is no doubt that total office occupation costs have depressed rental levels.

In conclusion, there is one other operating cost that cannot be ignored and that is depreciation. Even properly maintained buildings wear out, so that owners of property, whether for occupation or investment, need to respond to the fact that the capital invested in them is eroded. Depreciation arises through physical, functional and locational obsolescence. Little can be done about the latter, but the wearing out of the fabric and services and upgrading to improve suitability of use should be provided for through effective property management.

The useful life of an office building is normally put at 60-70 years, with major refurbishment after 20 years or less, when all elements of the building other than substructure and structure may need to be replaced. Table 4 indicates the useful life and capital values for the major elements of an office building.

The need to make suitable allowance for depreciation when considering office investments stems from growing concern about the need to refurbish a number of buildings constructed even as recently as the early seventies. The debate has only just started as to whether investment yields should be adjusted to reflect the true return, having allowed for depreciation, but it must lend support to the argument that an income and expenditure approach to investment appraisals should be adopted. This change would emphasise the need to gain a better understanding of operating costs and it would also re-affirm the importance of sound property management.

* Valuation and Development Appraisal edited by Clive Darlow published by The Estates Gazette Limited.

+ David Furbur ARICS is senior lecturer in the Department of Surveying at Liverpool Polytechnic

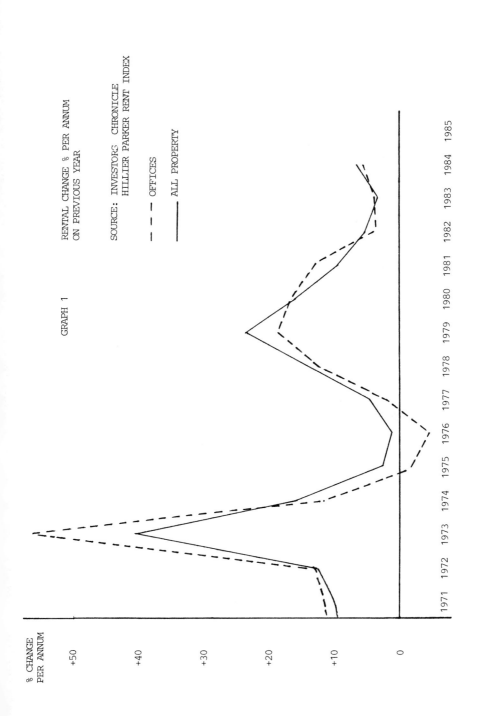

GRAPH 1

RENTAL CHANGE % PER ANNUM
ON PREVIOUS YEAR

SOURCE: INVESTORG CHRONICLE
HILLIER PARKER RENT INDEX

- - - OFFICES

——— ALL PROPERTY

% CHANGE
PER ANNUM

+50

+40

+30

+20

+10

0

1971 1972 1973 1974 1975 1976 1977 1978 1979 1980 1981 1982 1983 1984 1985

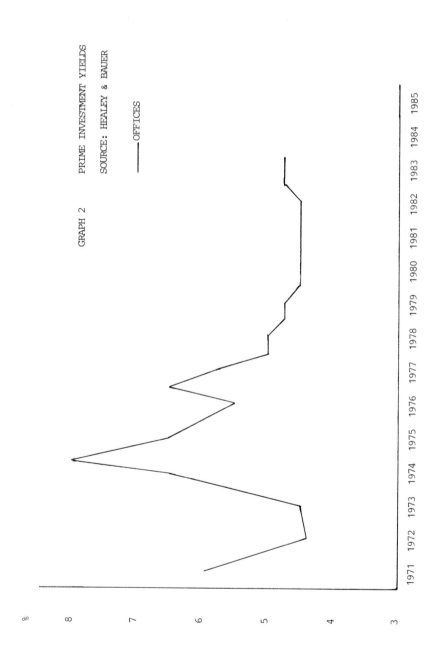

GRAPH 2   PRIME INVESTMENT YIELDS

SOURCE: HEALEY & BAUER

——— OFFICES

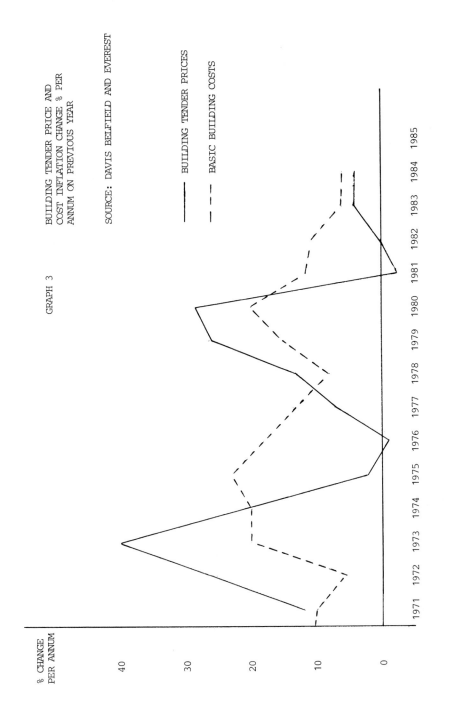

GRAPH 3

BUILDING TENDER PRICE AND
COST INFLATION CHANGE % PER
ANNUM ON PREVIOUS YEAR

SOURCE: DAVIS BELFIELD AND EVEREST

——— BUILDING TENDER PRICES

– – – BASIC BUILDING COSTS

% CHANGE
PER ANNUM

40

30

20

10

0

1971 1972 1973 1974 1975 1976 1977 1978 1979 1980 1981 1982 1983 1984 1985

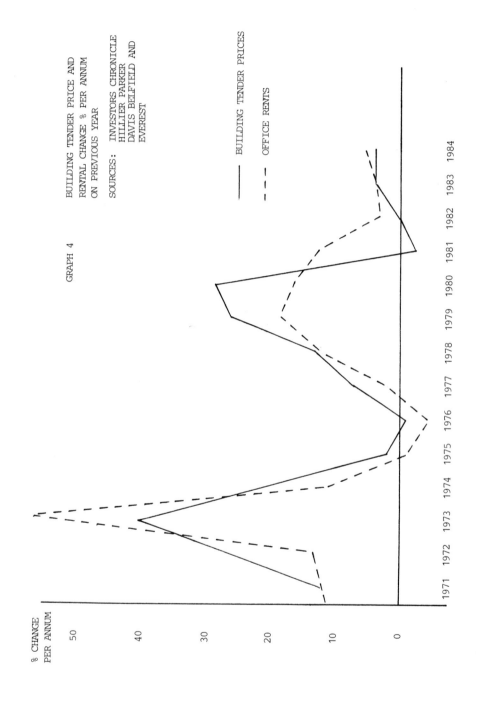

% CHANGE
PER ANNUM

50

40

30

20

10

0

1971 1972 1973 1974 1975 1976 1977 1978 1979 1980 1981 1982 1983 1984

GRAPH 4

BUILDING TENDER PRICE AND
RENTAL CHANGE % PER ANNUM
ON PREVIOUS YEAR

SOURCES: INVESTORS CHRONICLE
HILLIER PARKER
DAVIS BELFIELD AND
EVEREST

———— BUILDING TENDER PRICES

– – – OFFICE RENTS

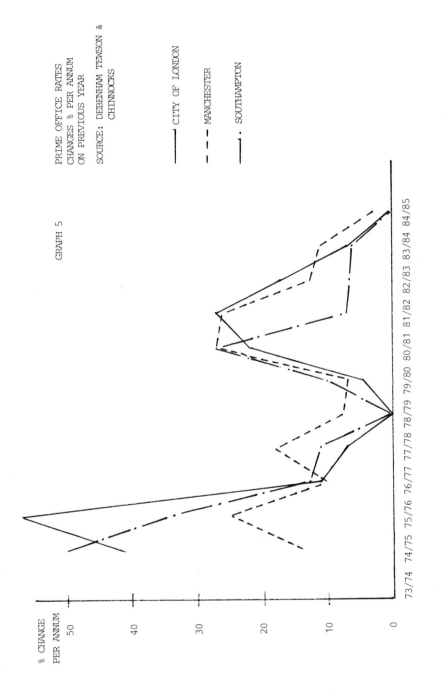

GRAPH 5

PRIME OFFICE RATES
CHANGES % PER ANNUM
ON PREVIOUS YEAR

SOURCE: DEBENHAM TEWSON &
        CHINNOCKS

————— CITY OF LONDON

– – – MANCHESTER

–·–·– SOUTHAMPTON

% CHANGE
PER ANNUM

50

40

30

20

10

0

73/74 74/75 75/76 76/77 77/78 78/79 79/80 80/81 81/82 82/83 83/84 84/85

TABLE 1   WORLD OFFICE OCCUPATION COSTS - MAY 1984

Business Centre	Rent	Rates/ Property Tax	Service Charge	Total Occupation Cost
	£/sq ft	£/sq ft	£/sq ft	£/sq ft
London - City	31.00	17.05	4.65	52.70
London - West End	22.00	8.80	3.96	34.76
Manchester	6.50	2.47	3.64	12.61
Glasgow	7.00	2.73	4.41	14.14
Brussels	4.99	1.45	0.55	6.99
Paris	16.17	4.04	0.81	21.02
Amsterdam	6.61	1.32	0.10	8.03
Frankfurt	9.54	2.38	0.10	12.02
Madrid	8.10	1.78	0.45	10.33
Barcelona	4.97	1.04	0.25	6.26
New York	34.83	4.18	6.62	45.63
Chicago	15.67	5.17	3.45	24.29
Los Angeles	20.39	3.67	1.22	25.28
San Francisco	23.19	5.10	1.39	29.68
Sao Paulo	6.21	1.92	0.25	8.38
Singapore	15.90	3.50	4.77	24.17
Hong Kong	19.25	2.89	2.11	24.25
Tokyo	36.79	4.05	3.68	44.52
Johannesburg	8.13	1.71	Not Quantifiable	9.84
Melbourne	11.01	2.64	1.32	14.97
Sydney	19.25	2.30	1.93	23.48
Perth	7.73	2.08	1.24	11.05

SOURCE: RICHARD ELLIS

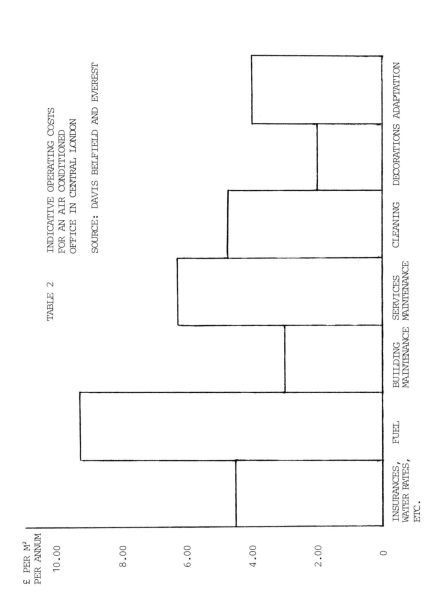

TABLE 2   INDICATIVE OPERATING COSTS
FOR AN AIR CONDITIONED
OFFICE IN CENTRAL LONDON

SOURCE: DAVIS BELFIELD AND EVEREST

TABLE 3   ANNUAL OPERATING COSTS FOR TEN OWNER OCCUPIED AIR CONDITIONED OFFICE BLOCKS

Building	Gross floor area	Depth of Office Medium 15 to 20m	Depth of Office Deep over 20m	Capital Cost Building	Operating Costs Fuel	Cleaning	Services Mainten-ance	Building Mainten-ance	Adaptations
	'000m²			£/m²	£/m²	£/m²	£/m²	£/m²	£/m²
A	24	●		500	7.50	-	-	-	-
B	22		●	408	5.75	-	-	-	-
C	8		●	-	14.20	1.10	9.15	1.50	2.55
D	19		●	421	4.00	-	-	-	-
E	18	●		460	5.75	3.35	7.20	0.30	0.65
F	3		●	502	8.30	5.30	-	3.45	-
G	5		●	378	4.30	4.60	3.40	1.75	-
H	20		●	668	6.95	2.60	1.80	-	2.55
I	19	●		620	7.50	3.20	0.45	0.25	-
J	25		●	431	7.00	-	-	-	-

SOURCE: DAVIS BELFIELD AND EVEREST

TABLE 4   INDICATIVE USEFUL LIFE AND CAPITAL VALUES
          FOR OFFICE ELEMENTS

	Element %	Total %
0 - 5 Years		
Decorations	1½	
	—	1½
5 - 10 Years		
Floor finishes	3	
Electrical fittings	2	
	—	5
10 - 15 Years		
Roof coverings	1	
Wall finishes	2	
Ceiling finishes	3	
Fittings	2	
	—	8
15 - 25 Years		
Partitions and internal doors	3	
Sanitary fittings and plumbing	2½	
Boilers	5	
Air conditioning installation	16	
Special services	2	
	—	28.5
25 - 40 Years		
Windows and external doors	10	
Heating, hot and cold water installation	3	
Electrical installation	8½	
Lifts	5	
	—	26.5
40 - 70 Years		
Substructure	8	
Structure	22.5	
		30.5
Total		100%

SOURCE: DAVIS BELFIELD AND EVEREST

DESIGN QUALITY AND COST IN EDUCATIONAL BUILDINGS

DR ALAN SPEDDING, Department of Surveying, Bristol Polytechnic

INTRODUCTION

Quality in design of buildings clearly bears some relation to, but
is not entirely governed by, cost. The word design will be used in
the context of those factors of the finished building which are
largely determined during the design process, and the term cost is
used to mean the price which the consumer has to pay for a building.
There have been many definitions of quality, for instance it has
been said to refer to the extent to which the properties of a product
fulfil the requirements or intentions of its use (1), or to be set
at levels which the community is prepared to accept (2).
Unfortunately there are a number of difficulties inherent in these
simple statements, because the assessment of quality is related to
the performance and aesthetics of the completed building. Whilst
some measures of performance can be devised, aesthetics is an area
where subjective judgement plays a great part, being very much
dependent upon factors such as the cultural, geographical and
socio-economic perspectives of the observer. It is beyond the
scope of this paper to discuss Contractors' workmanship as a quality
factor except to note that the Contractor who prices for a high
standard of quality in workmanship will probably not win a great
deal of work in competitive tendering situations.
    Satisfaction with design therefore relates to many factors,
including that of the expectations of the observer, and it is
doubtful that we can quantify satisfaction per se in respect of
buildings. However, as the majority of buildings with which we
have to work are part of our current stock of assets we can look
at what we hold already to establish some of the factors which
were important at the time when some of this stock was built, and
which have long-term implications for satisfaction with quality in
relation to cost.

THE CURRENT STOCK

The 1962 schools survey (3), found that approaching one-third of
maintained Primary Schools in England and Wales were built before
1875. Naturally, a large number of these were replaced in the
school building boom of the 1960s and early 1970s, but when one
considers that there was little school building in many rural
counties in the first half of the twentieth century, it is likely
that a significant number of schools still in use are over eighty
years old. Even assuming that these buildings are still giving
reasonable service for teaching purposes, they can hardly be
expected to last indefinitely, nor, without having had considerable
updating, is it likely that they are providing internal environmental
conditions which comply with current standards for new buildings.

The development of school building, particularly since the
beginnings of general education in the early 1800s has not
unexpectedly been heavily influenced by capital cost. Implicit
cost-limits based upon cost and area have held sway at various
times in various forms, both in the nineteenth and twentieth
centuries. Naturally, if these two factors are controlled in some
way at the design stage, then the value-for-money equation must
be affected by qualitative factors including functional and
environmental standards.

That the problem was recognised in Edwardian school building
days is evidenced by the Chief Architect of the London School
Board, who is reported to have said that the extra cost of school
buildings "of some architectural pretensions" was five percent over
the cost of the barest piece of accommodation (4). If one
measure of satisfaction is believed to relate to acceptable
standards of space, then we only have to look at the difference
between the finger-plan Primary schools of the late 1940s, which
provided over six square metres per pupil, and the schools of the
1970s at about three and a half square metres per pupil. The
standards changed dramatically, due mainly to capital cost
constraints, but who is to say what is the relative level of design
quality at either of these levels of provision of space? Naturally,
as the pedagogic methods employed have changed, so have the ways in
which buildings are used, and therefore the view of their
suitability for their purpose. If another measure of quality
relates to the criterion of initial cost, then this is, unfortunately,
not as helpful as one might hope, particularly when the use of new
materials, or materials in new combinations, is involved. The cost
of subsidies given to some of the prefabricated housing and high-
rise flat building systems after World War two, and their
subsequent performance may serve as illustrations of this.

The formation of the, now defunct, National Building Agency in
1964 (5), was intended to provide an advisory service to the
Construction Industry on system building, but the history of this
initiative indicates, amongst other things, that the problems of
deciding what is value for money in design terms are not easily
solved. The intention of one of the author's research projects,
therefore, was to try to expose some of the factors which operated

particularly in the mid 1960s to mid 1970s, in respect of the industy's response to the levels of demand then existing for schools.

RESPONSE TO DEMAND

It is not to be expected that we will experience in the forseeable future, demand for school buildings of the like of the 1960s and early 1970s. This was occasioned by the weight of pupil numbers, by geographical shifts in population and to an extent by the relatively slow rate of post-war building. However, the schools which were built in the post World War Two period are likely to be with us for a long time and we should be able to learn from our experience with them.

The problem faced by the Government was to ensure that buildings would be available when needed, not necessarily to provide them at a lower price, by using prefabrication. Evidence provided in the 1950s (6), and the author's research into school building costs of the 1960s and 1970s, suggested that, statistically, there was little significant difference in average capital cost between system and non-system built schools. Of course, there were parts of England and Wales where systems chould be shown to be more, or less, economical at certain times although the cost-limits operated by the Department of Education and Science could be expected to cause buildings of a similar function and apparent quality to cost something close to the price which the market would bear. Probably, most of the time, if a really cheap building was wanted, and time was not of the essence, then it was likely that traditional construction or an industrial shed would provide this, but how would the quality of such provision be measured? Although there was initially some concentration upon the external envelope of the building, bodies such as the Cleary Committee (7), had early recognised the importance of economy being sought in the total design. They also recognised that research should take place into the more intensive use of school premises by the community out of school hours, referred to as dual use.

The intention of stimulating prefabrication and standardisation was also to provide the opportunity for bulk-buying of components through consortia of Authorities. It must be said that the economies of scale anticipated from bulk-buying were not always clearly achieved (8), and there was some suggestion that increasing demand upon a manufacturer above an optimum figure could increase prices. Much of the response to demand involved the development of a wide range of building systems which included eight separate school building consortia in the public sector, ranging from the formation of CLASP in 1957 to ONWARD/MACE in 1966.

The author collected original design and cost data on over two thousand Primary schools built in England and Wales in the 1968 to 1976 period, which spanned a period of intense building activity and slump in demand, and these properties represent a significant part of the assets which now have to be managed. Figure 1 illustrates the relative movement into systems, and the change in the direction

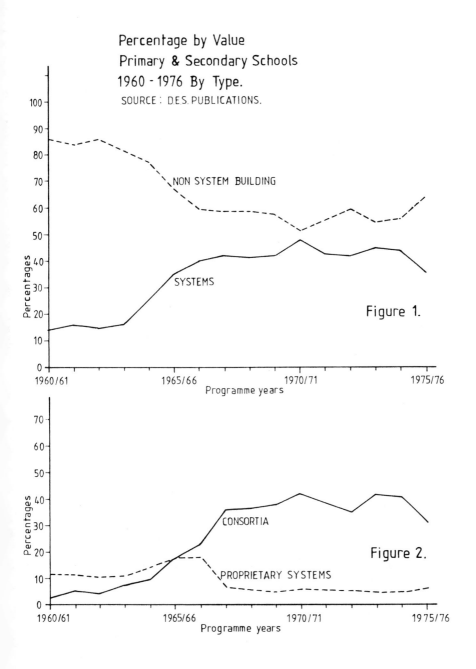

Percentage by Value
Primary & Secondary Schools
1960 - 1976 By Type.
SOURCE : D.E.S. PUBLICATIONS.

NON SYSTEM BUILDING

SYSTEMS

Figure 1.

Percentages

Programme years

CONSORTIA

PROPRIETARY SYSTEMS

Figure 2.

Percentages

Programme years

of the trend within this group when demand fell and other factors
began to take effect.

The school building consortia ranged in concept from attempts to
introduce as much prefabrication as possible, to rationalised
traditional building albeit usually relying upon a steel frame.
Many components were purpose-designed and some were virtually
standard off-the-shelf products, selected to fit the dimensional
discipline adopted. It was anticipated that design cost savings
would amount to at least one percent of capital cost of schools
due to the standardisation of detailing. The overall constraint
of designing within a limited catalogue of alternatives created
opposition in many quarters, although it was clear that the
ability of Architects to get the best out of a system varied even
within a County's Architectural Department. Thus, architecture
of quality depended upon individual skills, whether or not
designing was being undertaken in a relatively restricted
constructional and dimensional environment. The capital cost
relationship of system building to traditional building under
differing conditions of demand and cost inflation was found to
vary, illustrating the value of taking samples over a period of
time when examining relative costs. Systems working at economic
levels of demand could hold steadier when prices were rising,
quickly due to the built-in long term component supply agreements.
In times of falling demand they found, however, that fixed costs
were difficult to reduce quickly.

CONSTRUCTIONAL AND RELATED DESIGN FACTORS

A problem which caused difficulty with some systems was that the
structures chosen for buildings were initially designed for
economy in structural and production terms, and did not necessarily
achieve a good fit with the spaces to be provided. Planning grids
chosen were originally too large to cater satisfactorily for the
range of spans needed in classrooms, corridors, staircases and
toilet accommodation, and created problems in the fine-tuning of
space. Proprietary systems, particularly, within the non-traditional
schools group, see Figure 2 and Figure 1, had to cater for a wide
range of functions in order to achieve economical production levels.
The tendency to use steel frames in the school building systems,
together with the use of prefabricated components resulted in many
cases in problems of tolerances in fitting components together.
The original hope that unskilled workers could erect prefabricated
buildings was not generally borne out in practice because framed
buildings require precision jointing, lack of care with which has
led to making-up on site and maintenance problems. In additon,
costs of manufacturing equipment relative to site mechanisation, and
the wage differential between factory workers and construction
workers in Great Britain, was probably less favourable to the
economical use of prefabricated systems than in much of the rest of
the developed world.

Despite this the average price for system building over England and Wales become very competitive with traditional building in the early 1970s, particularly in the 240 to 320 pupil size schools. Table 1 gives average prices for a large number of single storey Primary schools in this group. Although the price per pupil was the basis for the cost-limit system, the cost per square metre has been given to illustrate the lack of significant difference in average price. The factor which cannot be revealed by quoting such figures is the quality achieved for the money and what short-cuts in design enabled building to be done within the cost limit when it was devalued by rapid cost inflation.

Table 1. Single Storey Primary Schools

Basic net cost at current average prices

Year	Non System Schools		System Schools	
	Number	$£/m^2$	Number	$£/m^2$
1969	59	54.61	74	56.14
1970	71	59.06	92	59.47
1971	107	67.77	136	67.44
1972	70	83.98	95	82.66
1973	81	99.71	108	98.09
1974	46	124.84	63	120.42
1975	71	143.57	61	141.41

The systems could achieve competitive prices, particularly if contractors were used who had experience of them, and if serial or repetitive projects were undertaken. In these conditions Contract planning with certain of the systems was aided by the documentation and the consequent discipline exerted on the design team. Of course, the use of frames for single-storey buildings speeded up roofing-in, but was not always as economical as it might have been, as the wall cladding used was often sufficient to carry imposed loads. If there was some duplication of structural function, and as the systems were, to some extent, competing against traditional construction on price, then economy had to be achieved elsewhere in the buildings. It may be that the service installations were economical in some systems due to standardisation of components and design efficiency, although perhaps the measure of efficiency can only be related to standards provided and consequent maintenance and running costs.

In these respects a significant factor in any school design is the lighting levels thought necessary. Whilst the psychological benefits of natural lighting can be considered as a quality factor which has some measurable cost, the benefits are not really quantifiable. The level of lighting desired in a school building is an important decision as it carries with it effects upon future costs as well as capital costs. Obviously, as the level of lighting deemed necessary rises, the ability of windows to provide this without causing glare and overheating in Summer reduces. Thus more reliance may have to be placed upon artificial lighting and ventilation. The single storey schools benefited from the development of efficient and economical rooflights, which were most suitable for the flat-roofed design solutions now eschewed by a great many Architects but multi-storey buildings and pitched roofs create problems. As windows are usually relatively expensive in the external envelope there is a heat/light balance to be achieved for any given size and shape of building, and for the hours and seasons of intended use in a given orientation. Lighting, therefore, also has effects upon the thermal and acoustic performance of the building and thus the construction of the envelope is also important, the characteristics of heavy traditional construction being quite different from lightweight construction.

Although the system versus traditional building argument has died down at present, the price of construction work is going up relative to the price of other manufactured goods. If there is an upsurge of demand for the products of the industry at a later date, say in inner-city areas, and a lack of skilled craftsmen, then the industry may have difficulty in coping with demand at reasonable prices and a renewal of interest in some forms of system-building may be inevitable. Of course, many recent buildings contain a high proportion of manufactured components, which represents a significant shift in Contractor's work from making to managing. This means that traditional construction is rapidly becoming more rationalised under competitive pressures, and one might hope that feedback on manufactured components should result in better performance and greater predictability in costs in use. Of course, money spent on expensive components and finishes may not necessarily produce lower running costs, and the idea of audits of finished designs before building has been mooted. Experience with NBA noted above may cause us to question, if we can't define what quality actually is, who is really fitted to audit someone else's design, and what responsibility will that party take for the audit?

FUTURE USE?

Although the initial cost of buildings in the public sector are generally expected to be amortised over a sixty-year period, it is clear that the essential fabric of a well constructed building will last considerably longer than this, and there are many colleges in Britain which are of considerable age. Therefore should the fundamental decision be made at the design stage about how long we really want a particular building to last in a particular location? It would seem to be important, and essential in public sector building, for this philosophy to be debated. We cannot forecast the precise nature of the teaching function which will take place in a building in sixty years' time, nor can we envisage where the population will be located in AD 2044. We do not even know that a school will even be a school, as we now know it, at that time. This might lead us to decide that such buildings must be capable of adaption and be flexible enough to allow for alternative uses.

The building of schools with dual educational and community use can be a desirable aim, making more intensive use of facilities and possibly reducing opportunity for vandalising empty buildings. Although some such schools have been built in recent years at what appears to have been reasonable prices, it is difficult to know what is the optimum cost/quality balance, particuarly if the dual use changes the design brief including the type of heating system required. Clearly the structural solution chosen for a school is important in capital cost terms, and also it has implications for future costs. For instance, rat-trad construction with load-bearing cross walls was used for some schools and gave an economical design solution. If, however, the spaces are not appropriate to new teaching methods in the future, it might be difficult to make alterations.

It has been suggested that designs should be tested against two major sources of possible change, i.e. class size/age groups, and changes in teaching and learning methods, (9). Whilst extra cost may be built in by such considerations, and an "ideal" design solution may not be attainable, it may make buildings more adaptable or flexible in use where it can reasonably be anticipated that change of use might occur. Possible structural solutions can be checked against alternative uses, and such studies can then reveal where load-bearing walls may be used without inhibiting change too seriously in the future. The maximisation of future potential is part of the long-life, low energy, loose-fit concept, (10). Probably it is in the Public Sector in particular that there should be opportunities made to examine these matters, due to the long payment period involved in financing development.

In future we will probably make more use of solar heat, ambient energy, insulation, and less traditional forms of construction, but only a total view of maintenance and running costs will tell us whether the money allocated in the initial design stage has proved to be well spent in such cases.

VALUE FOR MONEY

Many of the factors noted above lead one to question what was obtained for money spent on schools in recent years. Whilst innovations in design and construction helped to keep prices down, it also appears that, in many cases, materials and components were used with too little consideration for long-term performance. The financing of capital costs and maintenance costs out of separate budgets did not help here, as some authorities may have been tempted to get maximum area out of capital grants, and to deal with budget out of revenue when faults began to appear. One might expect, therefore, that many of the economies practised, if unsuitable construction ensued, will have resulted in excessive maintenance and failures. Therefore, it is likely that an indication of what was obtained for the money can be found in maintenance expenditure.

Naturally, the total cost of operating a school includes many items which are probably not directly affected by decisions on design alternatives at the constructional elemental level. A life-cycle view might take into account cost of land and buildings, providing mains services, fuel and other consumables; teaching, office, cleaning and caretaking staff, as well as maintenance and repair costs. A well known study found that, if building maintenance costs are taken together with fuel costs, they can amount to approximately half the total of running buildings, the other half being cleaning and caretaking costs, (11). Although all expenditure on buildings will be affected by policy decisions, probably the cost of maintenance is a factor worth close investigation when one tries to highlight value for money obtained from the design, in the light of standards current at the time the school was built, and the initial capital cost. Unfortunately, as was said before, quality in construction may not be achieved at the level intended by the design team. A BRE study, (12), indicated that lack of care on site, and unsatisfactory project information were major causes of significant quality deficiencies. Better communications and supervision would cost a little more but the co-operation would be very rewarding.

It is this aspect of communication of design teams' decisions and feedback on performance which should have become a positive feature of the school building systems. Such information is, no doubt, available inside the school building consortia in particular, but performance data need much more investigation in a coherent, co-ordinated manner, and designers need to interact with such investigations so that the less suitable design solutions are identified at the earliest stage possible. Building work, being labour-intensive is, as has been said above, becoming relatively expensive, and as maintenance and alteration work is even more labour-intensive than new construction, then we can expect the cost of maintenance to rise relative to new work, and the avoidance of maintenance and alteration work to assume greater importance. This trend is reinforced by the conservation lobby, although there could be a rapid swing away from an excess of conservation just as we have seen other apparently growing trends reversed by economic and other circumstances.

246

MAINTENANCE COSTS

The Maintenance Research Unit at Bristol Polytechnic has undertaken
a series of studies commissioned by various bodies, such as PSA and
BRE, on costs of maintenance of buildings. Studies of schools,
commissioned by the Department of Education and Science, have
investigated total maintenance costs and have attempted to
identify which elements of buildings require renewal.

An extensive study of maintenance expenditure in a large County
authority, (13), was undertaken and investigated costs of sixty
school buildings, incurred over a five-year period from 1977 to
1982. An histogram of costs for Primary schools is shown in figure
3, Secondary schools showing a somewhat similar general pattern.
From the histogram it will be seen that decoration is expensive,
despite County guidelines which attempted to virtually eliminate
such expenditure. Services installations and some external works
were also expensive, but the most significant element was roofing.
The conclusions reached on this study led us to believe that certain
design factors had heavily influenced costs of maintenance, but
also made us suspect that a study of even a much larger sample of
schools over such a period would tend to highlight policy, rather
than exposing the unavoidable cost implicit in the original
specification levels.

A subsequent study, (14), was undertaken for the Department of
Education and Science into the maintenance costs of a group of
fifty schools, half pre-1903 and the other half post-1946. The
study was designed to investigate costs of maintenance of these
schools from 1946 to 1982 inclusive. Therefore, we recorded the
maintenance costs for the whole life of the post-war group, and an
histogram for these schools' maintenance costs is given in figure 4.
Comparison of figures three and four shows that similarities exist
in the incidence of maintenance costs, but the cost of roofing is a
much lower proportion when looked at over the longer period. This
difference can be traced to the policy of the County concerned, in
that the late 1970s was a period when it was decided to replace all
troublesome roofs before they became completely uneconomic. Not
unexpectedly, we found also that the average costs per school for
building maintenance rose in the post-war sample as the schools
become older. Evidence from the pre-1903 group of schools led us
to speculate that the average costs of a large stock of schools
begin to fluctuate about some mean figure once the initial cost rise
has been experienced. The general level of costs thereafter depends
upon the sum of individual fialures, and policy. Thus it is in
the identification and classification of failures that potential
for more efficient maintenance cost budgeting, and asset management,
exists.

The author believes that a clear message emerges from these
recent studies, and that is that historical studies of schools
maintenance costs are of use only when the policy decisions which
gave rise to the level of costs can be identified, as well as
details of precisely what work was done in relation to the original
specification levels built into each school. This means that,

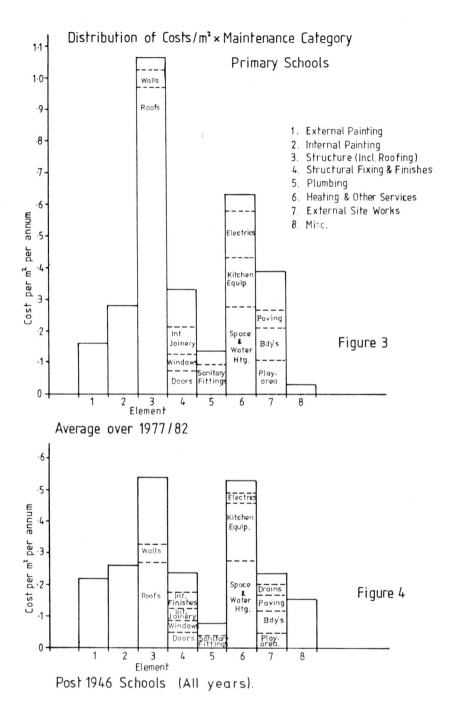

Distribution of Costs/m² × Maintenance Category

Primary Schools

1. External Painting
2. Internal Painting
3. Structure (Incl. Roofing)
4. Structural Fixing & Finishes
5. Plumbing
6. Heating & Other Services
7. External Site Works
8. Misc.

Figure 3

(Figure 3 bar chart — vertical axis: Cost per m² per annum, 0 to 1·1; horizontal axis: Element, 1 to 8)

Labels within bars: Walls, Roofs (Element 3); Int Joinery, Windows, Doors (Element 4); Sanitary Fittings (Element 5); Electrics, Kitchen Equip, Space & Water Htg. (Element 6); Paving, Bdy's, Play. area (Element 7)

Average over 1977/82

(Figure 4 bar chart — vertical axis: Cost per m² per annum, 0 to ·6; horizontal axis: Element, 1 to 8)

Labels within bars: Walls, Roofs (Element 3); Int. Finishes, Int. Joinery, Windows, Doors (Element 4); Sanitary Fittings (Element 5); Electrics, Kitchen Equip., Space & Water Htg. (Element 6); Drains, Paving, Bdy's, Play. area (Element 7)

Figure 4

Post 1946 Schools (All years).

ideally, the costs, together with their explanatory details, should be recorded and examined as they arise and such a strategy is made easier with a computerised recording system. The recording and monitoring of maintenance costs is one aspect of the problem, the other being the need for access to details of the stock held at any one time. Although it is recognised that not all features of design and specification can be economically recorded initially, the computer provides a means of storing and rapid updating of records.

Thus the computer can assist in drawing some aspects of the two strands of quality and cost together in a way which gives quicker and more effective feedback.

CONCLUSION

The estimation of quality in relation to cost inevitably involves subjective judgement. However, as the bulk of the stock of educational buildings which will be needed in the foreseeable future has already been built, we must learn what we can from it. We have seen that various pressures resulted in the wide range of constructional solutions already in being, ranging from traditional, proprietary systems, and consortia systems, to other specially designed solutions.

Some issues closely related to quality can be exposed by looking at initial and continuing costs, related to performance. In the case of initial provision, good property records can be computerised and then quickly updated, whilst continuing costs, particularly in relation to building maintenance can also with benefit be classified and recorded using a computer. This combination can enable us to move away from simply collecting historical costs to a more interactive method of decision making on maintenance work.

Thus, feedback to designers and managers of building assets can become more immediate, and links between design quality and cost can be made clearer.

REFERENCES

1.   J Sittig: Quality in Building: Build International
        Jan/Feb 1969.

2.   P A Stone: Planned economy in construction: Chartered
        Surveyor: August 1963.

3.   Department of Education and Science: The school building
        survey 1962: HMSO 1965.

4.   A Service: Edwardian architecture and its origins:
        Architectural Press: 1975.

5.   Government White Paper: A National Building Agency:
        Cmmnd 2228: HMSO 1963.

6.   M Nenk: School: Proceedings Building Research Congress:
        London 1951.

7.   Ministry of Education: Report of a Technical Working
        Party on School Construction: HMSO 1948.

8.   O Roskill: Local Authority Consortia; advantages and
        disadvantages: The Building Press: 1964.

9.   Programme on Education Building Team: Providing for
        future change: OECD: Paris 1976.

10.   A Gordon: The economics of the 3Ls concept: Chartered
        Surveyor B & QS Quarterly, Winter 1974/5.

11.   Clapp and Cullen: The maintenance and running costs of
        school buildings: BRS CP72/1968.

12.   M Bentley: Quality in traditional housing: BRE CP7/1981.

13.   A Spedding et al: Maintenance in schools: May 1983:
        Bristol Polytechnic.

14.   Asset management in schools: A Spedding et al:
        Bristol Polytechnic: May 1984.

Section VII
Case Studies Experience

THE IMPACT OF CHANGING OFFICE PRODUCTION PROCESSES ON OFFICE
QUALITY FOR THE USER

PETER ELLIS, Building Use Studies

1.   INTRODUCTION

This paper reports early findings from a study of office quality and
its relation to the office production processes, being conducted by
the authors and funded by the Anglo-German Foundation. One aim of
this study is to explore cultural differences in office quality
between Britain, West Germany and the USA. To this end the study
team includes researchers from all three countries. The results of
the cross-cultural study are not yet available, and this paper is
restricted to reporting work carried out in Britain during the first
phase of the project.
     The first task of the study was to define office quality. In
section 2 we examine this from the viewpoints of the office investor,
the corporate user and the individual office worker. Section 3 then
takes the four main office production processes - development, design,
procurement and management - and discuss how variation in these may
affect office quality. In the final section, certain crucial factors
are identified which will influence the achievement of office quality,
and in particular, the resolution of potential conflicts between
quality for the investor and for the various types of user.

2.   WHAT IS OFFICE QUALITY?

Inevitably office quality means different things to different people.
In our research we have focussed on three viewpoints, those of the
office investor, the corporate user and the individual user. In the
case of the very small office, these may all be one and the same
person. Our prime concern is with large offices, the majority of
which are leased from an absentee landlord. Investors' primary
concern is with the exchange value of buildings, as opposed to their
use value. Use value we characterise in terms of two corporate
criteria - organisational fit and building utility - and the primarily
individual criterion of habitability. Generally there is some
relation between exchange and use value, although there have been
times during post-war development booms when entirely useless buildings (measured by the extent to which they were occupied) have grown
in exchange value. Equally there are situations where a building with

253

high use value for a particular organisation at a particular time will
have little exchange value because it is unusable by anyone else.

Conflicts between exchange and use value are matched by conflicts
between different aspects of usability. Building utility is defined
in terms of the inherent properties of a building which render it
economic and convenient to design, build, manage and maintain
irrespective of how well it fits particular organisational needs.
A building with high utility value may be badly located or may convey
inappropriate imagery for many organisations. Many speculative
offices which meet such organisational criteria have a low utility
resulting from over-economy by the developer or sheer bad design.
Habitability too may conflict with corporate use criteria. For
example, the open-plan office tends to expose tensions between a
corporate desire for control and supervision, and individual needs
for privacy and self-management (1).

In our research we have explored the ways in which these different
criteria for quality and the conflicts between them are affected by
four major processes in office production, ie. development, design,
procurement and management. Before discussing these, we shall
consider in more detail the nature of office quality.

2.1 Quality for the Investor

(a) Estate factors

The need to secure good tenancy arrangements is essential to maximising return on investment and minimising risk. This leads to
emphasis on:-

sound leases, ideally for 25 year terms with regular rent reviews and
repair responsibility laid on the tenant;

location of the building in an area where demand will remain high;

'good covenant' with reputable tenants whose track record is known;

single tenancies where possible, to reduce management and maintenance
costs and tenant turnover.

(b) Quantity of lettable space

For highest profit the amount of lettable floor space must be maxim-
ised. The site will be developed to the maximum permitted plot
development ratio. Within the building, the ratio of lettable to
gross floor space will vary according to the designer's ability to:-

keep core areas, vertical access and non-office uses to a minimum;

use wall materials which do not encroach on floor slabs, like curtain
walling;

keep as many building services off the floor as possible.

(c) Lettable quality

Traditionally, lettable quality for the investor has meant a focus on
attractive packaging, and provision of general purpose accommodation
likely to suit the average user. Offices are fully fitted out before
letting to give the appearance of being ready for use. All parts of
the office space are equipped with uniform lighting and air servicing.
Floor plans are straightforward and of standard depth. Hall, lobbies
and reception areas are given a 'quality' finish with carpetting and
attractive decor.

Post-war developers in a buoyant and expanding market for office
space tended to take a short-term perspective, sacrificing building
utility to immediate lettability criteria. As financial institutions
took over more investment in new buildings following development
slumps in the 1960's, there was a shift to longer-term criteria of
providing offices with adaptable spaces and services which would
accommodate change in occupational needs (2). This trend has sharp-
ened recently as demand for office space has dropped. Investors have
been forced to pay more attention to building utility as a means of
attracting tenants. Designing for the accommodation of new forms of
office technology is a recent and particular example of this (3).

2.2 Corporate Quality

(a) Organisational fit

Using the term 'organisational fit' implies a definition of corporate
office quality not just in terms of the physical accommodation, but
of how well matched this is to the organisation. While building
utility makes an absolute contribution to office quality, there are
many variables which have no absolute value. For example, deep plan
office space suits some organisations not others. But the capacity
of a building plan to be changed from an open to a cellular arrangement as time demands is a feature of universal value, and therefore
an aspect of building utility.

It is important to distinguish between organisations' own perceptions of office quality, and those features of the office which
experience and research suggests do in fact contribute significantly
to corporate quality. Many organisations have low awareness of what
to look for in selecting office accommodation, and are strongly
influenced by values derived from agents involved in the market
exchange of office space, such as location, and image. A study by
Poyner (4) of 20 organisations in London showed that the majority
conception of office quality was of the office as an organisational
container, the aim being to get everyone under one roof and to
clarify the physical boundaries of the organisation. More sophisticated conceptions held by a minority included viewing the office as:-

a potential source of staff dissatisfaction

a medium for maintaining good staff relations

a means of improving organisational relationships.

Some managers on the other hand, had no notion of any organisational
implications, viewing the office merely as 'space'.

While many organisations are aware that the <u>quality</u> as well as the <u>quantity</u> of office space contributes to organisational fit, this is often conceived in purely economic terms. <u>The prime</u> advantages of open planning are perceived as economic use of space, and ease of change. Received ideas about improved communications, team spirit and equality sometimes accompany the economic aims, but few organisations are aware of the potential links between type of layout and work organisation; that open plan suits some kinds of work but not others; that corporate and individual productivity may be affected by the degree of privacy and control over the work environment. The significance of the physical environment as a medium for corporate communication (the quality of physical accommodation a company provides tells its workers a lot about what their managers think of them) is also often underrated.

Apart from the physical characteristics of office accommodation, organisational fit is also affected by terms of tenure. Office tenants are as much concerned with rent levels and lease terms as their landlords. In the slack market conditions of recent years, tenants have been in a better position than previously to negotiate a favourable lease, on such terms as frequency of rent reviews and whether these may be both 'up and down' as opposed to 'up only', rent-free periods, and responsibility for repairs. A minority of mainly larger organisations own and occupy their office premises. Often the motivation to own is to gain control. The desire of many smaller organisations for control over tenure and construction of their own offices is being met by various new forms of development which have recently emerged, using non-institutional capital and co-operative enterprise (5). It is thought that this trend will continue.

(b) Building utility

Although there may be a high degree of ignorance among user organisations as to what constitutes office quality, it is equally true that much of our office stock is of poor general quality, that is, it has low building utility. There are two aspects to this. Firstly there is the issue of every-day performance; how easily and economically the building may be maintained and run; whether the services function efficiently, and the fabric is in good repair. Secondly, building utility is to do with the inherent adaptability of the building to a range of use patterns over time. For example, some speculative office buildings of the 1960's and 70's were constructed with the service ducts embedded in the building fabric, making it difficult to change the services without affecting the whole building. Certain plan depths are more adaptable to a range of office uses than others. Very narrow or deep plans are limited in flexibility. A valuable concept in building adaptability is the distinction between long (shell), medium (services) and short (scenery) life elements of the building, and the need for these to be structurally independent (6).

Rapid organisational change and advances in office technology are two factors which have made building utility, and particularly adaptability, a more important issue recently. Office technology

tends to make organisations more aware of the need for efficient
servicing and the capability of building services to cope with
changing loads of equipment, associated wiring and cabling, heat output and electrical provision. In present market conditions, developers are responding more than in the past to user needs, and are
becoming increasingly aware that buildings with low utility may be
difficult if not impossible to let in certain locations.

2.3 Habitability: Quality for the Individual
There is a growing body of knowledge of individual user needs in the
office, and of the links between aspects of the office environment
and individual satisfaction and productivity. A brief summary
follows (7).

(a) Safe, healthy and comfortable environment
Does the environment provide thermal, auditory, and air qualities
that are both perceived as acceptable as well as meeting existing
health and safety standards whenever these exist? Is the environment
perceived as pleasant and comfortable? Does the furniture, equipment,
and space minimize fatigue, discomfort, strain, and stress?

(b) Social integration
Does the opportunity for meaningful small group contacts as well as
privacy exist? Can one control the amount, type, and degree of
contact with others? Is autonomy supported or discouraged?

(c) Adequate and timely feedback
Does the office facilitate direct feedback from supervisor as well as
indirect feedback gained from observing the results of one's actions
on others or on products and facilities? Does the arrangement of
space and equipment and furnishings promote task identity and significance?

(d) Adaptability
Does the setting allow the individual to influence the micro-environ-
ment of his workspace? Are the kinds of adjustments provided on
equipment and furniture considered meaningful by users? Are they
easy to use?

(e) Image
Is the office environment congruent with the worker's professional
identity and expectations? Does it support and enhance his sense of
dignity and the organization's standing with its staff, clients,
community, and competitors?

257

3. OFFICE PRODUCTION PROCESSES

3.1 The Development Process

The key issue for the achievement of office quality with respect to
the development process is the extent to which conflicts between
exchange and use value can be resolved. It is clear that office
developers have become increasingly professional over the last 20
years, but has this led to improvements in building utility?
Institutional developers are professionally advised by chartered
surveyors, architects and financial consultants, and market research
is more frequently used than in the past. Development briefs have
become more sophisticated, and techniques for maximising lettable
floor area have advanced. While these strategies are mainly aimed
at increasing exchange value, market pressures are tending to align
user priorities more with those of the investor, and some recent
new office developments exhibit a number of features likely to
enhance quality for the user. Three of these are notable:-

(a) Design for multi-occupation
Traditionally, developers have favoured single lettings for ease of
management, in spite of an increasing demand for smaller office units.
As a result, many buildings designed for a single occupant have
become multi-let, but do not meet the needs of multi-occupation in
terms of separate entrances to the building, divisible services,
and scope for design differentiation. Some recent office developments have been designed and marketed for multi-occupation by small
organisations.

(b) Accommodation of office technology
Many new developments are now being fitted with suspended floors or
other cable trunking facilities. Other building services like
lighting are being designed with the widespread use of VDU's in mind.

(c) Architectural quality
While the interiors of new developments continue to exhibit the
uniformity and banality characteristic of speculative buildings, more
attention is now paid to the 'architectural' quality of the exterior,
particularly in order to create a uniqueness and 'sense of place'.

Two further factors are likely to contribute to building utility.
The first is the growing proportion of re-development and refurbishment of existing buildings, due to the reduction in prime new development sites and to slackening demand. When a developer cannot rely on
prime location to guarantee lettability, he must focus on other
aspects of the building in order to increase its marketability.
Many of the buildings constructed during the development booms of the
1960's are now coming up for lease renewal, and thus to market
scrutiny as to their lettability. Secondly, few new developments are
now undertaken without the guarantee of a tenant for the whole
building, or an 'anchor tenant' in the case of multi-occupation,
early in the development process. This means that the tenant is in a

better position to negotiate utility in the building design, than in the case of a development which is completed before being put on the market.

In some ways, however, developers are reluctant to increase building utility. For instance, life cycle costing, and particularly energy costs, still take a relatively low priority. Nor is there any marked trend to let speculative buildings in an unfinished state, so that the tenant can organise his own fitting-out. Too frequently in the past, tenants have incurred the unnecessary expenditure of ripping out recently installed ceilings, services and other fittings which do not meet their needs.

The present state of the office market is undoubtedly favourable to a continuing trend to greater building utility. But there is no guarantee that this will not be reversed if a return to the speculative boom conditions of the 1960's occurs.

3.2 Planning and Design Processes

In recent years there has been a massive growth in the range of office planning and design services available to client organisations. Space planning has become professionalised, and is offered not just by architects and interior designers, but by management consultants, office planning consultants, and furniture suppliers and contractors. Tenants requiring a fitting-out service have the choice of using in-house specialists, employing and themselves integrating the services of a variety of specialists, or handing over project management entirely to a firm which offers a complete 'turnkey' service integrating all these functions. Our own research indicates that office quality depends as much on a client's selection of a service appropriate to his particular needs as it does on the physical design product.

The crucial factor here is the organisation's own competence in understanding its needs and developing a good design brief, and then in project management. Large organisations with their own in-house services should have the competence to achieve office quality through the selective use of specialist consultants. The competence of small organisations on the other hand is less predictable. An experienced office manager may be extremely efficient in briefing and integrating professional services with the minimum of outside help. A small firm with no experience may benefit most from a 'turnkey' service supplied by a contractor. Evidence of the growing awareness among clients of the range of new services available comes from the growing number of firms who approach furniture suppliers before they have acquired new office space. The planning and design service offered by these suppliers extends to advising clients on the type of office space best suited to their needs, and specification of services such as lighting as well as the furniture itself. Many of the newer office consultants and suppliers place much emphasis on designing for individual needs and ergonomic efficiency.

Increased competitiveness in the market for fitting-out services is of potential benefit to both corporate and individual office quality. But there is little information available to guide clients to the service most appropriate for them, and as office space and

259

equipment increase in cost, the dangers of a mismatch become more significant.

3.3   The Procurement Process

Whereas the development and design processes have most effect on that aspect of corporate office quality we have labelled 'building utility', the procurement process is central to achieving 'organisational fit'.   A bad fit can arise in three ways: firstly through the client's ignorance of his own office needs; secondly through a lack of information in the market place on where suitable office accommodation may be found; thirdly through the absence of available accommodation of the type required.   All of these are responsible for low office quality in terms both of physical configuration of office space, and of the more abstract factors of location, image and tenure.

We have noted the low level of awareness among many organisations of any functions of the office beyond that of an organisational container.   The corporate attitude to the office environment and its importance to organisational functioning appears crucial to efficient procurement.

This attitude will dictate the quality of the brief for new accommodation, the organisational status given to premises planning, the resources allocated to finding and fitting-out suitable space, and the sophistication of the search for new premises.

Market information comes mainly from estate and letting agents. While these agents usually have an intimate knowledge of rent levels and other quantative factors, they do not normally expect to advise clients on qualitative factors such as spatial configuration suitable to particular organisational needs.   Such information is increasingly being supplied by space planners, but is often not available to firms which either do not perceive the need for it, or cannot afford such a sophisticated service.

3.4   The Management Process

Traditionally, the management of office facilities is allocated low priority and status.   But this appears to be changing in many organisations for several reasons.   Firstly, office space, furniture and equipment are becoming increasingly expensive to acquire and maintain. Rents and rates in Central London are the highest in the world. Energy costs continue to rise.   Salary costs may well be equalled by the running and support costs of the building per head of staff.

Secondly, at a time of economic recession firms are under pressure to use all their resources - staff, equipment, plant - as effectively as possible.   Thirdly, office automation is increasing proportionate investment in equipment, and making the environment more critical to overall office effectiveness.

A new professional label - 'facilities manager' - has been imported from the USA, in recognition of the importance of integrating the acquisition and management of space, furniture, office equipment and even personnel.   There is a growing understanding, at least in the planning and design professions, that modern offices, particularly open or highly automated offices, require active management if effectiveness and economy are to be maintained. A constant process of tuning-up layouts and environmental services is needed if space is not to be used inefficiently or conditions to deteriorate.

Research is also showing the importance in this process of individual involvement or participation. User participation in office systems design is becoming established doctrine (8) and there is growing evidence of the benefits of involving users in decisions affecting their workplaces as a means both of making these more effective and of influencing staff attitudes and morale. Habitability is an aspect of office quality which has much to do with the individual's perception of his environment as with the actual physical conditions. Attempts to define office quality for the individual in absolute physical terms have been unsuccessful for this reason.

4. CONCLUSIONS: KEY FACTORS IN ACHIEVING OFFICE QUALITY

It is clear from the foregoing discussion that office quality is not a unitary concept definable through a set of technical criteria, but a socio-technical phenomenon which depends in part on the resolution of conflicting interests between the different parties to the office production process. Office quality for the user organisation may be in conflict on the one hand with the interests of the owner and investor, and on the other with those of the individual office user. There are certain key factors which affect the resolution of these conflicts and the optimisation of office quality as a whole. The first of these is the corporate attitude to the office environment as a means of achieving organisational and individual goals. An unsophisticated attitude which views office accommodation merely as an organisational container is unlikely to yield office quality either for the corporation or the individual. A poor brief and low priority and unprofessional resources allocated to procuring or managing office space, will probably produce a bad organisational fit and low building utility. Habitability may also suffer if the poor physical quality of the office fails to meet certain individual needs. Even if the physical quality is adequate, habitability may still be low if the corporate attitude fails to recognise individual needs for privacy, control and involvement. Office quality for the individual goes beyond hygiene; for most office workers physical luxury is inadequate compensation for psychological discomfort.

The second key factor is feedback from use. The low utility of much of our office building stock designed and constructed during the last 30 years has resulted in part from a lack of knowledge of user needs, both corporate and individual. This factor is largely independent of economics, since it is often no more costly to design and construct for high building utility than for low. The necessary design knowledge has to come from socio-technical research on whole buildings in use and not just from technical research. There has been a general lack of research of this type, particularly related to the office.

Feedback from use is important not only in the office development and design processes, but to procurement and management. There is a mutual interaction between feedback from use and the corporate attitude to the office environment. A more sophisticated attitude is most likely to employ feedback from use as an organisational tool, and it is through feedback from use that organisations become more aware of the significance of the environment in achieving organisational goals, Effective feedback from use will influence the quality of the brief and hence organisational fit. It should also enhance habitability in two ways; firstly by making the organisation more aware of individual accommodation needs, and hence more able to supply them; secondly, because user involvement is an industrial relations activity which involves negotiation between different interests and leads, when effective, to some resolution and an increased commitment to the outcome.

The final factor is economic, and concerned with supply and demand. Historically, it is clear that during the development boom conditions of the 1960's there was a disparity between exchange and use value which allowed the construction of many low quality buildings. As demand for office space has slackened, there has been a convergence between investor's and users' criteria for office quality. The office employment market is currently booming for employers. Growth in office employment has slowed substantially in recent years (9), and there is no shortage of office labour. In these conditions employers have no economic need to give a high priority to office quality for the individual office worker.

However, there are signs that office workers are becoming more articulate about office quality. Trades unions for example, are more interested than in the past in physical work conditions as a bargaining counter in industrial relations. In some European countries (Sweden, Germany) there is a statutory obligation for employers to consult their staff on decisions affecting their workplaces. As advanced information technology creates the demand for a more highly skilled labour force, this trend seems likely to continue.

REFERENCES

1. Ellis P. and Duffy F (1980), 'Lost office landscapes: evaluation of open plan offices'. Management Today, May.
2. Barras R. (1979), 'The returns from office development and investment'. Centre for Environmental Studies Research Series 35, London.
3. Duffy F. et al. (1983), 'The ORBIT report: office research into buildings and information technology'. Duffy Eley Giffone Worthington, London.
4. Poyner B. (undated), 'Office design and operational efficiency'. Tavistock Institute for Human Relations, London.
5. Cadnam D. (1983), 'He who pays the piper'. Architects Journal, 18 May.
6. Duffy F. et al. (1976), 'Planning office space'. Architectural Press.
7. Becker F.D. (1981), 'Workspace: creating environments in organisations'. Praeger, New York.
8. Mumford E. (1983), 'Designing participatively'. Manchester Business School, Manchester.

9. Bird E. (1980), 'Information technology in the office: the
    impact on women's jobs'. Equal Opportunities Commission,
    Manchester.

ACKNOWLEDGEMENTS

The author wishes to acknowledge the ideas contributed to this paper
by his colleagues Sheena Wilson and Francis Duffy of Building Use
Studies Ltd., Professor Peter Jockusch of University of Kassel, and
Professor Frank Becker of Cornell University. He would also like
to thank the Anglo-German Foundation for sponsoring the research on
which this paper is based.

263

THE REALITIES AND POTENTIAL PROBLEMS OF PRACTICE IN TRYING TO DESIGN
WITH QUALITY FOR VALUE

GABRIEL LOWES, Borough Architect, Darlington Borough Council

1.   INTRODUCTION

The objective of building is to convert the wishes of someone who may
know little about building but is expected to know what he wants from
his building, into a structure perhaps not just for his purpose but
also for, if anticipation is possible, a future use. That process
involves many people and at the design stage there is a frequent in-
tolerance of the difficulties which are not of its own making. The
resultant cash conscious compromise is all too frequently the outcome.
This paper considers eight existing areas of influence and one where
the future has a part to play.

The broader view of those affected by the design process should
acknowledge the known magnitude of the hurdles placed in its way.
Building Regulations and Planning Law are commonly criticised but
these are only two of some 600 pieces of legislation which have to be
taken into account in the building process. Building Contracts too
are now more numerous with all their attendant law cases and legal
opinions. And these are just a few of the many hurdles that affect
the discussions in this paper.

2.   ATTITUDES AND ABILITY

Perhaps the most essential ingredient in the equation is people,
their ability and attitude, for it is they who either ensure good relationships and in turn produce good results - value for money - or
destroy cost effectiveness and create conflict. The "drawing up of
battle lines" is an attitude of mind usually brought about by either
a fear of being taken advantage of or plain greed.

The incompatibility of people chosen on a cash basis (i.e. thrown
together by a tendering procedure) can not be measured in advance and
ability has much to do with establishing good or bad relationships.
It can be said that the designer is as good as his client allows and
this is certainly the case in some instances.

The failures in the field can also manifest themselves as a result
of outside influences on people. A building nearing completion where
a contractor has little forward work creates an attitude in the workforce to make the work spin out, creating delay. Likewise client

decisions may result from a political standpoint (which turns the argument back to legislation) or indeed it could be plain ignorance.

3. FINANCIAL

Setting aside profitability and cost effectiveness of the building process itself, the designer is most heavily influenced by the impact of the financial operations. Borrowed money creates a debt charge generating a revenue cost. That revenue cost must be offset and a profit situation created.

This does however, directly affect such issues as tender periods and that in turn will establish fixed or fluctuating tenders. Allocation of finance in whatever form for construction, depends upon cash availability and or ability to borrow. The Local Government sector of general public expenditure, for example, is allowed an "allocation to borrow" each year. That allocation is not usually known until two months or less before the financial year starts and is not available after the end of that financial year.

In these circumstances several things happen. If you are not ready to start, it is unlikely much can be done, but if you do start, the urge is to keep the tender period within twelve months. If it is within twelve months, it will be a fixed price contract (returning once again to legislation). There are over 300 smaller authorities in such a predicament and if you consider the pressure on the designer to produce to such a deadline (for if there is no expenditure in that year it is assumed you did not want it and the next allocation will be that much less), there is little time for the nicety of design/value. It is a case of design by fear.

This paints a gloomy picture, but is an aspect of design in one sector of the industry which has work to do but is denied the pre-planning which ensures a better value.

4. TIME

Time is something of which the designer is given little. The effectiveness of cost-in-use is part of the design process and ignoring that particular opportunity by not having time to consider it, reduces the effectiveness of value for money. It seems unfortunate that the Civic Trust has to check the structural quality of the buildings it endows with awards because some previous award winning buildings have failed as structures, creating an embarrassment.

It is appropriate under this heading to mention design by default. Even that is sometimes not achieved due to bad communications. The public authority architect or engineer has, unlike his counterparts in the private sector, to live with his buildings. Their organization maintains those buildings and inevitably the design team should include the maintenance divisions in an attempt to prevent design by default. There is little time for this to develop and the temptation to reuse that bill or that drawing again as a short cut is always there and sadly is encouraged by clients.

265

5. ACCOUNTABILITY

There is little commercial opportunism in the public sector. The public sector, as a major client for a large slice of the industrial cake in the UK, needs, not only to be fair with the use of public funds, but seen to be fair. Accountability to many people is wrongly interpreted as public participation. To design towards competition in such circumstances side-tracks the main issue of establishing effective value. The one question which is inevitable is "how do we get competitive tenders?"

The problems which arise from this issue generate from tendering methods and the design approach to them. This tends to exclude the co-operation and co-ordination from the contractors at a stage when the expertise of the contractors is most useful to the designer. This is implemented under the watchful eye of the auditor whose task is to ensure that the final cost, not just the construction but the provision of service, gives best value. The process contradicts itself since the best value it can achieve excludes a strategic part of the design process.

6. PROFITABILITY

The end product of design is two-fold, to achieve construction completion to the satisfaction of all and best use in an enhanced environment. To achieve this can present the team as a whole with more than just a job of work. Those who deal with the financial costs wish to either see a profit or possess something of value and it is possible that those processes are not so concerned about the contribution of the product to the environment or to best use.

Short cuts are therefore, inevitable and workmanship will suffer. No-one can be supervised one-to-one and the risk of failure is inevitable and that is increased where lump sum working is tolerated even though it is illegal. The most profitable builder may not be either the most efficient or productive and this is probably due to the attraction of profit above the requirement of the work to be done.

7. MANUFACTURING

This is part of the comments concerning Time. Suppliers who have to design their product to fit into a building, or as is more the current trend, start to manufacture when the order is placed, creates inevitable delay at some stage. These are delays which are known and accepted but are a product of the state of the industry. In a healthier market the delays seem to be the same. Either there seems to be too little or too much production and is a reflection of the barometric effect of the building industry and the effect on it of a national economy.

Availability of materials is only one side of the coin. The durability, quality, life expectancy and ability to be maintained are the other. Real value comes in part from the quality of product. The degree of quality inevitably depends upon the cost and the axiom, "you get what you pay for" is a reasonable though not infallible basis for assessment. It is a hard task for a designer to encourage quality in a building where an unconcerned client sees an enticing

266

profit. Public sector clients are perhaps learning the lessons, for it is they particularly who expect their own designers to live with their work.

8. UNFORESEEN CIRCUMSTANCES

This is a polite statement for the numerous vulgar expressions the industry applies to situations of this kind. If all is going well there is always something at the back of one's mind saying "why?" Nothing ever runs smoothly and if there is the slightest probability be sure it will not just go wrong, it will go wrong in such a way that no-one is conceivably responsible.

Many building failures are never black and white. They are always in shades of grey. To deal with them, two attitudes or stances must be taken. One is to deal with the problem to prevent a knock-on effect and the other is to establish a fair assessment of responsibility (or rather who pays) by agreeing quickly. To achieve this it is necessary to return to Attitudes and Ability (human error is always difficult to swallow) and to Profitability and Manufacturing. However, it is Time which is the most usual culprit. Inadequate forethought is usually not given or rather allowed to be given to the designer to prepare his information for the contractor. Nor is adequate time given to the contractor to prepare properly or work with manufacturers due to the tender periods which are geared to financial arrangements rather than effectiveness of the work.

9. SAFETY

The use of some materials and the construction site each have dangers to which little attention was given until relatively recently. Even so the Health and Safety Executive now places an obligation on the contractors of which the designer may know little, leaving the client somewhat on a limb when considering the cost of his development.

The building industry, because of the somewhat nomadic nature of a part of its workforce, is not easy to train and encourage to work in a safety conscious manner. The Unions wish to see uniform safety but the general effect at the moment is one of uncertainty, particularly at the tendering stage. Questions of, "to what extent has double manning been allowed for for safety?" or "what safety equipment is there within the price?" will return the argument to productivity and of course profit. It also highlights the statement that profitability of a contractor may not reflect efficiency or productivity.

10. THE FUTURE PROBLEMS

Inevitably the use of Information Technology among people not brought up and trained to use it, will create individual problems. It is far sighted to research on-site computer use but difficult unless the user can communicate. At the moment if, for example, the user is not familiar with a keyboard (and professionals still question why the keys are not in alphabetical order), the application is still very basic.

Perhaps the most obtuse effect on the processes of designing and building is the introduction of new ideas from clients about the buildings they need as life-styles and management techniques change, which the building industry dislikes because it is new and less profitable. Old practices die hard and it is these new ones which are being introduced which will produce the hurdles for the future. The increased time for leisure is just one example of a changing life-style and expectation which is affecting the design and versatility that is required from facilities, as they in turn too meet new demands.

The increasing cost of energy and the relative performance of building materials will change the effectiveness of our current building stock. Of the current stock 75% is expected to be still in use in the year 2000. The quality of buildings constructed in the last 30 years is unlikely perhaps to stand the tests of time to which the 18th and 19th C buildings have been subjected. The industry and professions must concern themselves with quality to ensure a heritage rather than short term profitability. In doing so, a more profitable longer term approach will be more cost effective.

To achieve this there is a need to either have more time or some form of easily applied technology to cater for the reducing time available for design. There is a well defined view that design is less like solving a logical puzzle and more like riding a bicycle whilst juggling. A fair statement, but it perhaps hides the considerable constraint by other agencies over and above the innovative design thought process.

THE ROLE OF QUALITY AND ITS ASSESSMENT IN DESIGN

JOHN RITTER, Property Services Agency

1. INTRODUCTION

This contribution to the Conference consists of observations from the
viewpoint of the Product Quality Unit in the Property Services Agency
(PSA). The Product Quality Unit is a small branch, concerned with
PSA's use of products in building work and with the special objective
of using PSA purchasing power and technical knowledge (derived from
its large construction and maintenance programmes) to promote more
competitive British building products, in terms of quality. Although
it is largely limited to products, the paper aims to show how important
the subject of quality is at the present time; and that in spite of
some special difficulties over the recognition of appropriate levels
of quality for building work, there are actions which a responsible
specifier can, and arguably should, take now.

2. IMPORTANCE OF QUALITY FOR PRODUCT MANUFACTURERS; THE SPECIFIER'S ROLE

The Property Services Agency, as every other responsible specifier or
purchaser for building work, must look for value for money continuous-
ly. This means not only using the right procedures for procuring
buildings effectively and fairly; but the specification of appropriate
levels of quality. Members of the Conference will be well aware of
the current National Quality Campaign, and will have seen, for example, newspaper advertisements aimed at manufacturing industry and
its managers in particular. Making products at the right levels of
quality is very likely to be the key to survival for a firm which is
competing for business in the 1980's. But in the construction industry perhaps more than in any other, the specifier has a crucially
important role, as well as the manufacturer. If the specifier, be
he an architect, a do-it-yourself householder, an engineer, a builder's buyer, or whoever, is not aware of what appropriate quality is,
and if he does not insist on it, then the very manufacturer who puts
most resource into improving his range will stand to lose the most
against competition from a cut-price product with hidden defects.

We are a trading nation; and we could not contemplate abandoning
the effort of competing with foreign products, even if the Treaty of

Rome and other obligations were to allow us to do so. If our imports of timber which cannot be produced in this country are left out of account, our exports and imports of building products are roughly equal at around 1.3 thousand million pounds per year. If our buying habits shifted by only 1% in favour of British products as opposed to foreign ones, we should save well over £10m per year in foreign exchange, with all that this means in terms of jobs, as well as of future supplies safeguarded. But if our buying habits became based on a sufficiently clear view of product quality, so that our home market became a better base for exports, and a further 1% export gain were achieved, then there would be a further £13m increase in foreign earnings as well. There are in fact actions which any professional specifier could take quite easily now, in order to be able to say that he is specifying responsibly: not only for his own sake and the sake of his clients, but for the industry itself.

3. PROBLEMS OF DETERMINING APPROPRIATE QUALITY IN BUILDING

This is all very well. But what is Quality? In this paper, I am treating quality as the ability to satisfy - to be fit for a purpose; even if that purpose is very broadly defined. This definition of quality does not imply a high level of sophistication or cost; it does not imply a Rolls Royce, as opposed to a Mini. It is a matter of appropriateness to requirement. But in determining that require- ment we come up against special problems in the building industry.

First, the life-span of a building is long. When a failure occurs it is often difficult for the owner or the occupier to appreciate the cause: the original design; installation; product manufacture; the gale last night; or the unexpected change of use last year. Feedback from experience leading to a clear view of required quality is difficult for building owners and occupiers to achieve, and can be clouded by old wives' tales. The average time which an owner-occupier spends in a given house is only about 7½ years. The expected life- span of the house is perhaps 60: and the owner's strategy for avoiding excessive expense in repairs or maintenance tends to be one of calculated risk-taking each time he moves home, rather than one of systematic planning.

The time-span of the individual building project is also long. A professional specifier may only deal with 30 or 40 projects in a life- time; and he may move jobs, perhaps every 3 or 4 years particularly in the earlier stages of his career. These early stages are the very ones in which, in many offices, he is most likely to deal with detailed matters or specification. If the average time-span of building projects is 5 or 6 years, the younger specifier may never appreciate the consequences of his own decisions. At any event, there is often room for improvement in the mechanisms whereby professionals in the building industry build up their bank of experience.

Thirdly, experience of quality levels is none too easily built up in the building trades themselves. The problems are legion, not only in coping with the increasing range of new materials and methods, and with an ever widening range of clients' requirements in terms of geo- graphical location and of new building types: but in the mobility of

270

labour. Building at a certain level of quality is not very difficult; so we have the 'cowboy' problem of people who come, and vanish again.

All these points mean that clear definitions of required quality in the building field are not easily achieved by the ordinary accumulation of experience, whether by users, professionals, or the builders themselves: simple market forces cannot be relied on to improve general quality in the long run, as might be the case with ordinary consumer durables. Sound recognition of appropriate quality levels would be difficult in the building industry for these reasons alone, even without the added complications of the aesthetic and psychological impacts of design. Yet, as I have said, it may be crucial to the survival of a manufacturing firm that it makes a better estimation of required quality levels than its competitor. To follow up this question of quality, I shall note those aspects which my Unit is (or is not) pursuing, and describe some features of PSA specification practice. From these some general suggestions may emerge.

4. ASPECTS OF QUALITY

In the current work of the Product Quality Unit, we are still treating quality rather negatively: more as the absence of defects, than the presence of delights. That is a serious omission, particularly in the context of promoting products that will compete among the international best. But the absence of defects in products is analogous to "hygiene factors" in personnel management. Where serious technical defects remain, the pursuit of delight is self indulgence. As soon as one can use today's materials, for today's purposes, with the same understanding and familiarity that a 19th century craftsman could possess in his limited range of materials and building types — then one can legitimately concentrate on designing for delight. There are, of course, many excellent product designers and manufacturers who can and do; and PSA policy is to make the most of them. However, my Unit's programme puts other questions first, and I am limiting my paper to these.

I am also limiting this paper in line with the Unit's present concentration on product design and manufacture, as opposed to overall building design, or the installation of those products; despite the fact that it is through design or installation faults that the majority of our failures occur. Partly, this is because of the import and export issue; although design and construction services do make a key contribution to our balance of payments. Partly however, it is because the quality of products, of product information, and even of product identification, are essential early stages in the achievement of quality in the building as a whole. And we must start somewhere.

The Product Quality Unit tends to concentrate on three areas. In choosing any item, the specifier's first concern is whether it has the right features for its job. If there is insufficient consensus on what these features ought to be for a given type of product, much effort can be saved by reference to Standards: Product Standards, Codes of Practice, Standard test methods and so on. The provision of sufficient appropriate and up-to-date Standards to support the building industry is however a major organisational and intellectual problem. There are the difficulties not only of achieving adequate user information, as I have already outlined, but of achieving consensus among legitimately differing interests. Timescale is of the essence: if in the five years that may be needed to achieve agreement on a Standard, technology has moved forward, the whole exercise and its considerable expense in manpower and supporting effort may have been wasted. The problems of cross-reference between the thousand and more Standards required by our industry are daunting. Yet particularly for basic materials and methods, Standards underpin almost all our other activity as specifiers. Standards provide the shorthand references without which the designer could never concentrate on those essential problems which by contrast need his individual attention.

It is however equally important for the specifier to achieve the right features in a product which, being innovatory or patented, is inappropriate for Standardisation. The authoritative certification or reviewing of innovative products is an activity perhaps as important as Standardisation itself. The library of standard Specification Clauses which PSA supplies for its designers has nearly as many blank sections for the designer to fill in with an individual manufacturer's name and product reference, as it has simple references to British Standards.

The third concern for the specifier is whether the product which he gets will actually be properly made and supplied. It is a disturbing fact that a large proportion of products do not actually comply with the Standards which lay down what their features ought to be. In 1983 the Building Research Establishment published an alarming report by Ken Fletcher on the non-compliance of goods. One response to this problem is third-party Quality Assurance: that is, the use of somebody independent, neither the supplier nor the buyer, to inspect and recheck suppliers' methods and output so as to reassure potential buyers about the quality management which the producer applies, and about the likely consistency of the product itself. I believe that far too many specifiers in the building industry take the claimed compliance of a product to a British Standard on trust, without further assurance.

5. THE PSA RESPONSE

The Property Services Agency responds to these aspects of the need for quality in a variety of ways. In particular, our Building Advisory Branch collects feedback on problems in the very large estate of property that we maintain. It then gives advice to our maintenance staff - but even more importantly, to our designers who are building new. The work of the Building Advisory Branch underpins all our building specification policy. Its feedback is of particular value, since there is no other British organisation which both designs and maintains its own buildings in such quantity, with such a range of building types, and with such a geographical spread, from Cornwall to Shetland; and its conclusions deserve to be applied systematically throughout the Agency. One mechanism for this is through the provision for PSA designers of a library of standard clauses which can be selected and printed out to make up an individual Job Specification. These standard clauses are revised each year, together with guidance notes, based on the feedback from the Advisory Branch and from the equally important Engineering Directorates, and thus ensure an automatic take-up of many of the lessons learned. A further mechanism is that in a few key instances of products such as windows, doorsets, or other items where costs or the risks from failure are high, PSA designers may use only components which have been specially approved and tested. These products are assessed against Performance Specifications, again based on our feedback; and subject to price, supply, and quality assurance arrangements they then go onto an approved list, usually for 3 years before reassessment. They are neither particularly up- nor down-market; they are assessed for value for money, and they can include a whole range of performance levels.

Both our Standard Specification and Procurement operations refer to British Standards wherever possible - around 300, at present, for building work and nearly as many again for Engineering Specifications. Wherever possible, these references are also linked to Third Party Quality Assurance schemes - so that a reference in our General Specification might be on the lines "Widgets to be to BS 1234, and supplied by a current British Standards Kitemark licensee" - the kitemark schemes being among the best known and most widespread quality assurance schemes for products. Rather similarly, where a product is to be specified by name and British Board of Agrement certified items are available and appropriate, PSA designers should name these as their first choice and only accept an alternative if it is offered with an equivalent level of independent assurance.

In all this, we try to keep in touch with manufacturers through extensive representation on committees at the British Standards Institution, through the help of the National Council of Building Material Producers, through Trade Associations and other organisations or directly with individual firms.

6. GENERAL CONCLUSION

I suggest that there are four key features of this PSA practice which may be of interest for this Conference.

    i. Feedback: A deliberate and systematic attempt to learn from the failures and successes in our building stock.

    ii. Application: The means of applying that feedback systematically as well as on an individual basis; in particular through the use of a standard approach to specification writing, with related guidance which not only incorporates our experience but saves designers trouble and worry at the same time; and through special purchasing initiatives where justified by the value of the products concerned.

    iii. Quality Assurance: Our reference to quality assurance schemes by independent authorities of good standing not only protects ourselves from inadequate products, but protects the 'quality' manufacturing industry from unfair competition by others whose output while superficially attractive and certainly cheap may all too often contain hidden defects.

iv. Contact with manufacturers: This is particularly valuable, and in particular is achieved through membership of committees at the British Standards Institution; but clear and useful communication remains difficult in a fragmented industry, for reasons including those outlined earlier in this paper.

The systematic application of feedback, and reference to quality assurance are both actions which can be taken by almost any responsible specifier. The establishment of effective feedback mechanisms, and of effective dialogue with the manufacturing industry are however areas where our scattered industry is at a disadvantage. The British Standards Institution in particular very much needs a better, and better co-ordinated, input from users of building products; but this can only be provided where their evidence is adequate and properly recorded, and where an idea can be given of the scale of purchasing involved. A British Standards committee cannot work to produce a better Standard, and thus to encourage manufacturers to invest heavily in a new product, if this proves only to be bought to a few exceptional instances.

I have tried to show in this paper that it is a special problem of the building industry for specifiers and users to focus their experience and to form views clearly on appropriate quality for the products which they will buy. I have also tried to show that their success in this affects not only themselves, but the continued competitiveness and prosperity of the manufacturers who supply them. While some appropriate actions can be taken readily by specifiers, difficulties remain; and I look forward to the opportunity of the Conference to pursue their solution.

CASE STUDIES OF ARCHITECTURE FOR INDUSTRY

MARTIN SYMES, The Bartlett School, University College London

1. INTRODUCTION

Given the importance that images of industrial buildings have had for the development of architectural style in this century, it is surely surprising that so little has been published in the way of critical research on their design. Pevsner's (1976) chapter is one of his briefest and least informative surveys. Winter's (1970) book, scarcely more informative on Nineteenth Century developments, does deal at greater length with more recent buildings, but his attitude is strongly coloured by the assumption that economic factors, properly considered, led to the development of appropriate aesthetic quality. In a similar vein, two monographs on the work of Albert Kahn (Hildebrandt, 1974 and Nelson, 1939) show the work of one especially prolific designer as making a major contribution to functionalist architectural design and give second place to his use of formal devices inherited from an earlier, cultural tradition of design determinants. Evans (1976) and Banham (1960) are two of the very few critics who have suggested approaching industrial buildings from cultural history and their references are to single buildings rather than the great corpus of construction which took place in the periods concerned.

The issue is also of some importance to society at large, as industry is changing throughout the Western world as well as in the Far East, and the existing stock of buildings may be in-appropriate for many of the new processes being developed, and become an inhibition on our adaptation to economic and social change.

The established approach to industrial building research seems to be that the prediction of a building's functional performance can be matched with possible clients' building requirements to form a basis for design specification. A further assumption is then made that future unsatisfactory performance would be followed by adaptation of the premises used, or the firm's relocation. The criteria employed in determining performance

275

are thus essentially economic and it can be argued, that an industrialist's choice of premises is like his choice of location: based on the cost at that place of the materials, labour and organisation necessary to carry on his business and compete profitably in the market place. Indeed, in architectural theory it has been repeatedly suggested (Steadman, 1979) that buildings survive and are valued in the long run if they fit precisely the needs of the task they are called on to perform, whereas for domestic or institutional buildings, the criteria of fit may include social or psychological factors, for industrial buildings and firms the costs of production are almost invariably assumed to predominate.

A critical observer might nonetheless take the view that other worthwhile questions can be approached using different methodologies. Architectural theory is widely understood to be concerned not only with the performance and use of buildings but also with their meanings, and although it may also be difficult to formulate research programmes concerning these issues (Groat, 1981) there is no fundamental reason why they should not be attempted. Writers such as Rapoport (1977) consider the built environment part of an expressive culture which holds intangible values constant over long periods of time. Social groups recognise objects as representing important features of social life and individual members receive 'psychic rewards' from this recognition. From this point of view both the amenities offered to users by an industrial building and the aesthetic properties of the building itself can come to symbolise its owner's or user's place in society and thus possess cultural qualitites which should be better understood if designs for industry are to be improved.

An important difficulty, both for theory and for research, seems to be that there is no agreement on the relative importance of economic factors and cultural qualities in determining design decisions. Indeed, common sense would suggest that designers make decisions which affect more than one aspect of a building at the same time and users probably also approach their choices in a holistic way. Banham showed one way in which functional and cultural aspects of architecture may interact with each other with his argument that the appearance of functionality has symbolic value during the emergence of modern design. Rapoport argued in much the same vein when describing the cultural values of a design as only latent functions while the economic aspects are manifest. In both cases the economic organisation is seen to come first and the cultural overtone to be acquired later. Barthes (1967) however suggests that the reverse might also be the case. Practical objects, which he calls sign-functions, having acquired meaning as a representation of a particular activity, can then be interpreted as indicators of where, when or how that activity can, will or should take place. Economic structures could in such a case follow from the perception of symbols in an environment, even though the original patterns of activity they represented were no longer optimal for the context concerned. Something like this probably happens from

time to time in Las Vegas (Venturi et al, 1967) where people
go to gamble because they see the signs of gambling halls are
there and the same process may sometimes happen with manufacturing industry as well. Unfortunately the research tools to
demonstrate whether this is the case at all and the extent
to which it occurs are in their infancy and have yet to be
applied to the topic at hand.

A final difficulty of research in this field, which has to be
mentioned, is that of the self-consciousness of designers and
clients in respect of research results which would suggest to
them how they might be expected to behave. It is clearly very
difficult to make independent controlled experiments where knowledge drawn from them can be fed back to the subjects concerned
at any time. Masser (1982) discusses the cyclical nature of
decision-making procedures in planning and his comments seem to
apply equally well to the making of design choices. He even
draws attention to the existence of research-based ideas to
which participants can say they adhere but may not always be
able to use in practice. These he terms "espoused theories".
One he mentions which has relevance to the discussion of industrial building design is that of economic rationality, a
theory which architects or potential clients may use to justify
decisions or choices which an independent observer might have
suggested were made on intuitive or other irrational grounds.
Alternatively it is of course possible that those making
design choices might sometimes believe they are following
cultural prejudices about design quality when in fact they are
using an economic logic to which they do not admit!

2.   APPROACH AND CONTEXT

The case studies in this paper represent an attempt to address
the problem of industrialists' motivation in their choice of
premises through reporting inverviews with leading executives
of three firms in the North East of the United States. They
were undertaken in 1980 as one of the preliminary steps in a
broader and as yet incomplete study of the imagery of industrial
estates, on which the author is engaged. They seem of interest
for the present discussion of design quality, costs and profits,
as they display quite vividly the various cultural qualities
which can be attributed to an industrial building and suggest
that the qualities sometimes override considerations of finance
in building choices.

The North East of the United States has not always been an
economic problem area. As recently as 1959, James Hund
reported that about 50% of the electronics industry was devoted
to military and industrial production and that in this sector
"both large and small firms (find) the (New York) region a
congenial place... The need for flexibility... in the design
and production of expensive equipment... (requires) large firms

277

(with) substantial resources... (and)... they... rely on many
small specialised plants to fill the gaps in their own
capabilities:.. constant liaison must be maintained,... design
and testing carefully meshed". Large plants tended to be in
suburban areas (IBM at Poughkeepsie, Bell at Murray Hill
and Holmdel, Fairchild on Long Island) and small ones in
New York City (Sperry Rand in Queens, Loral in the Bronx,
Skiatron on Manhattan itself). Hurd was forced, however, to
note a change in the market: "the substitution of computers for
clerical labour (was) proceeding at such a pace... that it was
the most important area of expansion". At the same time, he
wrote, "the advantages of the region are being gradually
duplicated elsewhere..." mentioning Florida, Colorado and
California, "where there are the attraction of educational
opportunities, research activities, and the natural climate
as well as an unusual 'esprit' in the business community".

A more recent study by the Port of New York and New Jersey
documents the increasing disadvantages of their area on the
first, second and fourth points:

    i.   Educational support: "contracts with four (universities)
         disclosed no special program to foster the transfer of
         faculty and student-generated science-technology innova-
         tions and no effort to support or facilitate scientific
         discovery transfer to the market place";

   ii.   Research activities: "are under serious threats from
         decreasing financial support, from ageing equipment, from
         changing demographies and from perceived decline in the
         opportunities for scientists and engineers";

  iii.   Business attitudes: here there are two problems. Firstly
         the large corporation may take its own development out
         of a region, and secondly the small company may find it
         difficult to gain support for the development of innovations.

As a result of such trends, and in the context of national
economic problems in the 1970's, the report concludes:

    "Historically, manufacturing formed the keystone of the
    (regions)... economic base, (so) the... loss of income
    was soon manifested in deteriorating and abandoned housing,
    in decaying commerical strips, in neglected public
    facilities such as parks and playgrounds, and more
    generally in the quality of life."

It was thus of some interest to discover from the New York
Times that a "high technology boom has shielded the New
England economy from the worst of the 1980 recession...
(partly by establishing) a Northern outpost in an old
red-brick shoe factory... barely 50 miles from the Canadian border."

278

One factor in this recent revival of business activity and
renewal of confidence may have been low labour costs, a cost
advantage which is sometimes important to new sophisticated
lines such as electronics, data processing and instruments before
they become established.

In other parts of the region, however, cooperation between industry, unions and government has been necessary to significatly
first change attitudes and then the trend. "Brooklyn" a
local paper reported "has an excellent labour force" but "the
image is the toughest one to fight... all you hear reported is
crime". Examples of successful intervention are given as the
Brooklyn Navy Yard, currently run by a public/private corporation,
the Bedford-Stuyversant Restoration Corporation, one of whose
vice presidents in quoted as saying "I'll be very happy if we
could duplicate the security and trust that the Navy Yard has
earned", and Williamsburg Industrial Development Enterprises
who have "the goal of establishing areas of cooperation
between the business sector and the government sector". Their
operation includes a Youth Employment Training Program on the
top floor of a seven - storey former building: a vertical industrial park. According to one of the tenants: "the factory is
clean and light... and the security is excellent".

Of course mortgage money, sometimes supported by Federal or City
Government grants or guarantees, have also been made available
to assist in the process of economic regeneration, but the neglect
of the environment in earlier years can still make it difficult
to show that an expenditure which is large in proportion to a
property's market value and devoted only to upgrading its
quality will necessarily justify itself and positive promotion
is undertaken as well.

According to the Port Authority of New York and New Jersey,
again, initiatives which have been taken include:

  i. The "New Jersey's got it" campaign, a multi-purpose
      promotion covering both tourism and the business scene.

 ii. The Philip Morris ad campaign which offered a guide to
      "home town New York".

iii. An attractive book describing Hoboken's history and assets
      is designed to elicit interest in the "square mile" city.

 iv. "Baltimore's allocation of $450,000 for construction
      purposes was, in effect, an image improvement mechanism".

The case studies which follow show just how important an
application of such qualitative factors can be for individual
"economic men" when choosing their place of work.

3. CASE STUDY ONE: A HISTORIC SETTING FOR HIGH TECHNOLOGY DEVELOPMENT

Today, Wang Office equipment is advertised world wide in the
daily newspapers and colour magazines. The firm is the creation
of Dr Wang who invented the magnetic core used in computer
memories in 1948 and started up his own firm to engineer one-off
products for special customers. A digital scoreboard was designed
for New York's Shea stadium when the firm only employed 6
engineers. Seven years later there was an exchange of patents
with IBM and in 1964 the firm began volume production of a desk-
top calculator. This line became the basis of considerable
expansion but was outmoded when competitors introduced smaller
machines using silicon chip technology. In 1972, the company
began producing word-processors and other office equipment.
According to Time magazine they currently have 35% of the world
market and have averaged 75% growth on profits for the last five
years. They are beginning to meet with more opposition but
are also developing new products. In 1978 employment seems to
have been about 3,500, but by 1980 it was over 10,000. Dr Wang
was reported as negotiating with the government of his homeland,
China, to set up a joint venture manufacturing small computers
there.

Wang studied at Harvard but his business started in an upstairs
room in Boston, moved to a house in the south side and then a
series of factory spaces ending up in Cambridge with 10,000
sq ft in 1955. They then moved to Needham and in 1962 built
a new building at Tuxbury. Wang purchased 80 acres of land, so
although the first structure was small, additions of 2-storey
assembly space were made in 1965 and 1969. The site sloped,
so, with a constant roof line, the 1973, 1975 and 1978 warehousing extensions could have 3, 4 and 5 storeys respectively.
In 1975 the firm had bought a disused modern missile manufacturing plant on the outskirts of Lowell and converted it to
office use. It is this building which has been designated
corporate headquarters.

Lowell is, indeed, one of the most interesting remains of America's
Industrial Revolution period. A canal had been built in the
1790's to circumvent the 30 foot drop in the Merrimack River at
the Pawtucket Rapids, and carry merchandise to the port of
Boston. The canal was a failure, for another, shorter route was
established, but it did prove the basis for developing in the
1820's a series of mill sites and around them the town of Lowell.
The proprietors of a Locks and Canals Company and of the Merrimack Manufacturing company developed here a model industrial city.
They were concerned that manufacturing in America should not
generate the slum conditions which existed in Manchester and for
twenty years or so seem to have run an exemplary company town.
Young women from the farms of New Hampshire were employed at
admittedly low wages but they were housed, fed and supervised
by matrons in boarding houses run by the company. Lectures

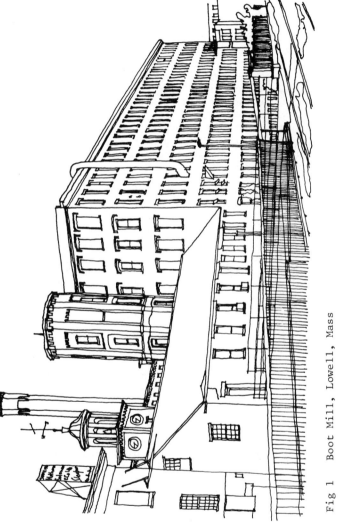

Fig 1   Boot Mill, Lowell, Mass

were arranged in the evenings and the women gained a kind of
independence they could not achieve elsewhere at the time.
By the mid-nineteenth century this idealism began to disappear,
Irish and later French Canadian men were employed and the boarding houses sold. Water power was replaced by steam and later
electricity. The canal as a means of transport was replaced
by the railway. A hundred years after its foundation Lowell
was an industrial town of around 120,000 people, but soon after
suffered from the general migration of cotton mills to the South.
Lowell's remarkable history has become an important focus of
recent attempts to revitalise its economy. The city has become
America's second urban National Historic Park (the other
surrounds Independence Hall in Philadelphia) and also receives
State aid for environmental improvement. Park wardens take
visitors on guided tours of the canal system, there are two
important museums and in 1979 15 active refurbishment projects.
The plan for developing this cultural areas has been published
as a model approach. Refurbishment attracts federal, state and
city funds, the educational system has established retraining
programmes for the unemployed and firms in service industries
have begun to move back into the town. All the officials who
were interviewed agreed that the Historic Park had changed the
image of the town and that this has affected morale at all levels.

In 1978 a former plastics factory in Lowell was acquired by Wang
and a federal grant obtained to build a 14-storey office tower
next to it. Back at Tuxbury, another 5-storey structure was
erected in 1979 and a warehouse with especially high ceilings
in 1980. This has enabled the company to reorganise its space
there, converting its earliest four buildings to a higher
standard office accommodation. There is a manufacturing
facility in Methuen, Mass, and overseas plants are in Puerto
Rico and Ireland. About a year ago, the city of Lowell asked
them to purchase part of one of its old mills, the Boot mill
No 6. This has been used by another electronics company which
was bought out by a conglomerate and then given new, but
inexperienced, management. Possibly for this reason the
company went bankrupt and the city feared the building would
be demolished for redevelopment. The city's industrial development office say Wang only bought the mill as a favour to them and
may never do anything with it, but Wang's construction manager
says they have begun to waterproof the building's shell and
evaluate the interior's potential use. He feels the city wanted
Wang as a lead tenant. The L-shaped structure constitutes
the whole of one side of the Boot mill's courtyard and half
another. It has a street frontage and also one on the canal.
It could accommodate up to 200 people with perhaps two floors ware
    housing for mail order division and another three floors of
office space to provide for soft-ware services and sales
support. A problem is the lack of on-site parking, but the city
may find a way of providing some nearby. A further unknown
is whether restaurants and shops will open up nearby to service

what for downtown Lowell will be a new class of employee.

There can be little doubt that Wang has made a shrewd business decision in choosing to redevelop his business in such an attractive city. In the context of issues raised by the introduction to this paper, he may be thought to have foreseen how expectations of profitability could be raised by the selection of a location with these qualities and not to have waited for the profits to appear first before using them to obtain such amenity. (See Figure 1)

4.   CASE STUDY TWO: RELIGIOUS ASSOCIATIONS FOR A DISUSED MILL

Enirex was established in 1958. The two partners had developed a new approach to heat transfer technolgoy and this is still the basis of the firm. Their work is mainly development work and results in short run manufacturing only. After some years a separate company was set up to deal with medical applications and this grew faster than Enirex. After about 20 years the two partners split up, one taking the medical equipment business and the other the more general work. It is this firm which is now reestablished with a new set of middle managers. Five younger men have been recruited for this level who have an active involvement in the evangelical work. The president believes that people who are creative in one sphere will be creative in others as welll. He certainly is as good as his word, having established two more companies in the last two years, one to manufacture and market switches, the other to undertake building work. Currently Enirex and its associates employ about 30 people, something less than they have in the past, but the firm expects to expand again very rapidly in the next few years.

The original firm was started in a friend's basement. After a year then moved to the current president's garage for a further year. During this time a 5,000 sq ft building was erected for they in Riverdale and over the years this was extended twice, once to 10,000 and later to 20,000 sq ft overall. The medical equipment company was in a separate building from its early days and went through a similar process of expansion. When the two companies split, the building in Riverdale was sold and Enirex moved to a 50 year old building in Paterson, New Jersey, one of America's oldest industrial cities.

Alexander Hamilton, the first secretary of the Federal Treasury and a leading figure in the Society of Useful Manufacturers, seems to have been instrumental in the commissioning of Pierre l'Enfant to plan Paterson as a new city around the Great Falls of the Passaic River in the 1790's. He and his successors constructed a series of millraces which provided power for the spinning and weaving, first of cotton and then of silk. Colt

283

revolvers were made there and later steam locomotives. By the
1880's the city was America's 15th largest city, but its industries all declined in the 20th Century. The mills became dilapi-
dated and some were demolished, while alternative employment was
established in, for example, aircraft engine manufacture, on
peripheral industrial estates. By the 1960's many of the city's
inhabitants were illegal immigrants from Spanish speaking countries. The turn round began in the late 1960's, when a plan to
demolish the mills and use their land for a highway extension was
resisted by a preservation group which included the mayor's wife
and a Columbia student of architecture. In 1971 the Historic
District was listed by the State and Federal Governments and in
1976 the President designated a National Historic Landmark.
About $15 million has been provided in assistance for a re-
vitalisation programme in which private citizens are also parti-
cipating. One of these is the president of Enirex, the case
study firm. He lives a few miles outside the city and has
been involved since 1971 with an evangelical movement to provide better education for children in minority groups. The
group has its headquarters in Michigan and initially raised
money for scholarships to send children to private schools.
In 1976 the Paterson members decided that their money would be
better spent on establishing their own school and the Enirex
president, in his own words,"practically took a year off work
in order to set this up". It started with about 60 children
and occupies another converted mill adjacent to the city's
newly established museum. The group is now working on job-
creation by establishing a craft workshop. The artists will
relate to groups of children in the school and will live in
a housing association nearby. The evangelical group does not
approve of subsidies and so is using the "sweat-equity" approach
to rehabilitation which they see as a complete contrast to the
city's own schemes. They believe there should be a coherent
philosophy underlying everything they do.

The mill was bought from a weaving company who will remain as
a tenant of the second floor for two years. The third (top)
floor is used for offices and for the switch making subsidiary.
This bench-type work will be moved out when a suitable small
mill can be found. The ground level houses general storage and
fabrication work, and an extension is used as a research lab.
This will go to Butler, where the researcher, who lives there,
has found space in a subdivided old mill, next year. The
extension will then be added to the production area. The
building subsidiary is in the process of refurbishing the
external walls, putting in new windows and increasing thermal
insulation at the same time. A new heating system has already
been installed. Further work is planned to the exterior
decorations and to the parking area. Similar work was done
to the interior of the other mill now used as a school, but
there the city insisted on refurbishing the exterior themselves.
That structure is much older (1891) and part of the Historic

District so the city actually bought and retain ownership of the walls and roof. The church group own only the interior.

In the long run, Enirex' president sees his mill as the production component of his restructured company and expects to have office space elsewhere, even though his personal office now has the best view in Paterson, directly overlooking the Great Falls. Thus starting from a religious motivation and exploiting his ability to see a social purpose in the disused mills, this entrepreneur aims to run a highly profitable business. By a strange reversal of conventional logic, he expects an environment he values mainly for its cultural quality to assist him in this aim. (See Figure 2)

5.   CASE STUDY THREE: POLITICAL CONNOTATIONS FOR A SCIENCE PARK

Digital Equipment Company, or DEC as it likes to be known, is considered the market leader in minicomputers. It was established in 1957 by two brothers who had helped build the Whirlwind computer at MIT. Their business idea was to market an "Inter active computer" which most of their competitors thought was decades away. From the beginning they refused to take government money and went out for profits immediately. During the 1960's they sold their machines through intermediaries and avoided both the need to develop software themselves and the creation of an elaborate dealer and after sales network. As the market began to change, DEC developed a wider range of products and adopted the micro-chip technology. They also found they needed to sell some software too. By 1972 the different parts of their $188 million organisation were competing with each other and the organisation structure was changed to one based on markets rather than products. After 1975, when sales were $534 million, DEC developed the distributed data processing concept and began to challenge IBM. The founders are committed to keeping individual groups within their firm small, and require each to bid for development funds from the Board.

DEC began in 8,500 sq ft of mill space in the former wool town of Maynard, Mass. As the firm expanded they gradually leased more of the mill and in 1974 were able to purchase the whole complex. They considered this cheap space and were prepared to set up a rolling programme of conversion and rehabilitation. The president in particular is committed to this environment. As the company has expanded they have established a policy of clustering up to 6 smallish (2,200 + workers) plants together in locations where they can establish either a market or a workforce. DEC manufacture now in Galway, Puerto Rico and Arizona as well as in New England. There is a small factory in "the black ghetto" of Springfield, there is one in a more modern empty factory in Malborough, Mass. and then there is the new unit in Roxbury, Mass. This community is one of the very poorest

285

areas on Boston's South Side. The City of Boston has achieved
an international reputation for the refurbishment for commercial
purposes of much of its waterfront and special attention has been
focussed on the Rouse Company's conversion of the old Quincy
Market for use as restaurants and boutiques. In the 1960's most
urban renewal programmes were concerned with housing problems,
but this programme could not lead to a permanent reduction of
deprivation without the support of employment and the jobs
were moving away just as they did from many other American
cities. The Community Development Corporation is now being
granted federal and city funds to attempt to re-establish
industrial employment in areas such as Roxbury. In the late
1960's the city had purchased 200 acres of land here for an
extension of the Freeway system, but that plan was abandoned
after community opposition, and replaced by proposals for an
arterial street on the ground. Some of the land now owned
by the city was thus available for other uses and a community
group negotiated for the right to participate in its redesign
and reuse. It was an area with a poor image and many of the
vacant structures had been demolished or destroyed by arson.
The new plan which emerged, and which the Community Development
Corporation is beginning to implement, included the use of land
beside the new road for new industrial buildings and the
refurbishment of the old buildings as incubator space. The
Cross Town Industrial Park, as it is to be called, has the
advantage of proximity to a residential area and to a new
hospital complex, but its disadvantages have seemed to outweigh
them. The disused structure in which the CDC is housed has
attracted no other tenants yet. It may do in future if a new
initiative succeeds. Control Data Corp., a computer manufacturer, has established a small Business and Technology
Centre in a remodelled 8-storey warehouse in St Paul and believe
they can do the same in Boston. Their aim is to make resources
available to "the small businesses of today (who) will be the
Control Datas of Tomorrow". If they are successful, the firms
attracted may also forge links with one of CDC's successes, the
new branch plant of Digital Electronics which has been established there.

The DEC factory is an elegant purpose-built single-storey
modernistic structure which stands out sharply from its
surroundings. The architect who designed it was black and so
will the manager who runs it be. DEC are certainly ploughing
their profits back into the community but the manner in which
they do this also seems to be a way of making forceful statements about the relationship between architectural and political
qualities. No doubt they expect these statements to feed
back in their turn to even greater profitability. (See
Figure 3)

286

6. CONCLUSIONS

In the introduction it was argued that architectural theory has
not yet satisfactorily described all the choices which have
to be made in industrial building selection and that a particular gap lies in the full description of possible relationships
between economic and qualitative factors which may influence
design choices. The paper suggested that some of the approaches
being made to the promotion of industrial development in the
North East of the United States assume that industrialists can
be expected to put qualitative matters before economic ones
in their assessment of the suitability of certain places for
the location of their firms. In the latter part of the paper
three case studies are detailed which appear to support this
suggestion. They seem to show that the industrialists'
theories about the value of an environment can vary quite widely,
however, even within this framework. In one case, the entrepreneur made a simple assumption that an attractive environment
would help the firm prosper; in a second case, the attraction
of the required environment was assumed to be strongly qualified
by religious and social objectives imposed upon it; in the third
case it appears that the developers of a new building have
assumed it will help them multiply the value of resources
generated first elsewhere. More work on the variety of
"espoused theories" which help link quality and profit in
industry is clearly needed before firm conclusions can be
drawn on the strength or general applicability of these observations, but it is hoped nonetheless that the case for further
investigation need not yet be rejected.

REFERENCES

1. Pevsner, N (1976), A History of Building Types, London,
     Thames and Hudson
2. Winter, J (1970), Industrial Architecture, A Survey of Factory
     Building, London, Studio Vista
3. Hildebrand, G (1974), Designing for Industry: The Architecture
     of Albert Kahn, Cambridge, Mass, MIT Press
4. Nelson, G (1939), Industrial Architecture of Albert Kahn Inc,
     New York, Architectural Book Publishing Co
5. Evans, R (1976), 'Regulation and Production', in Lotus
     International 12, Milan
6. Banham P R (1960), Theory and Design in the First Machine Age,
     London, The Architectural Press
7. Steadman, P (1979), The Evolution of Designs, Cambridge, The
     University Press
8. Groat, L (1981), 'Meaning in Architecture, New Directions and
     Sources', in The Journal of Environmental Psychology 1,1,
     London
9. Rapoport, A (1977), Human Aspects of Urban Form, Oxford,
     Pergamon Press

10. Barthes, R (1967), trans.Laver A and Smith C, Elements of
    Semiology, London, Cape
11. Venturi, R, Izenour, S, and Scott-Brown, D, (1977), Learning
    From Las Vegas, Cambridge, Mass, MIT Press
12. Masser, I (1982), 'The Analysis of Planning Processes: Some
    Methodological Considerations', in Environment and Planning B,
    9, 5-14
13. Hund, J (1959), 'Electronics', in Made in New York: Case Studies
    in Metropolitan Manufacturing, Cambridge, Mass, Harvard
    University Press, 241-325
14. Perilla, O (1979), Some Aspects of Technology in the Economy of
    the New York - New Jersey Region, Port Authority of New York
    and New Jersey, May
15. Cowan, E (1980),'New England's Economic Shift', in New York
    Times - Business Day, October 20
16. Levine, D L, (1980), 'A Great Place for Business', in North
    Brooklyn News, March 14-20

Fig 2   Enirex Mill, Patterson, NJ

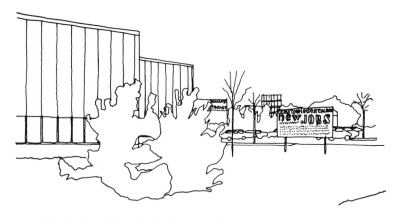

Fig 3   DEC Plant, Roxbury, Mass

THE RELATIONSHIP BETWEEN QUALITY AND COST IN THE DESIGN OF M&E
SERVICES

PATRICK VENNING, Davis Belfield and Everest

INTRODUCTION

When I was first given this brief many months ago, I accepted it
guardedly and reserved my position to adapt it perhaps severely to
fit my capabilities. It was a large enough subject to make me
nervous. M&E services are very complex and fast changing and
whilst the quantity surveyor has come to be listened to with respect
on cost - sometimes - it is a brave one who ventures into "design"
and "quality". But there is a need in the building industry for
the relationship to be examined and if environmental engineers dispute that, perhaps they will be prepared to accept that there is a
need for the relationship to be better communicated to clients and
other designers.
    Many papers start with definitions and this is no exception. I
will not write much by way of definition of "cost" but much needs to
be clarified regarding "quality".

QUALITY

"Quality" is not a precise, scientific and objective word. It can
mean many things to many people. Compare two cars - a Rolls Royce
and say a Fiat. Without wishing to libel a large Italian company,
most people would in preference award the word "quality" to the
Rolls Royce without much thought. Both cars and M&E services are
engineering products so the analogy may hold up quite well.
    I am suggesting five parameters by which quality can be judged:-

    1. Practicality - does it work?

    2. Life - how long will it last?

    3. Economy - is it efficient? What is the running cost?

    4. Appearance - does it look nice?

    5. Depreciation - does it hold its value?
In comparison between our two cars, we can with some confidence
make the following observations:-

289

(i)   The Rolls Royce, because of its high quality components and
       engineering, will function very well with minimal servicing
       and attention;
(ii)   it will have a very long life because of its quality metal,
       well engineered and well protected;
(iii) in appearance it will gain wide acceptance, and
(iv)   it will hold its market value - certainly after initial
       depreciation and if taken over a long period. However:
(v)   The Rolls Royce's economy is more questionable. Its miles
       per gallon will be low and its overall comparative running
       costs will be high.

   All these aspects of quality in the context of M&E services are
briefly discussed below.

COST

"Cost" means the initial capital outlay involved in the procurement
of the M&E services. They have a market procurement "price" in the
building industry. This paper is not the place to discuss the
relative merits of different forms of contract and sub-contract and
in what follows, the cost of services assumes that the price paid is
for efficiently and economically procured services elements including mark-ups, work-in-connection and the like. But the effect of
different services on the cost of building elements cannot be
ignored.

THE THEORETICAL RELATIONSHIP BETWEEN COST AND QUALITY

Before attempting to look at cost and quality for actual services
installations and services elements, it is worth first trying to
see if the relationship can theoretically be expressed in a common
unit. Discounted cash flow can be used to reconcile out-of-time
events so are we able, without discussing DCF and its weaknesses in
depth, to look at aspects of "quality" in terms of net present value
for numerical comparison with initial cost? Below, each of the five
suggested aspects of "quality" are discussed in turn.

   Practicality

Services that do not do what they are supposed to do (i.e. breakdown)
are a nuisance. It is a nuisance that cannot readily be given a
monetary value for the circumstances for different buildings and
situations and uses are complex. However, maintenance in terms of
labour costs and component renewal costs have a cost even if difficult to assess before the event. It would not be irresponsible to
assess a relatively higher allowance for maintenance for fan coil
units than for say a VAV air conditioning system. Any predicted
maintenance costs can be given a NPV.

   Life

The life of services is not to be confused with maintenance for at
the end of their life, we are faced with renewal and replacement

(not repair and maintenance). The life of services components and elements can be assessed from norms and again the costs involved can be given a NPV.

Economy

Services have running costs that of course include energy consumption and labour costs in operation, insurance etc. If these costs involved can be predicted, they can be given an NPV.

Appearance

Quantity Surveyors are often accused of wanting to apply numbers to everything. But you cannot put a price on aesthetics. It is suggested that this aspect of quality can be ignored, for function is the critical factor in choosing various services options.

Value

The purchaser of M&E services is not buying consumables. His purchase has a value and what is happening, is that monetary liquid asset is being exchanged for a less flexible asset. If that asset holds its market value close to the purchase price it is a better purchase than an option that does not. The quality of services in a building must affect the value of that building. A house with effective central heating will fetch a higher market price than the same house without central heating. Similarly an office block fitted out to high "bank" standards should be worth more than one specified to more basic speculative standards. However, it is doubtful whether the market is currently sufficiently equipped regarding this aspect to enable us to evaluate value as a factor of quality for direct comparison with cost.

In summary, therefore, whilst cost can be related to the NPV of maintenance (practicality), running (economy) and renewal (life) costs, appearance and value cannot be so treated.

A SIMPLIFIED AND THEORETICAL EXAMPLE TO EXPLORE THE RELATIONSHIP
BETWEEN COST AND QUALITY

In July 1983, the Quantity Surveyors Division of the Royal Institution of Chartered Surveyors published "Life Cycle Costing for Construction". This report, written by Roger Flanagan and George
Norman, had been commissioned of the Department of Construction
Management, University of Reading. It included the figure reproduced
below being the life cycle cost commitment for a small primary
school built for a UK Local Authority. Future costs are discounted
at 2% (net of inflation) and the school life was taken as 50 years.

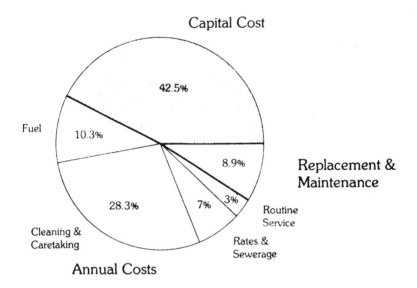

Local authority school life cycle cost
It makes the point that future annual, replacement and maintenance
costs cannot be dismissed as insignificant in comparison with
capital cost.
    That same report, in Chapter 9, included a worked example of a
life cycle cost plan for a 240 place primary school. This is reproduced below in full not because it is the last word in such a document but because it forms a convenient basis for showing how capital
cost changes may be related to the future cost (or quality) implications of those changes.

Worked example from "Life Cycle Costing for Construction" (©Royal
Institution of Chartered Surveyors July 1983)

The worked example detailed below gives the life cycle cost plan for
a particular design option for a 240 place primary school. No
attempt has been made to compare these costs with the costs of other

292

design options, the intention being to illustrate in some detail the application of the techniques.

If alternative designs are to be compared, similar calculations should be performed for each option. In doing so there will, of course, be many areas in which cost estimates are common to the various possibilities, thus saving computation effort.

The example has been provided by the Essex County Council using a base date of November 1982. A discount rate of 2% has been used which takes due account of inflation. This reflects the view of the County Council with respect to the difference between the cost of capital to the Council and forecast inflation rates.

The life cycle for the analysis is 30 years. Note that this is an estimate of the functional life of the building, not the physical life.

The estimated fuel use has been calculated in accordance with the model set out in the Department of Education and Science Design Note 17.

Life Cycle Cost Plan

General information

Project: 240 place primary school
Location: Essex
Date: November 1982
Discount rate: 2% (net)
Life cycle: 30 years

Description of proposed building

Steel and concrete framed single storey building, flat roof, low pressure hot water heating system. Associated external works.

Sources of data

1. Cleaning and caretaking - County formula
2. Grounds maintenance - County schedule of works values
3. County and district rates - Essex Local Government Finance
                              1982/3
4. Sewerage charges - Anglian Water Authority
5. Water charges - Essex Water Company
6. Insurances - County Insurance Officer
7. Energy - CIBS Guide, DES Design Note 17. Current Gas and
            Electricity Board Tariff.
8. Maintenance - Chief building surveyor's maintenance records.

Assumptions

Occupancy   - 8½ hours per day, 5 days per week excluding holi-
               days. No evening use.

Life cycle   - Use of building assumed indefinite but life cycle
               restricted to 30 years. No disposal costs or
               residual value included. Assumes current restri-

293

tion on internal decoration will be relaxed in the
near future.

Discount rate - Assumed that the major cost elements, viz energy
and labour, will inflate at approximately the same
rate. Therefore single discount factor used.

Measurement

Building data

Gross floor area	904m²
Circulation space area	80m²
Ground floor area	904m²
Volume	2,440m²
Upper floor area	- m²
Roof area	904m²
Wall area (net)	316m²
Windows and external door area (net)	106m²
Internal walls and partitions area	420m²
Internal doors area	70m²
Wall finishes area	840m²
Floor finishes area	830m².
Ceiling finishes area	813m²
Fittings and furnishings	- nr
Sanitary appliances	56nr
Water installation (draw off points)	93nr
Site works	10,000m²

Design data

No. of occupants	243nr
Occupants/m² gross floor area	0.27
Net/gross floor area	84%

Brief specification: concrete light aggregate wall panels, aluminium
windows, Durox and asphalt flat roof, fairface blockwork partitions
with sigmulta, generally carpet floor covering, low pressure hot
water heating system with steel radiators, fluorescent lighting.

Life cycle cost plan - worksheet

1. Capital cost

From Essex County Council Building Economics Unit Model for 240
place primary

Building 904m² @ £316.37 =	£286,000	
External works (from model)	£ 52,000	
	£338,000	(October 1982 firm
Design changes etc.	£ 60,000	price)
	£398,000	

294

2. Operation costs (annual)

Cost per annum

    2.1 Caretaking   £

Gfa 904 plus 28m² covered court = 932m²   5,200
From County Schedule, 1 caretaker required

    2.2 Cleaning

        (a) Floor and general cleaning

            From County Schedule total hours 39 per week
            Less caretaker contribution   20

            Part-time cleaner   19 hrs per wk

            Cleaner 19 hours @ £2.00 x 52 weeks = £1,976
            Materials   = £   250
            Equipment replacement and
            maintenance say   = £   200
                                     2,426

        (b) Window cleaning

            Twice annually @ £27 = £54   54

                     Total cleaning   £7,680

                               (£8.50/m²)

        Present value (30 years) = £172,000

                            (£190.27/m²)

    2.3 Rates

From National Rating Formula for Schools

Teaching area plus teaching storage   = 622m²
Less 10% for open plan   =   62m²

                             560m² ÷ 1.9

                    = 295 scholar places

Rating calculation: Gross
                                   Value

1. Basic price £16.25 plus £1.10 for high
   site value = £17.35 x 295   =   5,118
2. Kitchen   5.3m² x 42m²   =   223
3. Caretaker   =
4. Playing fields £150/ha x 1ha   =   150

                                  5,491

Rateable Value = $\dfrac{10\ (5,491 - 34)}{12}$ = 4,548

District rate : 4,548 @ 11.30p in £	=	514
County rate : 4,548 @ 125.90p in £	=	5726
Sewerage charge: 4,548 @ 18.65p in £		848
Water charges : 160,000 gals @ 84p/1000 gals		134
Standing water charge		29

       Total rates & water charges   =   £7251
       ($8.02/m^2$)

       Present value (30 years)   =   £162,400
       ($£179.65/m^2$)

2.4 Insurances
(Part fire insurance only)

Building cost £286,000
Premium rate: 80p/£1000

       Annual premium = £230
       ($0.25/m^2$)

       Present value (30 years)=£5150
       ($5.70/m^2$)

2.5 Staff

Administrative staff (establishment related only)
       = £7,500 p.a.   ($8.30m^2$)
       Present value (30 years) = £168,000
       ($£185.84/m^2$)

2.6 Energy

(Calculation follows the principles of DES Design Note 17)

Heat losses
a) Fabric

Element	Area	U Value	W/°C	Temp diff	Conv to kW	kW
Ground floor	904m² x	0.3	= 271.2			
Roof	904m² x	0.5	= 452.0			
Opaque wall	316m² x	1.3	= 410.8			
Windows	106m² x	5.6	= 593.6			

       1,727.6 x 19° ÷ 1000   = 32.82

b) Ventilation   Occupants   Air change/person

       243   x 30m³ x 0.33 x 19 ÷   1000   = 45.70
       Total loss = 78.52kW

c) Less gains

		kW
Occupants	:243 x 70W/person ÷ 1000 =	17.01
Lighting	:12W/m² (DES) x 904 ÷ 1000 =	10.85
Miscellaneous power:3W/m² x 904	÷ 1000 =	2.71 Less
		30.57

Net heat loss = kW 47.95

Annual energy cost

Natural gas	Total loss (kW)		Running hours		Calorific value	Price/ Therm	Cost
1. Heating	47.95	x	1731	÷	29.3	x 50.7p =	1,436
2. Hot water (kitchen)	12500			÷	29.3	x 50.7p =	216

Gas total =£1,652

Electricity

1. Lighting $\frac{52}{3.73}$ (kW/m²/a) x 904m² x 5.5p = 693

	W/m²	Hours/per annum		
2. Local hot water	2.00 x	400	=	800
3. Power	3.00 x	1,400	=	4,200
4. Pumps	2.00 x	1,120	=	2,240

7,240 x 904m²

÷ 1000 x 5.5p

= 360

Electricity total = £1,053

Energy total cost per annum = £2,705
(£2.99/m²)

Present value = £60,500
(£60.92/m²)

3. Maintenance costs

3.1 Annual

a) Building

	Cost/per annum £
1. Heating and hot water services plant (term contract)	300
2. Replacement of fluorescent tubes and bulbs 904m² @ 6p	54
3. Minor repairs 904m² @ 50p	452

Building total £806
(0.89m²)

b) External works

			£
Grounds maintenance for lha site	Labour	700
	Fuel and maintenance	150
	External works	850

Total annual maintenance £1,656
Present value (30 years) £37,100

3.2 Intermittent

Item	Quantity	Maintenance/ replacement cost (unit rate)	Interval (years)	Proportion replaced	Cost £	PVF	PV £
Roof covering							
Asphalt	936m²	£25/m²	30	100%	23400	0.5521	12920
External walls							
Panel joints	186m	£3.15/m	15	100%	586	1.2951	760
Windows & doors							
Anodised aluminium windows	92m²	£100	30	15%	1380	0.5521	760
Softwood doors	14m²	£160 (Rep.)	20	15%	1120	0.6730	750
	14m²	£3 (Redec)	6	100%	42	3.5505	150
Internal walls							
Sigmulta wall finishes	2074m²	£3.75/m²	15	100%	7778	1.2951	10080
Floor finishes							
Carpet (classrooms)	400m²	£4.75/m²	15	75%	1900	1.2951	2460
Carpet (circulation)	71m²	£4.75/m²	10	100%	337	2.0454	690
Carpet (admin)	48m²	£4.75/m²	20	50%	95	0.6730	60
Granwood	125m²	£3.00/m² (sanding)	10	100%	375	2.0454	770
Vinyl	186m²	£10.67/m²	30	75%	1488	0.5521	820
Ceilings							
Emulsion paint	813m²	£1.78/m²	10	90%	1300	2.0456	2660
Heating							
70kW gas boiler	1 Nr	£2750	30	100%	2750	0.5521	1520
Control equipment	1 Nr	£1500	20	50%	750	0.6730	500
Pumps	2 Nr	£300	30	100%	600	0.5521	330
Hot water							
Calorifier	1 Nr	£750	30	100%	750	0.5521	410
Pumps	2 Nr	£100	30	100%	200	0.5521	110
Local water heaters	6 Nr	£50	20	50%	150	0.6730	100
Lighting & power							
Allow contingency	–	£1500	15	100%	1500	1.2951	1940
Building present value = 37790							
Present value/m² = £41.80/m²							
External works							
Playgrounds & road surfaces	2220m²	£350	20	90%	6993	0.6730	4700
Fences & gates	–	£500	30	100%	500	0.5521	270
External works present value							**£4970**

Building present value   = £37790
External works present value   = £4970
                                    _____
                                    £42760
Building and engineering surveyors charges   -   £5140
                                    _____
                                    £47900

Intermittent maintenance total present value say £48000

298

Life cycle cost plan summary

Project: 240 Place primary school
Location: Essex
Date: November 1982

Costs	Estimated target costs	Present Value
1. Capital costs		
Building	336,750	336,750
External works	61,250	61,250
Total capital costs	398,000	398,000
2. Running costs		
2.1 Operation costs		
Energy	2,705	60,500
Caretaking and cleaning	7,680	172,000
Rates	7,251	162,400
Insurances	230	5,150
Staff	7,500	168,000
Total operations costs	25,366	568,050
2.2 Maintenance costs (annual)		
Building	806	18,050
External works	850	19,050
Total maintenance costs (annual)	1,656	37,100
2.3 Maintenance/replacement/alteration costs (intermittent)		
Building	–	42,300
External works	–	5,700
Total maintenance/replacement/ alteration costs (intermittent)	–	48,000
2.4 Sundries	250	5,600
Total sundries	250	5,600
Total running costs	–	658,750
3. Additional tax allowances		NIL
4. Salvage & residuals		NIL
5. Occupancy costs (excluded)		–

```
                 Total present value of
                 life cycle costs =              £1,056,750

                 Annual equivalent value of
                 life cycle costs =              £   47,200
```

Full year effect costs

Note: All costs at November 1982 prices.

```
                                                 Extl
1.  Capital repayment      Building  (£/m²)    Works    Total

    £398,000 debt charges    47,200  (52.21)   8,600    55,800

2.  Annual costs (from
    LCCP summary)
    2.1  Operations costs    25,366  (28.06)      -     25,366
    2.2 Annual maintenance      806   (0.89)     850     1,656
    2.3 Sundries                250   (0.28)      -        250

2.  Intermittent costs (from
    LCCP summary)
    Present value   : £48,000
    Annual equivalent: £48,000

              ÷ 22.3965*       1,890   (2.09)     253     2,143
    * (YP for 30 years 2%)
                             £75,512  (83.53) £9,703  £85,215


    Full year effect costs
    (year 1 at current prices)                       £85,215
```

(End of worked example from "Life Cycle Costing for Construction"-
ⓒRICS July 1983)
   This example is, as is stated, based on:
(i) low pressure hot water heating with steel radiators and
(ii) fluorescent lighting
   Let us now examine what could happen to the relevant figures if
low pressure hot water heating was changed to a warm air heating
system - and if fluorescent lighting was changed to recessed fluorescent luminaires in a suspended ceiling.

Item 1 Capital Cost would read:

```
Building 904m² (including part false ceiling,
warm air system and recessed fluorescent
luminaires)                                      £322,000

External works                                     52,000
                                                  374,000 (October
Design charges etc.                                66,000 1982 firm
                                                 £440,000 price)
```

Item 2.4 Insurances would read.

Building cost: £322,000
Premium rate :   80p/£1000   Annual premium =   £   258
                                          (0.285/m²)   ———
                            Present value (30 years) =   £5,780
                                                         ———

Item 2.6 Energy would read:

Heat losses

a) Fabric element   kW
   as before   32.82

b) Ventilation occupants

   904m² x 2.5m height x 4 air changes
   x 0.33 x 19° ÷ 1000   =   56.68
                                    ———
                   Total loss =   89.50 kW

c) Less gains
   as before   30.57
                                    ———
            Net heat loss =   58.93kW
                              ———

Annual energy cost

Natural Gas	Total loss kW		Running hours		Calorific Value		Price/ Therm		Cost
1. Heating	58.93	x	1731	÷	29.3	x	50.7p	=	1,765
2. Hot water (kitchen)	12500			÷	29.3	x	50.7p	=	216
							Gas total	=	£1,981

Electricity

1. Lighting as before   = 693
                              W/m²   Hours/per annum
2. Local hot water   2.00   x   400   =   800
3. Power   3.00   x   1,400   =   4,200
4. Pumps and fans   3.00   x   1,120   =   3,360
                                                       ———
                                                    8,360 x
                                                    904m²
                                                    ÷ 1000
                                                    x 5.5p = 416

                                 Electricity total £1109
                                                    ———
                    Energy total cost per annum £3090
                                 Present value £69110

301

Item 3 Maintenance costs would read:

3.1 Annual

   a) Building

	Cost per annum £
1. Heating and hot water services plant (term contract)	350
2. Replacement of tubes and bulbs as before	54
3. Minor repairs as before	452

Building total = £856

   b) External Works   as before

External works = £850

Total annual maintenance = £1706

Present value (30 years) = £38200

Item	Qty	Maintenance replacement cost (unit rate)	Interval years	Proportion re-placed	Cost £	PVF	PV £
Building present value							37790
Ceilings Omit Emulsion paint	813m²	£1.78/m²	10	90%	1300	2.0456	(2660)
Add Suspended ceilings	813m²	£15.0/m²	20	100%	12200	0.6730	8210
Heating Add Air handling units & fans	1Nr	£3500	30	100%	3500	0.5521	1930

Building present value = £45,270

Present value/m² = £50.07/m²

External works present value   as before   £4970
                                                                 ─────

                        Building present value   =   45,270
                    External works present value   =   4,970
                                                         ──────
                                                          50,240
        Building & engineering surveyors charges   =   6,040
                                                         ──────
                                                         £56,280

        Intermittent maintenance total present value say £56,300
                                                         ──────

Life cycle cost plan summary:

This would be:

Costs	Estimated target costs	Present value
1 Capital costs - Building	378,750	378,750
- External works	61,250	61,250
Total capital costs	440,000	440,000
2 Running costs		
2.1 Operation costs		
Energy	3,090	69,110
Caretaking & cleaning	7,680	172,000
Rates	7,251	162,400
Insurances	258	5,780
Staff	7,500	168,000
Total operations costs	25,779	577,290
2.2 Maintenance costs (annual)		
Building	856	19,150
External works	850	19,050
Total maintenance costs (annual)	1,706	38,200
2.3 Maintenance/replacement/alteration costs (intermittent)		
Building	–	50,600
External works	–	5,700
Total maintenance/replacement/alteration costs (intermittent)	–	56,300
2.4 Sundries	250	5,600
Total sundries	250	5,600
Total running costs	–	677,390

Life cycle cost plan summary contd.

Costs	Estimated target costs	Present value
3. Additional tax allowances		Nil
4. Salvage and residuals		Nil
5. Occupancy costs (excluded)		−
Total occupancy costs		−

Total present value of
life cycle costs   £1,117,390

Annual equivalent value
of life cycle costs   £   49,900

Full year effect costs (all costs at November 1982 prices)

This would read:

		Building	(£/m²)	Extn Works	Total
1.	Capital repayment £440,000 debt charges:	53,100	(58.73)	8,600	61,700
2.	Annual costs (from LCCP summary)				
2.1	Operations costs	25,779	(28.51)	−	25,779
2.2	Annual maintenance	856	(0.95)	850	1,706
2.3	Sundries	250	(0.28)	−	250

2. Intermittent costs (from
    LCCP summary)
    Present value   £56,300
    Annual equivalent: £56,300 ÷

		Building	(£/m²)	Extn Works	Total
	22.3965*	2,261	(2.50)	253	2,514
*(YP for 30 yrs 2%)					
		£82,246	(90.97)	£9,703	£91,949

It must be stressed that the above example and what is drawn from
it cannot be taken as the source of precise data for application
to other projects. There is an inbuilt approximation within a
life cycle cost estimate for many reasons and it must be understood that such exercises have more validity for evaluating the
differences amongst options that as a prediction technique for
single solutions.

What the example does show is how the running, maintenance and replacement costs can be related to initial cost. In the former life cycle cost plan, £55,800 (in full year effect) is spent on initial cost in comparison with £29,415 on future costs. In the latter adjusted life cycle cost plan, the initial cost commitment is £61,700 (an additional £5,900) for future costs equivalent to £30,249. There is thus no detectable benefit from the additional initial cost, for the future costs have gone up and the total life cycle full year effect costs have increased.

SOURCES OF DATA FOR FUTURE (QUALITY) COSTS

Despite the recent surge of interest in future costs (largely moti-vated by inflation and substantial increases in the cost of energy) it is very difficult to locate useful ready-to-use data. This is further discussed in the conclusion to the paper.

SUGGESTED SERVICES ELEMENT UNIT RATES FOR OFFICES AND THE QUALITY
RANGE INHERENT IN THEM

In May 1983 the Cost Research section of the author's practice produced the tables of element costs per m² of QS gross floor area
reproduced below. These costs were based on the following assumptions:-

1. June 1983 price level (Davis Belfield & Everest Tender Index
    200).

2. Outer London.

3. Contract value over £½ million.

4. Competitive tender.

In addition further assumptions were made for the purpose of
deriving a reasonable range of services costs:-

5. Air conditioned/comfort cooled buildings are double glazed.

6. Fairly high level of wall and roof insulation.

7. Window to wall ratio not exceeding 50%.

8. Normal exposure.

9. Floor to floor height not exceeding 4 metres.

These were fed into an office estimating programme for computer
application.

COST PARAMETERS OF MECHANICAL SERVICES - OFFICES (£s per m² of floor area)

	SPECULATIVE			HIGH QUALITY SPECULATIVE			OWNER OCCUPIED		
	Up to 1000m²	1000-5000m²	Over 5000m²	Up to 1000m²	1000-5000m²	Over 5000m²	Up to 1000m²	1000-5000m²	Over 5000m²
Plumbing and Sanitary Fittings	3.00-10.00	7.50-12.00	7.50-12.00	3.00-10.00	7.50-12.00	7.50-12.00	4.00-11.00	9.00-13.00	9.00-13.00
Hot & Cold Services	4.50-9.00	5.50-11.00	5.50-11.00	4.50-9.00	5.50-11.00	5.50-11.00	4.50-9.00	5.50-11.00	5.50-11.00
Hose Reels & Dry Riser				2.00-4.50					
Sprinklers (Hazard Ordinary)	6.50-9.00	6.50-10.00	6.50-10.00	6.50-9.00	6.50-10.00	6.50-10.00	6.50-9.00	6.50-10.00	6.50-10.00
(Extra for Storage Tank)	1.00-2.00	1.00-2.00	1.00-2.00	1.00-2.00	1.00-2.00	1.00-2.00	1.00-2.00	1.00-2.00	1.00-2.00
LPHW Radiator heating	20.00-27.50	24.00-38.50	27.50-38.50	21.00-31.00	26.50-42.00	31.00-42.00	22.00-34.00	26.50-44.00	31.00-46.00
LPHW Radiant panel heating	23.00-31.00	27.50-42.00	31.00-42.00	24.00-34.00	30.00-45.00	34.00-45.00	26.50-38.50	31.00-48.50	35.00-51.00
Ventilation - Supply & extract	38.50-55.00	44.00-60.00	44.00-60.00	44.00-60.00	44.00-66.00	44.00-66.00	44.00-66.00	44.00-71.00	44.00-71.00
COMFORT COOLING									
Fan Coil units) / Induction Units)	60.00-77.00	64.00-82.50	64.00-82.50	66.00-82.50	68.00-88.00	68.00-88.00	70.50-84.00	72.50-93.50	72.50-93.50
VAV Ductwork	60.00-88.00	64.00-93.50	64.00-93.50	68.00-92.50	71.50-99.00	71.50-99.00	71.50-101.00	75.00-110.00	75.00-110.00
AIR CONDITIONING									
Fan Coil Units) / Induction Units)	71.50-99.00	77.00-110.00	77.00-110.00	77.00-104.50	82.50-115.50	82.50-115.50	82.50-121.00	88.00-132.00	88.00-132.00
VAV Ductwork	75.00-110.00	77.00-115.00	77.00-115.50	79.00-121.00	86.00-121.00	86.00-121.00	86.00-127.00	90.00-132.00	93.50-132.00
Car park extract	←—————————————— 16.50-27.50 ——————————————→								
Toilet extract	←—————————————— 1.25-3.50 ——————————————→								

Includes: Main Contractors profit and overheads and general builders work in connection

Excludes: Gas Board charges, Energy Management, Restaurant Equipment and Supplies

COST PARAMETERS OF ELECTRICAL SERVICES - OFFICES (£s per m² of floor area)

	SPECULATIVE			HIGH QUALITY SPECULATIVE			OWNER OCCUPIED		
	2-5000m²	5-15000m²	15-30000m²	2-5000m²	5-15000m²	15-30000m²	2-5000m²	5-15000m²	15-30000m²
Sub-station equipment	—	—	—	—	—	2.00-2.50	—	3.00-3.50	2.00-2.50
Standby equipment	0.-2.00	1.00-2.00	1.00-2.00	0.-4.50	1.50-4.50	1.50-4.50	0.-5.50	3.00-9.00	1.50-7.50
Mains & Sub-mains	7.00-10.00	5.50-9.00	4.50-6.50	8.00-11.00	6.50-10.00	5.50-7.50	9.00-13.00	6.50-11.00	5.50-9.00
Power	5.50-10.00	5.50-10.00	5.50-10.00	9.00-14.00	9.00-14.00	9.00-14.00	10.00-14.00	10.00-14.00	10.00-14.00
Lighting	3.00-5.00	3.00-5.00	3.00-5.00	3.00-5.00	3.00-5.00	3.00-5.00	3.00-5.00	3.00-5.00	3.00-5.00
Emergency Lighting	0.50-1.00	0.50-1.00	0.50-1.00	0.50-1.00	0.50-1.00	0.50-1.00	0.50-1.00	0.50-1.00	0.50-1.00
Luminaries	11.00-16.50	11.00-16.50	11.00-16.50	11.00-22.00	11.00-22.00	11.00-22.00	13.00-27.50	13.00-27.50	13.00-27.50
Electrical for Mechanical Services	7.00-10.00	5.50-10.00	4.50-9.00	6.50-11.00	5.50-10.00	4.50-9.00	7.50-11.00	6.50-10.00	5.50-9.00
Telephone Wireways	0.50-1.50	0.50-1.00	0.50-1.00	0.50-1.50	0.50-1.00	0.50-1.00	0.50-1.50	0.50-1.00	0.50-1.00
Clocks	—	—	—	—	—	—	0.10-0.20	0.10-0.20	0.10-0.20
Public Address	—	—	—	—	—	—	1.00-2.00	1.00-2.00	1.00-2.00
Television	—	—	—	0.-0.25	0.10-0.25	0.10-0.25	0.-0.25	0.10-0.25	0.10-0.25
Fire Alarms	3.00-3.50	3.00-3.50	2.00-3.00	3.00-3.50	3.00-3.50	2.00-3.00	3.00-5.50	3.00-4.50	2.00-3.50
Security	—	—	—	1.00-3.50	1.00-3.50	1.00-3.50	1.50-6.50	1.50-6.50	1.50-6.50
Lightning Protection	0.50-1.00	0.50-1.00	0.50-1.00	0.50-1.00	0.50-1.00	0.50-1.00	0.30-0.50	0.30-0.50	0.30-0.50
External Lighting	0.-0.50	0.20-0.35	0.20-0.30	0.-0.50	0.20-0.35	0.20-0.30	0.-0.50	0.20-0.35	0.20-0.30

Includes: Main Contractors profit and overheads and general builders work in connection

Excludes: Electricity Board Charges, Telephone Wiring Equipment & Connection Charges, Auto-Controls for Lighting, Computer Supplies, Energy Management, Restaurant Equipment and Supplies

Mechanical Services

Sanitary Fittings

At the lower end of the scale these would be standard white fittings
with standard chrome plated brassware. At the upper end these would
be non-standard coloured with special fittings and coatings.

Plumbing

Plumbing would range from UPVC soil and waste pipes in a single
stack system to cast iron with copper wastes in a one pipe system.

Hot and Cold Services

At the lower end of the scale would be envisaged plastic pipework,
local storage and local water heaters, minimum permitted thickness
of insulation, traditional water useage and untreated mains water.
At the upper end would be envisaged copper pipework, central storage,
maximum economic thickness of insulation, economiser taps, timed
flushing of appliances and on site water softeners and treatment.

LPHW Heating

Between the lower and upper ends of the scale the range would be:-
(i)   from basic pipework systems to reverse return systems,
(ii)   from black mild steel pipework to copper pipework
(iii) from standard finishes to heat emitters to special or stove
       enamelled finishes
(iv)   from compact distribution (i.e. high rise offices) to very
       extensive distribution (i.e. single storey)
(v)   from single plant items to duplicated or standby plant, and
(vi)   from simple controls to extensive controls

Ventilation and Air Conditioning

Between the lower and upper ends of the scale the range would be:-
(i)   from warm air ventilation to comfort cooling ventilation
(ii)   from comfort cooling to full air conditioning
(iii) from basic systems with no allowance for up-grading to space
       and facilities built in system for future enhancement or
       enlargement of capabilities
(iv)   from single plant items to duplicated sets
(v)   from economic duct sizes to special duct sizes to suit space
       limitations
(vi)   from basic ductwork construction to enhanced duct thicknesses
        and more complex jointing and sealing provisions
(vii) from no pressure testing on low velocity systems to pressure
       testing
(viii) from standard finish grilles or equipment to special or stove
       enamelled finishes

310

(ix) from minimum permitted thickness of insulation to maximum economic thickness of insulation
(x) from basic design of system to heat reclaim facilities
(xi) from basic simple controls to extensive and integrated controls linked to energy management systems
(xii) from no control of energy usage to energy management system with data recording
(xiii) from selection of plant for duty only to plant selected for economy of maintenance in addition to duty
(xiv) from fixed open plan coverage to cellular and full flexibility
(xv) from single fuel to multiple fuel provision, and
(xvi) from full maintenance programme with own staff to avoidance or minimisation of maintenance programme or staff

Electrical Services

Sub-Station Equipment

For most office blocks the supply is usually taken at low voltage. Where the size or location of the building dictates the provision of a sub-station, the equipment would normally be provided by the Electricity Board and the connection charges adjusted accordingly.

Occasionally the building owner/developer will decide to take the supply at high voltage, where the size of the building warrants this and the owner/developer has staff capable of operating and maintaining high voltage switchgear and transformers (or is prepared to pay the Electricity Board for so doing).

Standby Equipment

At the lower end of the scale, standby supplies would only cover emergency services such as, car park ventilation, sprinkler pumps and hose reel pumps.

At the upper end of the scale maintenance of near normal working conditions in limited areas of the building would also be covered.

Where more extensive standby is required for larger operational areas and/or computer installations the cost of standby will go even higher. [e.g. a large bank development (28,000m²) - £19.50/m² or a Broadcasting Centre (54,000m²) - £11.50/m²]

Mains and Sub-Mains

Distribution costs increase with "quality" because electrical load density increases with:-
(a) Greater use of current consuming office equipment
(b) Larger proportion of areas with higher than average illumination levels or lighting techniques giving a higher load density
(c) Increase in central mechanical plant loads to cater for (a) and (b) and to give more generous operating margins on the plant.
The impact on larger buildings is less because costs are not directly proportional to area.

311

Power

At the lower end of the scale a simple skirting trunking and/or
widely spaced under floor trunking system might be provided.
    At the top end the installation might include flush floor trunking
on a grid of 3m or less with outlet boxes at intervals of about 2.5m.

Lighting (Emergency lighting and luminaires)

All the figures are based on general lighting to provide 500 to 600
lux in office areas and emergency lighting to escape routes only.
    At the lower end of the scale the general lighting might be pro-
vided by 1500mm or 1800mm fluorescent luminaires spaced at maximum
permissible centres with little or no consideration to the flexibility of possible partition locations.
    At the upper end of the scale a great number of smaller and more
expensive luminaires might be used (say 1200mm), set out on a grid
which permits a high degree of flexibility in the location of partitions.

Electrical for Mechanical Services (Power and controls cabling
                                                            only)

Lower quality buildings would have heating and some ventilation to
toilets.
    Higher quality buildings would be air conditioned. Therefore
there would be more plant and controls wiring.

Telephone Wireways

This element deals with cable trays and trunkings from the intake
position to the equipment room and from the equipment room to suitable locations on each floor.
    Wireways within each office floor are covered by the skirting
and/or floor trunking systems dealt with under power.
    The configuration of the building is the main factor in deter-
mining the costs for this element. The lower end of the scale
represents a simple compact building with perhaps a single riser
position. The upper end represents a building with a large plan
area and several risers.

Fire Alarms

A basic system would comprise manual alarms generally with automatic detection in plant rooms and ventilation systems.
    A more elaborate system would include more extensive automatic
detection to a larger proportion of areas.

Security

At the lower end of the scale this element might include contacts
on all external doors and ground floor windows connected to a

312

central alarm panel.
   At the top of the scale there might be any or all of the following
(a)   Movement detection
(b)   Simple or graded card access control and monitoring
(c)   Closed circuit television monitoring of key areas inside and/or
      outside the building

   Lightning Protection

This element is mainly dependent on building configuration and type
of construction.

   External Lighting

In its simplest form this element might comprise canopy lighting to
the main entrance.
   At its most elaborate it might include access road lighting, car
park lighting and flood-lighting to the building facade.

CONCLUSION

The arithmetical relationships between initial cost and future cost
have been known for very many years. "Cost-in-use" and "Life cycle
costs" are phrases that are not new. Why is it therefore that cost/
quality relationships are not offered to building industry clients
as a normal consultancy service? The following reasons are
suggested:
1.   The relationship between the pure cost of money and inflation is
     not understood. Thus the choice of appropriate discount rate
     becomes difficult and if as a result, a wide range of discount
     rates is encountered, a wide range of net present values for
     future costs results, which tends to down-value the relevance
     of the exercise.
2.   There is still insufficient feedback to those who design build-
     ings of data as to how existing buildings behave in cost terms.
     The position is slowly improving but the users and operators of
     buildings do need to be persuaded that monitoring running costs
     and publishing them will ultimately result in better future
     building industry products.
3.   No matter how compelling the case for further initial expendi-
     ture in future cost saving terms, if that extra capital is
     simply not available, it will not be committed.
4.   Cost-in-use and life-cycle studies commit professional time
     which costs money. We have not arrived at the point where such
     work is seen as value-for-money regarding the fees involved.
   This paper has only touched the very surface of only few of the
aspects of a subject that could be sufficient material for a large
book. If it, in a small way, encourages people to address themselves to the above problems, it will have achieved something.

313

THE ROLE OF THE DESIGNER IN A CHANGING CONTEXT

MERVIN PERKINS, Hampshire County Council

1. CONTEXT

Perhaps the most challenging problem faced by architects in local authorities during the last decade has been the need to maintain and sometimes improve design standards, with ever decreasing resources. At the same time, the focus of building programmes has switched from capital expenditure on new work to the rationalisation of existing resources. This has forced architects into new roles and to adopt new design strategies. These may well derive from building maintenance and estate management considerations.

In Hampshire, this period has been marked by a radical change in design emphasis – away from standardised products and towards individual design solutions.

Between 1960-75, Hampshire was the main user of the SCOLA building system and constructed some five hundred school buildings of this type. The solution to the problems of volume construction was not only to employ system building techniques, but to produce standard designs which could be repeated on several sites. It was not uncommon for as many as fifteen identical primary schools to be constructed from a single set of design drawings. Similar logic was extended to layouts of school sites – with building orientation dictated by service access points rather than climatic or environmental reasons; planting schemes geared to machine maintenance and chain-link fencing to separate the building from its community.

Today the results of this approach seem utilitarian and lacking in character. Moreover the legacy of such an expansive construction programme is now major expenditure on replacing roofs, cladding and heating systems. Hampshire's current design philosophy (1) represents a conscious attempt to produce solutions which are unique to their situation and draw inspiration from their context. It is based on attaining broadly acceptable and functional levels of design rather than narrow optimum solutions founded on abstract standards. Obviously there is a danger of producing 'one-off'

314

designs which are wasteful of resources and based on untried methods. In practice, however, there is much common ground between seemingly different designs and efforts are made to evaluate and learn from previous work.

By its very nature, this approach is wide-ranging and often multi-disciplinary. For example, it may involve architectural research to investigte contextual factors or evaluate prototype solutions. Alternatively, estate and building management practices may be employed to generate financial resource. Often the finished building represents the tangible product of a total design and economic appraisal in which the architect may have acted as entrepreneur in initiating the project as well as designing it.

In this respect, the paper will describe two recent case studies: a new school and the redevelopment of an existing college, in which the designs have been influenced by contextual, economic and management factors.

2. CASE STUDIES

The following examples are intended to illustrate two areas in which local authority architects are becoming increasingly involved. In the first instance, a more efficient use of resources has resulted from the re-evaluation of past designs. The second demonstrates the role of the design in generating the initial financial resource.

Yateley Newlands (2) is a one form entry school for 280 pupils in the five to eleven age group. It was completed in 1980 and serves an area of recent housing development.

Prior to this it had been Hampshire's policy to purchase sites of ten acres to accommodate two schools, although the initial requirement was often for a single school. In many cases no further building has occurred due to falling school roles in recent years. At Yateley the involvement of the design team at inception esulted in the purchased site area being reduced from ten acres to three. The consequent saving was more than the cost of the new school building. In design terms, however, it resulted in the need to preserve the essential character of the small tree-covered site whilst exploiting its whole area for educational uses.

Sectional Perspective   Yately Newlands School

Atrium

Yately Newlands School

At the time of design, research studies were being undertaken with Portsmouth Polytechnic School of Architecture. These aimed to investigate aspects of environmental control in primary schools, through the use of automatic monitoring, fuel metering and behavioural mapping techniques (3 & 4).

Although the main objective was to assess user influences on fuel expenditure, the study evaluated this in the context of the educational environment. It was considered necessary for such research to be broad-based and multi-disciplinary in order for any

316

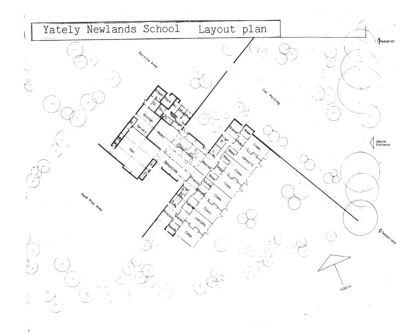

conclusions to have a practical application. In this case, a personal involvement in both the research and the design for Yateley, meant the study results had a direct and immedite influence.

The initial findings of this research work stressed the qualitative aspects of the educational environment and in particular the significance of scale and variety. It suggested a passive educational approach enabling a range of different organisational patterns and individual teacher preferences.

Thus the Yateley design aims to achieve a variety of educational environments. Externally these range from the exposed games pitch to sheltered south-facing sitting areas. Wherever possible, birch and oak saplings have been retained and landscaping costs reduced by transplanting others to form natural study areas and provide shelter from the elements. To achieve a gradual transition from the outside, external walls and paving continue inside the building.

Internally the south facing teaching block contains class bases, tutorial rooms and practical areas under one large roof volume.

Overhead service ducts at an intermediate level reduce scale, join together the variety of spaces and prevent the need for costly underfloor service ducts. The northern block contains shared, administrative and ancillary accommodation, such as the hall, kitchen and staff areas.

317

View from Playground   Yately Newlands School

A most important and popular element in this design is the central glazed courtyard which not only provides additional external
teaching, but also allows landscaping to be brought into the heart
of the building. Much of the planting in this area was, in fact,
provided by the parents under the direction of the design team.

The conservatory reduces internal circulation and allows a teaching area 20% above the minimum. A further benefit is that it has
been shown to reduce fuel consumption for heating by approximately
10% (5).

In the field of energy use, in particular, the design has been
shaped by the parallel research. The research studies indicted
that traditional methods of environmental control avoiding mechanical systems were more acceptable to the users of this type of
building and generally produced lower running and maintenance costs.
Thus this design aims to give the occupants a high degree of individual control whilst providing efficient plant and automatic systems
where appropriate. Heating is by means of radiators fitted with individual thermostatic valves but these are served by a gas-fired
modular boiler linked to an optimum start controller.

Likewise the form of the building attempts to combine the educational benefits at a deep-plan layout with a selective approach to
environmental control - making use of daylighting, natural ventilation and solar gains. In this respect, the small area of roof
glazing gives an even level of daylighting and cross ventilation in

318

summer. Neither feature was included in the standard SCOLA designs studied. Similarly, the Yateley design has roof overhangs, rather than clipped eaves, to prevent summer overheating, whilst allowing the penetration of the low-angled winter sun.

In practice, the design has proved most successful in educational and environmental terms whilst having low running costs. Furthermore, the internal organisation of the County Architect's Department has enabled the design team to retain direct links with the school in use. For example, the reduction in the demand for school meals has recently allowed part of the kitchen to be converted into a specialist pottery area. The initiative for this came from the users and the design team rather than the client body.

Such situations are becoming increasingly common where architects have a direct involvement with building management duties. This role and the associated expertise can mean that they are better placed to propose corporate measures than the specialist client bodies that exist within local authorities. The second case study clearly illustrates this view.

****

In 1982, Farnborough College of Technology accommodated some 1,400 full-time and 4,500 part-time students on a wide range of full-time, sandwich and part-time day and evening courses.

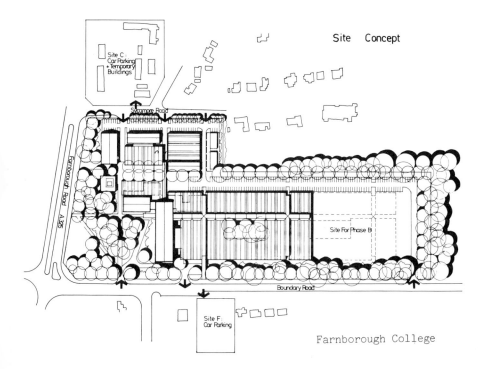

Farnborough College

NORTH-SOUTH SECTION   Farnborough College

Its campus consisted of six adjacent sites near the town centre.
In addition, three of the eight departments were located in a former
secondary school at Aldershot. This was intended as a short-term
occupation.

At this time, the County Architect's Department were briefed to
provide extensions with a budget cost of £900,000. However, due to
the nature of this particular establishment, a full rationalisation
study was initiated by the design team (6). This highlighted the
following problems:-

. the disadvantages of a campus divided by major roads

. a shortage of accommodation amounting to almost 10,000m^2 when
  compared with current Department of Education and Science standards

Perspective View from Carpark   Farnborough College

320

- a high proportion of general teaching accommodation in temporary buildings

- the general use of facilities which had not been designed for their present needs

- most accommodation being in a poor state of repair and some likely to need major maintenance within five years

Overall there appeared to be a surplus of land occupied by the College, but a substantial shortfall of suitable permanent accommodation to meet further needs. Certain sites appeared to be under-utilised and not fully exploited in terms of building provision.

The objective of the study was, therefore, to examine the existing facilities, in the light of future requirements, in order to identify a priority list of building needs and to formulate a long-term strategy for development. Inherent in this strategy was the proposal to finance later phases of construction through the disposal of surplus land. This, in turn, allowed for the relocation of College facilities and the consolidation of existing and new on a smaller site area.

The study involved the design team in finding alternative commercial use for the surplus sites and a change in financial regulations to enable the capital receipts to be retained for use in the redevelopment.

The outcome was that the scope of the project changed from one of piecemeal extensions to the major redevelopment project, now under construction at a cost of six million pounds. The greater part of this latter sum is being realised from land sales.

The design itself is very much a response to the particular problems of the College and the financial strategy for its redevelopment. Its stated objectives were:

- to unite both existing and new accommodation to form a coherent overall development

- to provide accommodation capable of meeting changing requirements during the design period and following occupation

- to provide a design of architectural and environmental quality which established a dynamic image suited to the changing role of such a College in the 1980's.

The resulting construction is based very much on prefabricated components, which can be varied according to detailed requirements, without affecting the overall building structure and its envelope. It involves the use of a single storey steel frame located on a repetitive grid. This is integrated with an overhead services grid and roof glazing to give natural lighting to deep-plan areas.

321

An overall planning strategy was developed for the whole site which involved the zoning of facilities. Car parking and service access is provided on the north side of the site with adjoining communal facilities such as the library, lecture theatre and refectory.

The teaching facilities of individual departments are positive to the south of this around landscaped courtyards. Within this area there is a similar zoning between highly serviced areas such as laboratories and general teaching space. In this way a planning and servicing matrix is developed which can accommodate a high degree of change.

Important elements of the design are the glazed streets with link accommodation. These are essentially unheated circulation spaces which give protection from rain and wind. As such these are developments from previous projects such as the Yateley Newland's conservatory.

The glazed malls also act as buffer zones to reduce heating requirements for adjacent space and external noise intrusion. In particular, their solar gains are employed to pre-heat ventilation losses. The design of malls and the associated ventilation system was actually influenced by fire requirements as much as environmental or other considerations. In this respect, all the malls have automatic smoke ventilation and the main east-west mall is treated as a fire-protected compartment. Thus ventilation air is taken

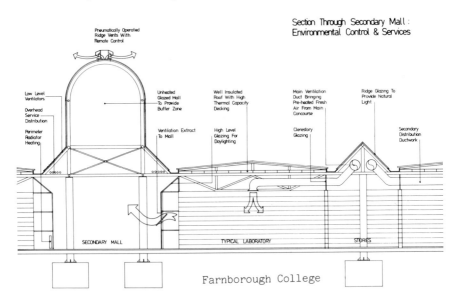

Section Through Secondary Mall :
Environmental Control & Services

Pneumatically Operated Ridge Vents With Remote Control

Low Level Ventilators

Overhead Service Distribution

Perimeter Radiator Heating

Unheated Glazed Mall To Provide Buffer Zone

Ventilation Extract To Mall

Well Insulated Roof With High Thermal Capacity Decking

High Level Glazing For Daylighting

Main Ventilation Duct Bringing Pre-heated Fresh Air From Main Concourse

Clerestory Glazing

Ridge Glazing To Provide Natural Light

Secondary Distribution Ductwork

SECONDARY MALL   TYPICAL LABORATORY   STORES

Farnborough College

from this mall into the building, with discharge via the secondary malls. The required smoke ventilators are also employed for environmental control. Detailed environmental design analysis of this aspect of the design has been carried out using the E.S.P. model at the University of Strathclyde. It suggests worthwhile fuel savings from this provision.

There is a strong design similarity in the use of such malls with recent shopping developments. This was a conscious decision to encourage greater public and commercial use of the college facilities, and avoid the institutional character of many similar educational establishments.

In practice, the planning strategy would allow for alternative commercial usage of certain zones should this become desirable in the event of falling student numbers.

3. CONCLUSIONS

The increasing awareness of commercial pressures is just one of the ways in which the traditional views are changing within local authorities. The opportunity now exists for architects to play a greater role both at the inception of projects and after their completion, by working more closely with building users in the fields of estate management and maintenance. However, it does require the designer to broaden his understanding of other professions and their objectives.

The problems now facing local authorities require their designers to adopt changing roles. No longer are their major concerns with the need for new buildings which can be designed to standard briefing requirements on green field sites.

Through the increased involvement of the designer, it is hope that environmental standards can be improved and resources utilised more effectively.

REFERENCES

1. Le Cuyer, A., 'Public Panache", The Architects' Journal, 16 April, pp.761-775, 1980

2. Hawkes, D., 'Yateley Newlands Primary School, Hampshire', The Architects' Journal, 24 June, pp.1199-1214, 1981

3. Powell, J.A. et al, 'Energy Utilisation in Primary Schools and Improved Design', Portsmouth Polytechnic Occasional Paper, 1982

4. Perkins, M. and Powell, J., 'The Assessment of Energy Utilisation in Primary School Building (in Hampshire), Portsmouth Polytechnic,

Occasional Paper, 1979

5. Hawkes, D., 'Atria and Conservatories 1' The Architects' Journal, 11 May, pp.67-74, 1983

6. Hampshire County Architects Department, 'Farnborough College of Technology Redevelopment - Design Report', Internal Report, 1983

Section VIII
Management

TOWARDS THE CLIENT'S OBJECTIVE

DAVID ALLEN, FRICS, Building Design Partnership

1. INTRODUCTION

There is a saying in Yorkshire : 'Nobody does owt for nowt!' It
serves here to highlight that, in every field of human endeavour,
there is usually some driving motivation - some objective!
    It is said that the objective of this Conference is to 'discuss
the relationship between the benefits derived from building projects
and the resources required to achieve them'. It is towards that
objective that my endeavours have been, and this paper is, directed.
Hopefully, it will also act as 'a stimulus to improve the quality,
efficiency and effectiveness of building design and construction.'
    To be effective, building design and construction has to be
directed towards achieving the client's objective. As we all know,
building design and construction is a complex business. It
comprises many specialised inter-related, but not necessarily co-
ordinated - or, more particularly, integrated - parts.
Potentially, each part is motivated by its own objectives. A
client's project, therefore, can be - and often is - the end-
product of a combination of disparate individual objectives.
    In order to transcend this tendency, it is essential that the
client's objective, in commissioning a building project, is fully
understood. Furthermore, it is necessary that the entire efforts
of all parties are directed towards the achievement of that
objective.
    The remainder of this paper is directed towards :

(a) understanding clients' objectives; and
(b) exploring the building industry's response, in satisfying
them.

2. UNDERSTANDING CLIENTS' OBJECTIVES

Returning to the saying : 'Nobody does owt for nowt!' and applying
it to the business of building, it is true that nobody builds for no
reason. Building is an expensive business and is the physical
result of somebody's driving motivation.

Whilst a great deal, and a growing proportion, of building is
driven by profit motivation, a significant amount, particularly in
the public sector, is driven by other forms of motivation. An
expression, which perhaps sums up all these forms of motivation, is
the 'perceived benefits to be derived from building projects'.
Those benefits will vary, depending upon the nature of the client
and the project he seeks to have built.

Benefits, in the public sector, will normally relate to some form
of service to the public-at-large. In the private sector, they will
normally relate to profit, except perhaps in the domestic context.
Here, benefits may be viewed as the quality of life enjoyed by the
individual and the family; although there are those who, even in the
domestic context, perceive their residence as a vehicle for profit.

Having recognised that, in general terms, it is perceived
benefits which provide the driving motivation for building projects,
it is important to identify, in the case of each client and each
project, the nature of those perceived benefits. Given that
understanding, the client's objective becomes apparent.

Paraphrasing the introductory sentence of the Conference
brochure, the challenge for the building industry, in playing its
part in Britain's survival as an advanced nation, is to become more
effective and inventive at utilising the nation's scarce resources
in satisfying its clients' objectives. In so doing, the industry
will also be demonstrating its response to the Department of
Industry's campaign, 'Design for Profit'.

The question remains : 'How?'

3. THE BUILDING INDUSTRY'S RESPONSE

Traditionally, I submit, the building industry's response to a
client's needs has been heavily constrained by a whole series of
pre-conditions, which the industry has developed over time. If it
is to meet the challenge posed by this Conference, the industry must
review every one of its practices and consider whether, individually
and collectively, they are effective in utilising the nation's
scarce resources in satisfying its client's objectives.

The scale of such a review is considerable and goes beyond both
my remit for this paper and the specific title of the Conference. I
intend, therefore, to limit my exploration to the topics of
'design', 'quality', 'cost' and 'profit'.

4. DESIGN

According to the Encyclopaedia Britannica: 'design is the process of
developing plans or schemes of action. The resultant plan or scheme
provides the pattern for making a product. It indicates, primarily,
an interrelation of parts intended to produce a coherent and
effective whole. Ordinarily, it bears four limiting factors in
mind:

(a)   capacities of materials employed
(b)   influence of methods of adapting these materials to their work
(c)   impingement of parts within the whole; and
(d)   effect of whole on those who see it, use it or become involved in it.

In fine arts, design is the creative process per se.   In engineering, it may mean a concise record of the embodiment of appropriate concepts and experiences.   Whilst in architecture, the artistic and engineering aspects of design tend to merge - thus an architect cannot design according to formulas alone.'

Given such an understanding, design needs direction if the 'pattern' which it provides is to result in an end-product. Furthermore, in the context of building design, that direction must take full account of :

(a)   the client's objective
(b)   the context in which he operates
(c)   the constraints to which his operation is subjected
(d)   the particular context in which the building is to be constructed; and
(e)   the particular constraints to which the design of the building and its construction are to be subjected.

I submit that, in most building projects, insufficient attention is given to these fundamental aspects, before design commences.   In many, I doubt whether they are ever clearly established - and yet they are so fundamental to every action and decision that ensues.

By observation, the normal approach to most building projects is that :

(a)   the client has already decided that he needs a building and has already identified his site
(b)   he draws up a schedule of his requirements for that building, both in spatial and performance terms.   Sometimes, a layout of his spatial requirements is also provided; and
(c)   the design team produces (maybe alternative) outline proposals, in response to the client's requirements, complete with an indication of specification, programme and cost.   This may, or may not, meet with the client's approval.   If it does, the project proceeds through its remaining stages.   If it doesn't meet with approval, it may have to be modified, to varying degrees, or started all over again.

The entire process starts from the assumption that the client has:

(a)   identified, in his own operational terms, the best means of achieving his objective
(b)   carefully analysed his spatial requirements, as they relate to those means; and
(c)   similarly analysed his performance requirements.

In many, if not most, instances, I submit, the client has done
none, or at best only some, of these things. This leads to the
inevitable conclusion that the design of many building projects is
mis-directed, with the inevitable result that those designs are
inappropriate to the clients' objectives.

Some examples, of where such possible mis-direction was arrested,
will serve to highlight the point :

Example 1 : the client who approached his design team, with the
intention of extending an existing laboratory complex in two places;
continuing its tendency to 'grow like topsy'. Upon investigation,
it became apparent that the client had given little consideration to
his future needs; both in terms of expansion and his day-to-day
operations. The result of the design team's initiative was an
overall development plan, involving a rolling series of building
phases. Some of these included refurbishment of the existing
buildings, to accommodate changing occupational patterns as the
development plan unfolded. In operational terms, the entire
laboratory complex became much more effective and efficient.
Inevitably, this would lead to relatively greater profits for the
client - which, after all, was his original objective.

Had the design team adopted the more traditional approach to this
particular client's initial brief, the two extensions proposed would
have only served to further compound the relative ineffectiveness
and inefficiency of the existing complex.

Example 2 : the major vehicle manufacturer who approached his
design team, with the intention of building a new engineering centre
for the whole of his car design activities. The client's initial
intentions were to build a separate office block, workshop block,
styling centre, etc., to house his existing departments. Adopting
the approach I am advocating, the design team sought answers to the
question : 'How are (or should) motor cars (be) designed?' The
building design, which emerged from this process, was substantially
a single building, with each of the functions involved in designing
a motor car sensibly related. The other consequence of the process
was that the entire organisational structure of the company's
engineering centre would change, to respect the clarity which the
briefing process had produced.

Again, this is an example of the design team seizing the
initiative, in the interest of satisfying the client's real
objective. In this instance, the effect of that initiative extended
far beyond their immediate function as a design team.

Example 3 : by way of contrast, there was the Polytechnic,
which, in spite of stringent cash limits, sought and achieved a
similar level of service to the public as was achieved elsewhere,
where cash was not so limited. The design team's response was to
give careful consideration to the use of space. By careful
time-tabling and the resultant greater use of each space, the total
amount of space required was significantly reduced. This,
inevitably, reduced the capital investment required, without loss of
service to the community.

This serves as an example of where the client had a very clear objective and had the wisdom to involve the design team in identifying the most economic means by which that objective could be achieved.

The key to success, in each example, was the initiative and perception of the particular design team or, in the latter case, the client. In all three examples, had a more traditional approach been adopted, it is extremely unlikely that the client's needs would have been met by the resultant designs.

Mis-direction of design not only results in non-achievement of client's objectives; it also results in poor utilisation of the nation's scarce resources. No matter how efficiently and sparingly those resources are applied in executing the resultant designs, if they are not achieving the clients' objectives, the inevitable result is waste! – and, possibly, on a large scale.

Given appropriate direction, the design team's task becomes one of achieving the client's objective, with efficiency and economy of means in mind. Such efficiency and economy, again, needs to recognise the client's objective. It must take into account the time-related and life-cycle expectations of the building, implicit (or explicit) in that objective.

Apparent cheapest first cost can be, and often is, misleading in this regard. The use of intrinsically cheap materials can have a direct and immediate effect on the duration and capital cost of other aspects of the building process. It can also result in significant recurring costs of maintenance, replacement and resultant disruption, during the life of the building.

Similar effects can result from apparently cheap methods of detailing and assembling the components of the building.

For design to be effective, it is paramount that time-related and life-cycle expectations of the building are clearly established at the outset, as part of understanding, in absolute terms, the client's objective. Thereafter, it is for the design team to achieve that objective with efficiency.

5. QUALITY

The resultant quality of a building project is determined by a combination of :

(a) the quality of thought embodied in all aspects of the building's design

(b) the intrinsic quality of the materials employed in its construction; and

(c) the effectiveness of the quality control, exercised during construction.

Many of the disasters of the '60's bear witness to the inadequacy of the attention paid by the industry to this crucial aspect of its products. In part, it relates to the inadequacy of the education and training of its personnel; in part to the 'cheapest first cost' attitudes of the industry and the majority of its clients; and, in part, to the fragmentation of the design process.

331

Design is an all-embracing process, involving the selection of all materials and components of which the building is to be constructed. In view of the many specialisms involved in the process, the achievement of a coherent and effective whole necessitates teamwork of the highest order. As with all other forms of teamwork, direction, co-ordination and integration are a pre-requisite and, ultimately, this has to be achieved by one person.

Fragmentation of the design process into its separate specialist interests leads, inevitably, to self-interest and separate driving motivation of each of the parties involved. This militates against teamwork and, therefore, against the quality of the end-product. This tendency can be further exacerbated where design work forms part of the activities of the contractor or his sub-contractors; unless adequately controlled.

The education and training of each of the separate disciplines further contributes to the potential for disintegration within the design team. There is no unifying policy and the objectives of the separate disciplines tend to be somewhat disparate. This tendency is further exacerbated by the fee scales of the separate professional bodies. Whilst architect, and to a lesser degree quantity surveyor, fees are based on the cost of the entire project, those of each of the specialist designers are based on the cost of those aspects of a project which each specialist designs. Whilst in many ways this seems eminently sensible, it does have the disadvantage of promoting yet further partisan attitudes within the design team.

'Cheapest first cost' attitudes are perhaps the most prevalent reason for absence of quality in the end-products of the UK building industry. They not only reflect in the intrinsic quality of the materials employed, but also in the extent to which professionals are employed in the design and quality control of buildings.

Whilst such attitudes, on face value, may appear attractive to clients, they tend to be short-sighted. Notwithstanding their effect on the design of the building, as witnessed by those who see it, use it or become involved in it, they often militate directly against the client's objective. This may be reflected in relatively higher operating and replacement costs, reduced rental income, reduced lettability and/or reduced asset value. The combination of these disadvantages, taken over time, is likely to be detrimental to profit-orientated clients; and to involve unnecessary recurring outgoings to service-orientated clients.

For design to be effective, it must embrace appropriate quality of thought, materials and control in its execution. It is of paramount importance, therefore, that the quality expectations of the building are :

(a) clearly established at the outset, as part of understanding the client's objective; and
(b) embodied, thereafter, in all aspects of the building's design and construction.

332

6. COST

Cost can be defined as the monetary value of goods and services that producers and consumers find it necessary or desirable to purchase.

Before embarking upon a building project, it is essential that there is a common understanding as to what constitutes 'cost', for the purpose of that particular project. Such understanding is as important between the members of the design team as it is between the client and the design team.

It is important to establish, at the outset, the costs to be taken into account, in progressing towards achieving the client's objective. The building, including its engineering services, fittings and external works, tends to be an obvious starting point; but what about land costs, site investigation, preparatory works, furniture, furnishings, plant, machinery, equipment, site supervision, professional fees and expenses (including legal, accountancy, estates, etc.), off-site costs, client artisan and management costs, overall contingency provision and VAT? It is too easy to assume that the client will separately budget and account for them. Furthermore, whilst the client may well account for them, separately, it may be that such costs should be reflected in design team deliberations as to the most economic means of achieving the client's objective.

There is also the matter of credits; tax allowances, development and other grants, fire damage and other insurance proceeds, sale of land, salvage or re-sale value of the products of demolition. Their inclusion, or otherwise, in the design team's deliberations can influence the overall design solution.

Costs will also vary, depending upon their time datum, and whether they are to include for future inflation and interest on capital employed, prior to building completion or occupation. It is, again, important to establish, at the outset, a clear definition as to how each is to be reflected (or not reflected) in the design team's considerations.

Having established what constitutes 'cost', for the purpose of a particular project, it is then important to understand any aspects of the client's operation or objective which will affect those costs. These may include procedural, capital, cash-flow and programme constraints.

It is also necessary to ascertain, from the client, his criteria as to the relative importance of :

    (a)   time and cost
    (b)   cost and value (in terms of both the resultant building and any residual parcel of saleable land)
    (c)   capital cost and recurring cost; and
    (d)   design, quality and cost.

This process of establishing an absolute understanding of the term 'cost', as seen by the client, can usefully be thought of as obtaining the 'cost brief'.

Given such a 'cost brief', the design team can proceed towards achieving its client's objective, in the certain knowledge that it has a clear understanding of :

    (a)   the components of cost which it has to take into account
    (b)   the time datum of those costs
    (c)   client constraints which will influence costs; and
    (d)   the relative importance of (capital) cost and, respectively, time, value, recurring cost, design and quality,

for that particular client on that particular project.

In the absence of such a brief, at the outset of a project, the entire design team's efforts can be mis-directed. In consequence, the capital costs, recurring costs and value may, individually and/or collectively, prove incompatible with the client's short- and long-term objectives.

I submit that such rigorous cost briefing rarely takes place, with the result that much abortive effort, loss of calendar time and unnecessarily high recurring costs and low values ensue. This is unlikely to be compatible with the client's objectives and, inevitably, militates against the effective and efficient use of the nation's scarce resources.

7. PROFIT

In general business usage, the term 'profit' is usually ascribed to the excess of total revenue over total cost, during a specific period of time. It is the private sector's normal driving force and, therefore, tends to become its prime objective.

The public sector, on the other hand, tends to replace 'profit' with 'service' (to the public-at-large), as its objective.

As stated earlier in this Paper, both these objectives may be summed up by the expression 'perceived benefits to be derived from building projects'.

A design team's perception of the benefits which a client obtains from his buildings may be fundamentally different to those which the client perceives. As such, it is vitally important, at the outset of a project, that the client's perception of the benefits the building will provide are diligently sought and understood by the design team. Omission of such understanding can cause the resultant design to be totally mis-directed.

Such mis-direction, I submit, is a regular occurrence in the building design process. It arises from the 'folk-lore' which prevails within the industry and between the industry and its clients. That 'folk-lore' relates, significantly, to the notion that the prime vehicle for communication of a client's needs is floor area. It permeates the industry and provides the normal base for relative expressions of cost.

334

Cost comparisons, adopting the notion of 'cost per square metre (or foot)', can be wholly misleading, when considering the relative cost of satisfying a client's objective. Alternative design team responses, to an identical client's objective, may produce designs which contain significantly dissimilar floor areas and yet both satisfy the same objective. They will result in significantly different 'costs per square metre' of floor area and yet the total cost of achieving the client's objective may be very similar.

An example of this very situation relates to two 'high-tech' laboratory buildings, for the same client. The client's objectives were very similar in both cases. Arising from the physical context in which each was to be located, the resultant design of the two buildings was totally different. The one was a six-storey development, with interstitial service floors. The other was a single-storey development, with a service void over. Adopting the 'cost per square metre' notion, the cost of the one was 50 per cent greater than the other.

The prime function of each building was 'research and development' and, in each case, the remainder of the building was wholly servant to a number of R&D laboratory modules. In terms of the client's objective, his vehicle for profit was the output of the laboratory modules, and not the floor space provided within the building. When the costs were analysed relative to the laboratory modules, the comparative costs were found to be virtually identical.

From this experience, it would seem that there would be considerable benefit, to the building industry and its clients, if building costs were analysed relative to clients' objectives. I would not see this as replacing the 'cost per square metre' notion, but supplementing it.

Without the analysis having been done, it is difficult to say; but it may be that analysing costs in this way would result in greater consistency of costs, when expressed relative to client objectives. Even if such greater consistency isn't to be found, such expressions of cost should certainly provide a cost communication vehicle compatible with client objectives. Such a vehicle would enable the client, in the initial briefing stages, to relate costs of construction directly with the benefits he anticipates will follow. Recurring costs and value could be similarly analysed, with further benefit to design teams' communication with clients, at the outset of their projects.

Apart from the direct and immediate benefit which such analyses would provide, they would also serve to focus the attention of design teams and clients, alike, on the most important aspect of the brief - the client's objective.

Another benefit which could arise would be the greater freedom of expression which such an approach would tend to give to design teams. There is a tendency for the present method of cost analysis to inhibit design, particularly from a 'form' point of view.. This arises from the use of 'cost per square metre' as a measure of relative extravagance and economy in building design. For those who tend towards 'cheapest first cost' attitudes, there is an attraction in the lower end of the 'cost per square metre' range. What isn't always appreciated is that this can significantly inhibit the design team's response, without necessarily saving the client any money. It can also inhibit the design team, in terms of their producing a design which will satisfy the client's objective.

In speculative office developments, for instance, there can be little doubt that the client's objective is profit. It is also well known that the measure of that profit relates to the difference between the cost of the development and the income received from rents and sales, over time. The vehicle for that income is the unit of lettable (as distinct from built) floor space. However, the level of that income, although primarily established by the building's location, is also influenced,amongst other things, by the 'usability' of its lettable floor space. The true objective of the office developer, therefore, is usable (as distinct from lettable) floor space. In this regard, dimensions of spaces, positions of columns and the use of single-glazing can all serve to reduce the amount of lettable space which is usable. Given the objective of maximising usable space, as the vehicle for profit, it is quite possible that the additional cost of increasing the usability of the lettable space can be accommodated within the parameters of the client's cost brief.

The vehicle for profit varies considerably between clients. It is frequently misunderstood or misinterpreted. What is needed is a standardised system of analysis and classification, to suit all client and project types. Given such a system, all buildings would need to be analysed accordingly, in addition to the present method of cost per square metre of built floor area.

From the resultant data bank, the building industry would be far better placed, than now, to relate to its clients, directly, in cost terms compatible with each client's own objectives. This, I believe, would be of great benefit to both the industry and its clients.

8. CONCLUSION

A client's project can be (and often is) the end-product of a combination of disparate individual objectives.

This situation arises from a combination of :

(a) fragmentation of the industry into a wide variety of specialised inter-related, but not necessarily co-ordinated (or, more particularly, integrated) parts

(b) a whole series of pre-conditions, which the industry has developed over time

(c)   insufficient attention being given, before design commences, to fully understanding the client's objective and operational context and constraints; as well as the context and constraints within which the building is to be designed and constructed

(d)   lack of initiative and perception by design teams in exploring, with the client, his real, as distinct from apparent, needs, towards achieving his objective

(e)   the 'cheapest first cost' attitudes which permeate the industry

(f)   inadequate attention to quality of thought, materials and control in the design and construction of its buildings

(g)   inadequate cost briefing at the outset of projects

(h)   the notion, which again permeates the industry, that the prime vehicle for communication of a client's needs is floor area; and

(j) more importantly, the notion of relating cost, primarily, to built floor area.

If the building industry is to respond, positively, to the Department of Industry's campaign, 'Design for Profit' (and I believe it should), it needs to :

(a)   review every one of its practices and consider whether, individually and collectively, they are effective in utilising the nation's scarce resources in satisfying its clients' objectives

(b)   recognise that 'nobody does owt for nowt' and seek absolute clarity, at the outset of each project, as to the client's objective

(c)   understand, fully, the context and constraints within which its clients have to operate and their buildings have to be designed and constructed

(d)   direct its entire integrated efforts towards its clients' objectives, and within their given context and constraints

(e)   employ greater initiative and perception in exploring, with clients, their real needs, towards achieving their objectives

(f)   take into account the time-related and life-cycle expectations, implicit (or explicit) in clients' objectives, when designing their buildings

(g)   pay greater attention to the quality of thought, materials and control employed in all aspects of each building's design and construction

(h)   establish, at the outset of a project, a clear and concise cost brief

(j)   establish, as part of understanding clients' objectives, their perception of the benefits which the completed building will provide; and

(k)   analyse building costs relative to clients' objectives, as well as to built floor area.

Such a response by the industry, I submit, would lead to a significant improvement in :

(a)   the quality, efficiency and effectiveness of building design and construction; and

(b)   the relationship between the benefits derived from building projects and the resources required to achieve them.

337

The objective of this Conference was to discuss and, hopefully, stimulate improvement in both these aspects of the building industry's performance. I trust that this Paper has met that objective and that those who read it will profit from its deliberations.

MANAGING FOR QUALITY IN BUILDING

PROFESSOR JOHN BENNETT, University of Reading

> Quality is never an accident; it is always the result of
> intelligent effort.
> John Ruskin

Conventional wisdom tells us that higher quality costs more than
lower quality. This is commonly reflected in the pricing strategies
adopted in both private and public enterprise.

In a highly competitive international market like motor car
manufacture the relationship between quality and price is fairly
straightforward. A Rolls-Royce is a piece of very high quality
engineering. A Lamborghini is a piece of very high quality aesthetic
design. Both are expensive and I suspect that they would not be so
highly regarded if they were not. We expect to pay a high price for
high quality.

At the level of the popular mass produced cars the situation is
not so straightforward. Here manufacturers must design and build
down to a price. For many years British cars were thought to provide
poor quality. Indeed it is clear that one of Sir Michael Edwardes'
main tasks at British Leyland was to improve the quality of his cars.
Fortunately for British industry he had some success. The improvement in Jaguar is already legendary and amongst popular cars the
Metro and Maestro are generally well thought of.

The interesting point is that in this difficult competitive
business of providing cars for a mass market, the same price can buy
different qualities. The Japanese motor car industry illustrates
this point most clearly. It achieves high quality and high productivity and therefore low costs; and of course depending on market
conditions Japanese car manufacturers have the still profitable
option of selling a good quality product for a low price. This
option is not available to less productive manufacturers.

So we have something of a paradox. In broad terms, Rolls-Royce v
the Metro, higher quality equates with higher price. In the detail
of a specific market sector higher quality and lower price can coexist depending on levels of productivity and as is being increasingly recognised, depending on management competence.

Building of course is not an international highly competitive market to anything like the same extent as motor car manufacture. As a consequence questions of quality, cost and value appear to be less clear. The position I take is that as far as these questions are concerned the general principles derived from manufacturing industry hold good for building. Therefore I assume that in broad strategic terms higher quality means higher costs. However at a more detailed tactical level higher quality, depending on management competence, may mean either higher or lower costs.

Fairly obviously, if these statements are true there must be at least two kinds of building costs which behave differently from each other. One of which varies directly with quality and the other varies inversely with quality.

Before beginning the search for these two kinds of building costs it is perhaps sensible to consider whether there is any evidence that quality and costs behave anywhere in building in the same way as I suggest they do in the Japanese car industry. In other words are there examples of high quality building co-existing with high productivity.

Several years ago I made a short study tour of West Germany with Dr Roger Flanagan. In the brief report of that visit (1) we said:

"In Stuttgart we visited the head office of a medium sized building contractor. The entrance foyer would have graced a five star hotel. The staff swimming pool was of a higher quality than any pool we had seen in the UK. The same standards are employed in the lavish new offices of the local ratio station which we visited later the same day. As an example the doors to the offices had rebated edges, are filled with sand, the hardwood frames are fitted with plastic inserts to ensure a tight fit all round and the ironmongery is heavy and solid. The result is an impression of weight and quality as one closes the door safe in the knowledge that one's conversation will not be interrupted by noises or draughts from the corridor."

We did get a consistent impression of high technical quality in the general run of buildings in West Germany. Turning now to levels of productivity, as a coincidence the best figures I have available suggest that West Germany has the most productive building industry in the world. The figures are given in Table 1 and are taken from the August 1982 National Institute Economic Review.

Table 1: Output per employed worker-year in construction in 1973 $ thousand

	1973	1980
West Germany	14.5	17.3
France	11.2	13.2
USA	11.6	11.0
UK	10.7	9.5
Japan	7.6	7.1

source: Labour Productivity in 1980: an International Comparison by
A D Roy in National Institute Economic Review No.101 Aug 1982.

It therefore does not seem unreasonable to assume that in building as in motor car manufacture high quality and high productivity, and therefore low costs and prices, can coexist. Having provided this reassurance we can return to the search for two kinds of building costs, one of which varies directly and one inversely with quality.

My most recent research into building costs (2) arises from Science and Engineering Research Council supported work aimed at producing a computer model to simulate specific construction projects. Amongst other things the simulator attempts to predict the range of likely costs and times for specific construction projects.

The simulator takes the form of a computer system which enables users to model specific construction projects in the form of an interactive bar chart. The user enters a series of bars graphically against a selected time scale which may be in days, weeks or months and draws in the logical links between the bars.

The user can enter details of holiday periods which occur during the project and these are automatically taken into account. The costs associated with each bar can be entered. Other facilities exist which are not central to the argument in this paper, including entering and using resource data and weather data.

The simulator is based on a hierarchy of bar charts to reflect the way successful managers are seen to plan in practice. Thus the user is presented with a choice of levels of model and so can consider increasingly complex questions as schemes and strategies for a project develop.

The first hierarchy level normally comprises 12 to 15 major work packages. This is illustrated in Fig. 1. Each primary work package may be analysed by the user into secondary work packages. This is illustrated in Fig. 2, which provides a detailed bar chart at the level of a good contract construction programme.

Preliminary costs or those costs for establishing and managing the construction site as a productive unit can be entered on to the primary bar chart. This is shown in Fig. 3.

The simulator runs the project a number of times, usually 100, selecting values from within the range of data entered by the user for each iteration. The results are presented as a graph with a frequency distribution and a cumulative plot. Separate graphs are available for cost and duration results as shown in Figs. 4 and 5. The graphs indicate the likely worst and best cases, the pattern of possible results and the likely chance of success for any particular value.

Four case studies have been completed using data from live projects which represent the range of work of the building industry. They are a low rise housing project, a speculative office block, a new hospital and a refurbishment project.

To continue the search for two distinct types of costs, it is necessary to consider the types of data needed to model the live case studies successfully. It was necessary to express costs in two parts for each work package. These are one fixed cost and one cost which varies with time. In gathering data for the case studies it was found that very broadly material costs can be adequately modelled if

341

they are regarded as fixed while labour and plant costs are best modelled if they are allowed to vary with the time taken to complete the work.

Time is also expressed in two parts. These are a basic time assuming perfect conditions and an additional range of probable times depending on a number of factors including variations in productivity, external interference and the interactions and dependencies between different gangs of operatives. These factors can usefully be grouped together as a measure of the overall uncertainty of the work.

Thus at a simple level the model of costs which has emerged from the research takes the following form:

$$C = m + l + lt \qquad\qquad (1)$$

where C   direct construction costs of a given quantity and quality of building

     m   material costs

     l   basic labour and plant costs

     t   factor which relates the variable time element due to uncertainty to the basic labour and plant costs.

There are other costs which contribute to the total cost of construction. They include fixed overheads, profit and of most interest for the purposes of this paper, management.

I take the view that the purpose of management is to reduce uncertainty. That view probably requires some elaboration. If we consider the basic requirements of well managed construction, we would probably agree that management's aim is something like the following.

Managers seek to ensure that each trade or specialist arrives on site at a time when their workplace is available and accessible, the necessary materials, components, tools and equipment are at hand and the operatives know what they are to build, to what quality standards and within what time-frame. It is also important that this work is within the normal competence of the operatives. Also they must be motivated to complete the work properly. None of this happens by chance; building work must be managed. It does not seem to do an injustice to characterize the above aims as seeking to reduce the uncertainty faced by operatives.

This point is put very directly by Reg Revans (3) one of Britain's most brilliant management researchers:

> "an environment that appears to a man to be haphazard, that is brought about by a management that is slipshod and, above all, that wastes his time is something that he will refuse indefinitely to tolerate. Whether this refusal takes the form of deliberate absenteeism, of a calculated strike, or of an industrial accident depends on the local circumstances."

Our simple equation must now be elaborated to incorporate management:

$$T = m + l + lt + p - ltn \qquad\qquad (2)$$

where T   total management and direct construction costs of a given quantity and quality of building

     p   management costs

        n   factor which relates variable labour and plant costs to the
             reduction in uncertainty due to management

and the other terms are defined as in equation (1)

Equation (2) accounts for the proposition that at a strategic
level an increase in quality leads to an increase in price. It will
account for the proposition that at a tactical level higher quality
may mean lower costs depending on management competence if we can
assume that better management leads to higher quality and in equation
(2) ltn exceeds p.

Let us take each of these assumptions in turn. Does better management lead to higher quality. The most compelling and direct evidence is provided by the Building Research Establishment study
"Quality Control on Building Sites" (4). It concludes:

      "...quality depends on the ability of site staff to create
      an environment where good work could and was likely to take
      place. Sites with acceptable quality standards tended to be
      characterized by a 'consultative' approach to problem-solving
      - anyone on site could raise questions and many individuals
      could contribute to solutions - whereas sites with low
      quality standards tended to be 'non-consultative' with only
      the clerk of works really concerned with quality matters."

This conclusion is based on observations on 27 building sites. It
really leaves little room for doubt about the link between quality
and management.

The second necessary assumption is that the term ltn exceeds p in
equation (2). What this means is that the management applied to the
work results in a reduction of total costs. It is very likely in a
well managed project that this is the case since if it were not so it
would be cheaper to do without the management. So we only have to
assume rational behaviour to accept that ltn will normally exceed p.

While it has required some effort to piece together the evidence
it does seem to be extremely likely that the effects we noticed in
motor car manufacturing also exist in building. What this means is
that once a strategic decision on the broad range of acceptable
quality standards has been made, the actual quality achieved within
that range depends on management competence.

It is interesting to notice that this is also thought to be the
case in the Japanese car industry. The Nissan Motor Company have
produced a very fine book called 'The Dawns of Tradition'. In
describing motor car manufacture it says:

      "To Japanese managers attitude is the most important intangible
      of quality. Improving production technology and heightening
      productivity demand good equipment and machines, but to an
      equal degree they need excellent relations between labour and
      management. The latter factor has close ties with workmanship."

What are the practical implications of these relationships?

In beginning to answer this question we need to recognize that
performance including the quality of the end product are determined
by the craftsmen and specialists who actually build. Reg Revans (3)
provides much evidence to support the following propositions:

343

1.	the intelligibility of the working task is the prime motivator to its enthusiastic and successful discharge

2.	the readiness of managers to listen to their subordinates - thereby enabling both manager and subordinate to learn how the working task may be more successfully discharged - is the principal influence in making that task intelligible

3.	group leaders tend to pass on to their subordinates the treatment they perceive themselves to get at the hands of their own superiors

4.	that among managers of all ranks, the most important condition for professional success is that they shall understand how and by whom their targets are originally set

5.	when things happen at the work place that are either unintelligible in the first place, or that do not yield to correction on appeal, there is likely to be a failure of morale; this will manifest itself in ways such as: spoiled work, unmet schedules, absenteeism, accidents and disputes.

These propositions tell us quite simply that if we want good quality buildings at sensible prices produced quickly we must ensure that the craftsmen and specialists understand what they are to build and accept it as sensible. On the other hand it means that if designers use novel materials in unusual or intricate construction details, if they change their mind and write variation orders and do not take the trouble to explain their decisions to the people directly influenced by them, they will get and they will deserve poor quality, high costs and slow completion.

Creating the circumstances in which high performance can be achieved requires management throughout the client:design:construction phases of projects. In the initial client based phase of projects key strategic decisions must be made. As far as quality is concerned these include first of all a decision on the overall level of quality to be aimed for. This in combination with overall project characteristics of size, complexity, repetition and speed, the likely nature of the detailed construction work, and the influence of the project environment helps to establish the general level of costs and times likely to be experienced and equally importantly helps to establish the likely variability in performance around that general level. The strategic task is to bring these factors into balance. Essentially this means that if the client's budget and timetable imply a second hand Metro from the local garage it is foolish to set off looking for a new Rolls-Royce or Lamborghini. On the other hand if the client wants the equivalent of a customized Lamborghini then he must understand and accept the costs, times and other consequences.

Essentially the strategic choice is between three options. First, to use standard answers. This is essentially the Metro or Maestro answer. Their equivalents in building are the products of the speculative volume house builders or standard warehouses, factories, and other similar buildings which can be bought from a catalogue.

The second option is to use traditional construction. That is any

344

method of construction in which designers know the performance they can expect from the end product, the builders know the sequence of trades or specialists, the nature of the work and the kind of plant and equipment they will need and the necessary materials and components are readily available locally. In other words the details are essentially predetermined and well-known and understood by the local industry but the overall form of the building is original.

The third option is to seek an original answer. That is the method of construction and the overall form of the building is original or at least is likely to be original.

The choice among these three options has direct effects on performance, costs and times. The second key strategic decision for the initial client based phase is to match the general form of the project organization to the choice of end product. Standard buildings tend to be produced by permanent organizations: Barratt, Wimpey, Conder are obvious examples. In selecting a standard building we are nearly always also selecting the matching project organization.

Traditional construction requires a traditional organization. This does not mean the UK orthodox approach as described in the RIBA Plan of Work and implied in over full detail in JCT80. It means an approach in which the designer designs and then organizes on behalf of the client direct contracts with the separate trades and specialists required by his design. Traditional organization existed in England until the early years of the last century, in Scotland until just after World War II and still exists today in much of West Germany. It is no coincidence that West Germany enjoys high productivity and high quality in its buildings. In this country of course Ray Moxley with his Alternative Method of Management (5) is rediscovering the benefits of the traditional approach. Traditional construction requires nothing more complicated than this and it is because we have strayed away from well established forms of construction that our orthodox approach in the UK has become so cumbersome.

Original, innovative designs and constructions require problem solving organizations. Essentially this means that the client, design and management responsibilities are each clearly and separately allocated. These three separate areas of responsibility provide respectively the project objectives, a model of the end product and a model of the required organization expressed in terms of costs and times. The teams responsible for the separate areas of responsibility interact in a manner which is problem led.

Much of what is currently known about problem solving and its organizational implications is described by Ben Heirs and Gordon Pehrson (6). Organizations need to tackle problems in four stages:

1. the question - develop a carefully defined question

2. the alternatives - develop alternative answers by using individuals skilled at assembling the needed information and creating or formulating the required alternatives or options

3. the consequences - predict using the same or different people the consequences of acting on each alternative

345

4. the decision - determine which of the alternatives - or which combination of thoughts among the alternatives presented - should be chosen to be acted on or tested, or what further thinking effort needs to be made in earlier stages of the decision-thinking process before a final decision can be made.

Having matched the strategic description of the end product and the organization with the client's objectives design can be begun. Here the main requirement of high performance is consistency. That is consistency between the strategic decisions and design decisions. It is all too easy to spoil an excellent strategic decision to use a standard building by introducing modifications and specials. Or to spoil a decision to use traditional construction by straying outside the comfortable range of the local craftsmen and specialists. There is often a great temptation to introduce improvements but if they are outside the normal understanding of the work people who will build they are likely to lead to failure. Innovation requires very careful attention to management and is always expensive. It may in the long run lead to lower costs but the experience of all industry, and building is no exception, is that prototypes are expensive.

Achieving the required consistency on original, innovative projects requires those responsible for management to monitor the design decisions against agreed budgets and programmes expressed in terms of clearly defined work packages. They must check that the contents of each work package are consistent with the general level of costs and times agreed at the strategic stage. They must also give careful attention to the boundaries between work packages. Their aim being to eliminate interference between the teams subsequently responsible for building each work package and to ensure that each work package can be carried out as one continuous operation. The managers also need to ensure that the description of the required construction work is sufficient to be intelligible to, and within the competence of, construction teams likely to be available to build the project. If it is not and the design cannot be modified they must look outside the construction industry for the required skills and knowledge, provide for additional supervision, provide training, or in some other way seek to match the work to the workers' competence.

The construction requirements of high performance are that the craftsmen and specialists find their work intelligible and they are allowed to carry out their work without interruption. The design phase should provide clearly defined work packages. The construction phase requires each of these to be matched to teams competent to carry out the work. Where either the work or the workers is such that performance is likely to be unpredictable that work package should be insulated by means of time buffers so that it does not corrupt other more predictable work packages.

In all three main phases management should seek to protect the project from interference from the environment. In the main this means anticipating likely sources of problems and making sure that the project organization is strong in those areas. If planning permission is likely to be difficult to obtain use designers who

346

understand the planning officers' aims and prejudices. If material
supplies are likely to be difficult use an experienced buyer who
knows the local industry better than anyone else and give him the
resources needed to negotiate the best possible deals. If local
trades unions are militant buy the best local site management, have
the most senior manager negotiate all industrial relations matters
directly with the local union bosses and encourage the workforce to
join the unions and take part in its affairs.

The end result of good management is likely to be high quality
and high productivity. Clients of the building industry are entitled
to expect nothing less.

REFERENCES

1. Bennett, J and Flanagan, R (1976), 'Impressions of the German
        construction industry'. Chartered Surveyor, Building and
        Quantity Surveying Quarterly. vol. 4, no. 1, pp 6-7.
2. Bennett, J and Ormerod, R (1984), 'Construction project simulat-
        or; Final Report on SERC Research Grant'. Department of
        Construction Management, University of Reading, Occasional
        Paper No. 12.
3. Revans, Reginald W (1982), The Origins and Growth of Action
        Learning. Chartwell-Bratt, Bromley.
4. Bentley, M J C (1981), 'Quality Control on Building Sites'.
        Building Research Establishment CP 7/81.
5. Moxley, R (1978), 'Alternative Method of Management', in Archi-
        tects Journal, 22 February.
6. Heirs, B and Pehrson, G (1982), The Mind of the Organization,
        Harper & Row.

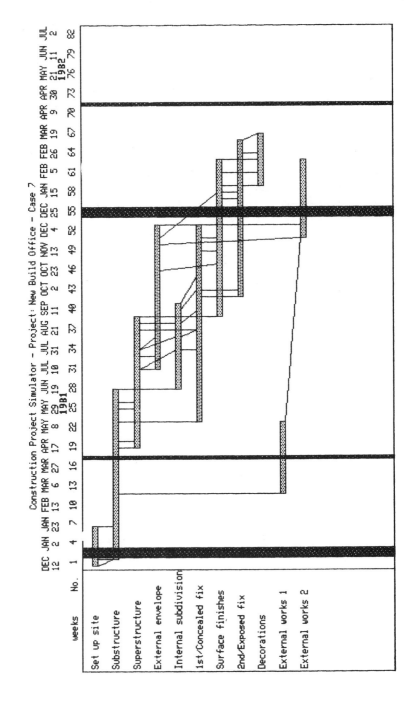

Fig. 1. Bar Chart; primary

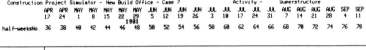

Construction Project Simulator - New Build Office - Case 7   Activity -   Superstructure

	APR	APR	MAY	MAY	MAY	MAY	MAY	JUN	JUN	JUN	JUN	JUL	JUL	JUL	JUL	JUL	AUG	AUG	AUG	AUG	SEP	SEP
	17	24	1	8	15	22	29	5	12	19	26	3	10	17	24	31	7	14	21	28	4	11
							1981															
half-weeksNo.	36	38	40	42	44	46	48	50	52	54	56	58	60	62	64	66	68	70	72	74	76	78

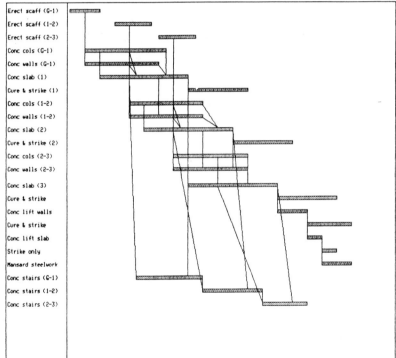

Fig. 2. Bar chart; secondary

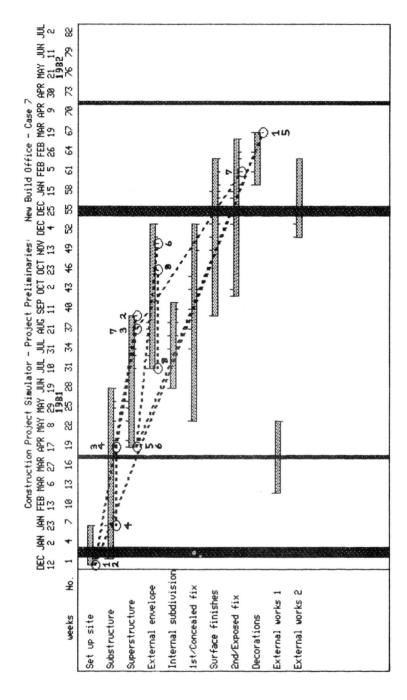

Fig. 3. Bar chart; with preliminaries

Construction Project Simulator - Preliminaries Descriptions and Costs 04-09-1984

1.	Set up site	(0)	to	Decorations
2.	Set up site	(0)	to	Superstructure
3.	Substructure	(15)	to	Superstructure
4.	Substructure	(4)	to	Substructure
5.	Superstructure	(15)	to	Decorations
6.	Superstructure	(15)	to	External envelope
7.	Superstructure	(33)	to	2nd/Exposed fix
8.	External envelope	(27)	to	External envelope

Description
(61) Agent,huts,temp.services
(35) Engineers
(33) Tower crane
(15) Mobile crane
(61) Assistant agent
(46) Bricklayer foreman
(55) Hoist
(42) Scaffold

Fig. 3. Preliminaries categories

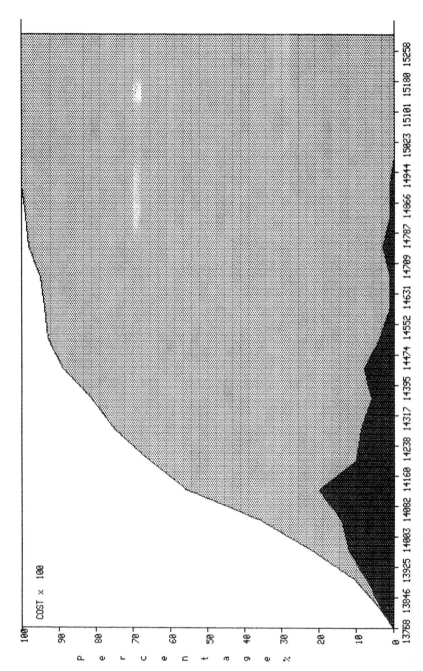

Fig. 4. Cost prediction

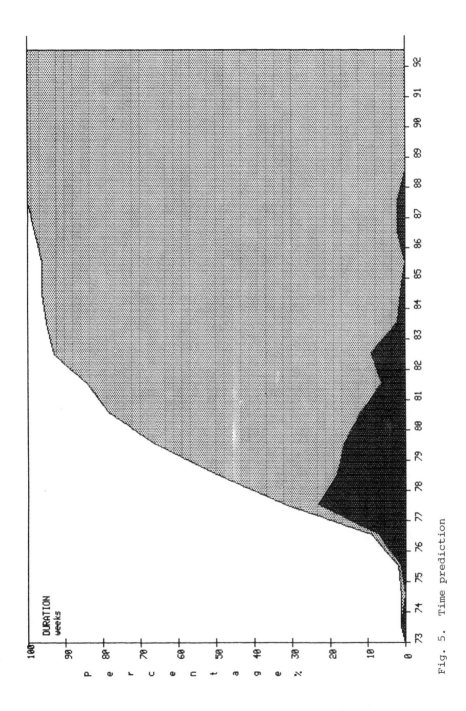

Fig. 5. Time prediction

RELATIONSHIP OF COST AND TIME TO QUALITY ASSURANCE

PETER F. MILLER   MSc FCIOB   J SAINSBURY PLC

Definitions

Cost Price (to be)   paid for a thing; expenditure of time, labour
law expenses especially those allowed by court
against losing party; prime cost - price at
which a manufacturer produces.
cost price - the price at which a merchant buys.
past tense : be aquirable at, involve expenditure of, result in the loss of estimate or fix
cost of production.

Time   -   indefinite continuous duration regarded as dimension in which a sequence of events takes
place.
Finite duration as distinct from eternity; more
or less definite portion of this associated
with the particular events or circumstances,
historical or other period.

-   allotted or available portion of time, time at
ones disposal.

-   portion of time destined or suitable for a purpose
-   period of gestation
-   terms of imprisonment
-   (payment for) amount of time worked

-   time as recognised by conventional standards
-   period of enjoyment

-   time bomb - bomb with time fuse or other
device so adjusted it will explode after
predetermined interval.

-   time constant - mathematical expression
giving indication of delay involved in heating, charging, moving etc., when the process
produces an opposing effect proportional to
the temperature, charge, speed etc.,

time honoured - respected on account of its
antiquity.

time server - one who on grounds of self interest adapts himself to opinions of the
times or persons in power.

                    time table - tabular list or schedule of
                    times of classes in school, arrival and
                    departure of trains, boats, buses.

Timeless   -   unending, eternal.

Quality   -   degree of excellence, relative nature or kind
                    or character.

            -       class or grade of thing.

            -       characteristic trait

            -       high rank or social standing

            -       (logic of proposition) being affirmative or
                    negative.

Assurance   -   positive assertion, self confidence, impudence.

Assure   -   make safe, tell (a person) confidently.

Confident   -   trusting
                    and although not in the title;

Profit   -   excess of returns over outlay.

Argument

Time   -   calendar

            -       elapsed

            -       available

            -       cummulative

Relationship of cost and time

1) Time expended at a time unit rate;

a) resources at time cost labour - working capital interest.

b) resources at quantity cost - materials.

2) Costs (similar graphs) for

a) staff take on

b) equipment procure and store

c) materials procure and store

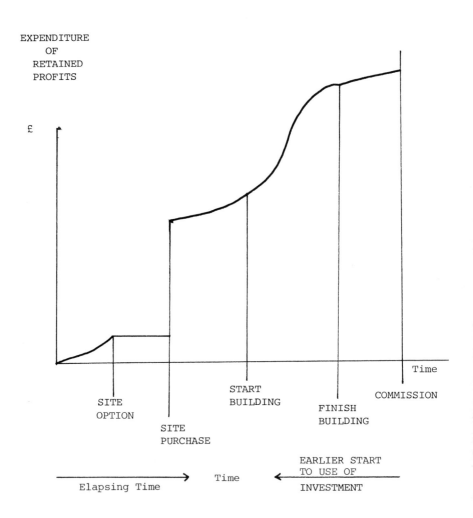

DIAGRAM 1

3) Time Pressure

The cost effect of compressing elapseable time is = the saving of
sum of the interest charges on cumulative expenditure during the
compressible period less the inefficiency/works generated in the use
of time paid resources. Benefit from applying time pressure is conditional upon the comparison resulting in the whole of the subsequent series of events moving together to equally earlier dates, i.e.
a coordinated move. If this does not arise interest costs rise due
to earlier ineffective expenditure.

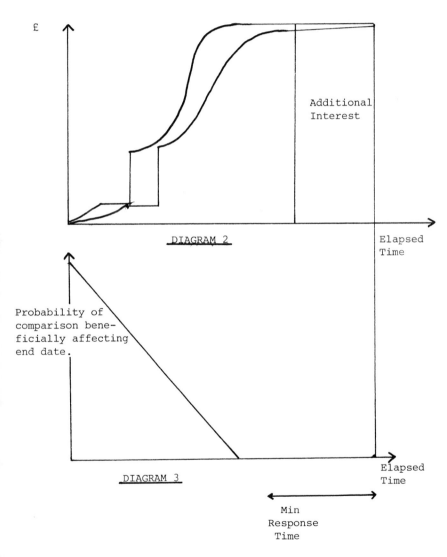

DIAGRAM 2

DIAGRAM 3

357

The above appreciation is based on elapsed (calendar) time. The principal use of elapsed time (in construction projects) is in the decision making processes of the project team, this includes those with a decisive influence upon the project's advancement.

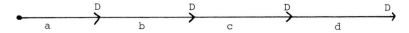

DIAGRAM 4

A project advancement is "a sequence of events", these events are decision events.

If a decision is to be made properly it is necessary to provide the decision maker (who has criteria) with data, preferably data related to these criteria. (A decision maker without criteria is a guesser).

Data preparation for any decision depends on

    1.   a definition of need   -   I need to know the sites boundaries.

                                     -   I need to know the cost of the service road.

    2.   a preparation of data -   a site survey and drawing.

                                       -   a drawing, same quantities, same unit costs, an estimate or a tender.

    3.   criteria   -   the site has to be 4 acres plus.

                                       -   the cost may/should not exceed £200,000.

358

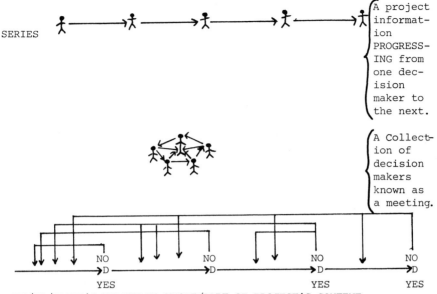

SERIES

A project informat-ion PROGRESS-ING from one dec-ision maker to the next.

A Collect-ion of decision makers known as a meeting.

NO   NO   NO   NO
→D ───────────→D ─────────────────→D ──────────────→D
YES   YES   YES   YES

YES/NO/MAYBE'S RELATE TO WHOLE/PART OF PROJECT'S CONTENT
[DECISIONS]

DIAGRAM 5

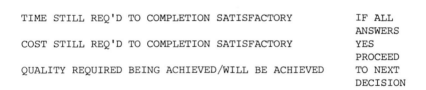

TIME STILL REQ'D TO COMPLETION SATISFACTORY   IF ALL
                                                     ANSWERS
COST STILL REQ'D TO COMPLETION SATISFACTORY   YES
                                                     PROCEED
QUALITY REQUIRED BEING ACHIEVED/WILL BE ACHIEVED   TO NEXT
                                                     DECISION

DIAGRAM 6

1.05   1.10   1.16   1.22   1.28

DIAGRAM 7

So if each decision maker in a series errs 5% (in appreciation of
data received combined with correct communication of results) the
geometric error is about 30%; a SIGNIFICANT ERROR

359

Using present day techniques the project's events assume a serial form and their individual times are of course cumulative, with many inactive time spaces between where data is sent, received and then waiting to be used.

If the truth is that the probability of compression of elapsable time   beneficially affecting the end data diminishes with the time that has elapsed so far on a project it can be seen for example that to speed up site aquisition is a straight gain to the later end user and hardly affects the intermediate users of the time in between.

This work is however beset by others who have a decisive say in the process, lawyers, neighbours, planners et al, but no responsibility for or interest in achieving the projects objectives. Once the site is procured (except perhaps for minor snags) the design process starts.

Seen from the end users/owners point of view, there is now a "sequence of events" almost unparalleled in our society, there we have a series of people "professionals" "time honoured" on account of their antiquity and often "time-serving" in adopting the opinion of those in power. These people proceed to communicate. The processes are too well known to need description and they are often thought to be adequate. As they are not however perfect or optimum it is not too impertinent to consider that they may be improved.

To restructure the present arrangement it is necessary to redesign the data flow and the decision making process from a series one to a parallel one.

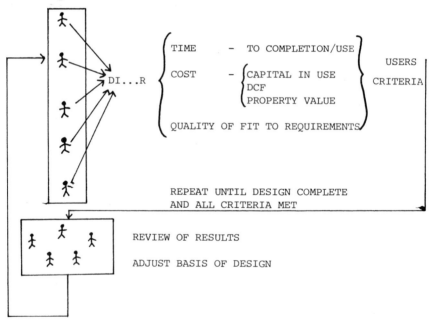

360

What is the correlation between the time/costs/quality values, the clients criteria and the design process.

How do you improve the discounted return on capital employed, usually the end users key criterion :-

1)   build it smaller - exact fit!

2)   change the configuration! - for example single storey public use buildings do not need lifts with their capital cost plus insurance, maintenance and power, capital cost now, less tax allowance, includes all space taken up including ingress/egress areas.

Insurance - probably increasing with RPI.

Maintenance - probably increasing above RPI.

Power - certainly increasing faster than RPI say 20% pa compound.

Let alone any cost in user work related to the presence of lifts.

3)   design it quicker - time pressure benefit with high probability of benefit.

4)   reconsider criteria used for each of the sub-systems of which the entity is to be composed to obtain economic use of materials (non time related).

5)   Interactive efficiency of the sub-systems of the designed buildings.

Before proceeding we must review the nature of the processes involved and the techniques in use and becoming generally available.

1.   Spacial definition and arrangement, usually drawn in outline.

2.   Technical solutions to provide the space, its enclosure and its environmental protection/control - always drawn and in detail, rarely capable of radical change say to overall grid once sufficient detail has been applied.

3.   Measurement of quantities and specification of quantities, increasingly speeded up by computors.

361

4.   Estimating and capital costing using calculators at least.

Each technical step is usually quite separate, and each step seems
to depend upon the other.

SPATIAL	TECHNICAL	MEASUREMENT	ESTIMATING	CAPITAL
DESIGN	SOLUTIONS			COST £
(Outline Drawings)	(Working Drawings)	(B.Q.)		

It is also "time honoured" wisdom that more detail increases the
accuracy of the forecast. This is doubtful.
     If anything increases the possible accuracy of a forecast it is
not the inclusion of more variables of one type but the inclusion
of all the significant types of variable and their balanced correlation.
     The evolution of computer manipulated data bases facilitates the
rapid comparison of many similar projects. This is usually a fairly
useless occupation unless the cost and other data is structured in
the first place with this comparison approach in mind.
     It is a truism that all projects are unique, it is also true that
similar bits and pieces are strung together to obtain these unique
results.
     The structure to be described is in use now and its effect is to
encourage the thought that parallel working is attainable.

Structure of Cost Data

A hierarchy best used from top to bottom but most readily explained
from detail to collective.

Resource-model	–	a set of resources (labour, plant mater-ials etc) whose employment results in a product which is incorporated into the finished works.
Spend-unit (S-U)	– –	The unit of product of a resource model.

Construction-planning-unit   (C-P-U)

                              –   a set of spend-units which provide a
                                  functional system, e.g. structural frame
                                  or rainwater disposal, which achieves a
                                  constructional purpose and physically
                                  expresses a design function.

362

Sector	-	a collection of C-P-U'S which provide a user facility.
Sub-projects	-	the set of sectors related to a commercial interest in a project.
Project	-	the complete set of works comprising the investment entity.

The taxation categories appear singly at S-U level and higher levels are a separable mixture of these.

Studies of cost kept in these hierarchies have shown that at project level no useful regression models can be found but a few characteristics only need be expressed in order to provide regression models of acceptable accuracy for certain sectors such as building shell and certain aspects of finishing and installing M & E services.

It is interesting to note that the more systematic or disciplined the 'system' used is, the more readily the models can be detected and the higher up the hierarchy they appear.

It seems that :

1. codes of practice or rules and regulations force solutions to technical problems into set patterns.

2. market forces keep the costs of the comparable sets of high/medium/ordinary quality style/appearance.

3. combinations of choices to meet the requirements of performance of a system tend to narrow the spread of costs for finished work.

Work on these analyses continues and the data base grows, the significant thing is that the pattern becomes more sharply defined and easier to see.

Of particular interest is the emergence of these small models with few (5 or so) variables. The problem of modelling used to be thought to be the core size and speed of the computer, the real difficulty is now realised to be in obtaining values for the variables particularly at the earliest stages of design.

Studies of extracting quantified data from computer designs on a VDU show that this is no real problem and that direct extraction with minimal keyed in specification details is available now.

The objective of this work has always been the statement of capital cost directly from the design at the earliest stage. Rather more exciting possibilities are now apparent :-

1.   the variables can be quantified and with the use of the models
     the direction of movement the values in it to reduce cost and
     the cost effect of such a move can be calculated.

2.   the options using alternative models and their iteration to
     obtain the best result, one of which is lowest cost (in the
     full cost-in-use overall sense) involves little human effort.

Possible design solutions can be generally evaluated in anticipation
of all the details being drawn out. One example of interest is the
surface car park;

1.   define area available

2.   define site levels

3.   express criteria for circulation, individual parking, ped-
     estrian routes

4.   test - surface
           - partial/whole decking
           - multi-storey
           - benefit of additional site area (determine value of
              more site if at all available)

5.   relate results of 4 in terms of capital, cost-in-use etc.,

6.   accept option which meets cost, appearance and planning crit-
     eria.

To conclude; if the design team were to be a parallel working set
of decision makers (as well as option creators) backed by suitable
models of the functional systems from which they have to choose in
order to convert their performance criteria into constructable entities, it is possible to see (in the near future) a methodology
which provides an analysis of a great number of options in capital,
running, dcf, terms which are reliable to known limits.
    The cost would be known, the elapsed time would be brief, the
quality of the result assured and the real options available to the
client early enough to be taken.
    The time bomb of client dissatisfaction, so evident in reports in
circulation now could be defused with the time constant concept
replacing it.

PROJECT ORGANIZATION AND IT'S IMPACT ON QUALITY

ANTHONY WALKER, Liverpool Polytechnic
WILLIAM HUGHES, Liverpool Polytechnic

Introduction

At its broadest level, quality in building is the degree
to which a completed project satisfies the requirements of
a client. The way in which the building professions and
industry organise and manage the process of producing
buildings has a fundamental effect upon the achievement of
clients objectives and hence on the quality of the
completed project measured in these terms.

This paper examines the organization and management of
three building projects which resulted in the levels of
client satisfaction shown in table 1 (1).

TABLE 1: Comparison of client satifaction

Project	%Satisfaction
1	68
2	68
3	47

Whilst the method of measuring client satisfaction was
crude (2), and there remains great scope for developing
approaches to this problem, nevertheless the relative
degree of satisfaction is a sufficient guide when
considering the organization and management approach which
produced them.

The method of analysing the effectiveness of project
organizations being undertaken at Liverpool Polytechnic's
Department of Surveying analyses the performance of the
construction project management team as a whole to
identify organisational causes of deficiencies in the
project outcome. These deficiencies and client's overall
response to their satisfaction with the completed project
form the basis of the measurements of client satisfaction
referred to above.

The analysis requires a systematic method of filtering
out information on relationships, decisions and actions
from the records and documentation of a project.   The

systematic approach is provided by the technique of Linear
Responsibility Analysis (3). The model upon which the
technique is based is described below, the main
propositions of which can be summarised as:

(i) The building process is divided into the systems of
Conception, Inception and Realisation at Primary Decision
points and into Sub-systems at the Key Decision and
Operational Decision points.
(ii) The differentiation of the system should be
matched by the provision of a corresponding level of
integration effort.
(iii) The managing system and the operating system
should be differentiated from each other.
(iv) The managing system itself should be
undifferentiated.
(v) The client and the building process should be
integrated.

The technique of Linear Responsibility Analysis (LRA) can
be used to apply the model because it offers a systematic
approach to defining the operating and managing systems of
a particular project. Using this technique it is possible
to identify relationships between contributors, where the
responsibility lies for decisions, in what capacity any
decision was taken, and most importantly, how effective a
management structure has been. The measure of the
effectiveness of a building project organisation used is
the level of client satisfaction with the project outcome.
This paper shows the application of the technique to three
projects as structured post-mortems and draws conclusions
about them based on quantitative comparisons of the
effectiveness of the management of the different projects.

Analytical Framework*

The propositions mentioned above arise from Walker's model
of the process of building provision (4). Within the
framework offered by Walker, decisions on a project are
classified, or ranked, according to the degree to which
they commit the client to given courses of action or
commitment of finance. The highest ranked decisions are
called primary decisions. In the context of a building
project primary decisions are common to all projects viz:

(a) The decision that the client needs to acquire real
property to achieve his objectives.
(b) Deciding the form that the real property acquired
should take.

* A glossary of terms is given in Table 2

366

TABLE 2: Glossary of terms

Term	Definition
Differentiation	The attribute which serves to define the boundary between systems, sub-systems or people.
Discontinuity	The interruption of the managing system betwen tasks.
Environmental Forces	See Environmental Influences.
Environmental influences	Events which are not part of the system under consideration but which have an effect upon the system.
Integration	The coordinating activity which overcomes the effects of differentiation or discontinuity.
Key Decisions	Decision points which provide the client with regions of control and differentiate the sub-systems of activity.
Linear Responsibility presenting Analysis (LRA)	The technique of collecting, collating and the data collected from an analysis of a project so that an objective assessment can be made of the project's organizational structure.
Managing system	The requisite activities, skills, technology and facilities to keep the Operating system going.
Operating system	The requisite skills, technology and facilities for actually carrying out work.
Operational Decisions	The decisions normally taken by the project team which provide the boundaries between one task sub-system and another.
Primary Decisions	The highest rank of decision in the client's organisation affecting the project which provide the boundaries between the major systems.
Project Conception	The system which results in the decision that the client needs to acquire real property to achieve his objectives.
Project Inception	The system of activity which results in the client deciding upon the form that the real property - which is required to achieve his objectives - should take.
Project Realization	The system of activity which encompasses the design and construction of a building.
Reciprocal Interdependency	A relationship between Task sub-systems such that each is mutually dependant on the output of the other.
Sequential Interdependency	A relationship between Task sub-systems such that the execution of a Task sub-system relies upon the output from a preceeding Task sub-system before it can act.
Sub-system of activity	A sequence of decisions and activities which take place between two Key decision points.
System of activity	A sequence of decisions and activities which take place between two Primary decision points.
Task sub-system	A sequence of activities which take place between two Operational decision points.

These primary decisions form the boundary to the three major systems of activity on any building project as shown in Table 3, i.e. the systems of Project Conception, Project Inception and Project Realisation.

TABLE 3:   Primary Decisions vs. Systems of Activity

PRIMARY DECISIONS:	SYSTEMS:
Adaptation to external influences	
	Conception
Provision of a performance through the acquisition of real property	
	Inception
Identification and construction of a new building	
	Realization
Final Completion	

The system of Project Realisation contains what is commonly understood to be the design process and the construction of the project. For the projects studied in this paper, the first two systems of activity were not recorded to a sufficient level of detail for analysis.

The next level of decision-making is the key decision level. Key decisions are determined by the client as a result of the client's internal procedures for expenditure and similar approvals, and will be strongly influenced by environmental influences upon the client's activities. They range from, for example, approval of design and budget proposals and decisions to delay the project to decisions to change the nature of the project. If the client's organisation is not responsive to environmental forces, key decision points may be innappropriately identified to the detriment of project outcome. They act as major feedback opportunities within the client's firm and also for the process of building provision.

The third level of decision-making lies at the operational level. Operational decisions contribute to key decisions and are constrained by them. They will mainly be concerned with implementation of procedural aspects of building project organisation and will move the project incrementally towards a key decision and provide secondary feedback opportunities.

Key and operational decision points cannot be universally prescribed for all projects and need to be uniquely identified for each project analysed.

Thus, each system created by primary decisions consists of a number of sub-systems created by key decisions, which in turn consist of a number of task sub-systems created by operational decisions.

The sub-systems of activity are sequentially interdependent, that is, succeeding sub-systems rely upon the output from preceeding sub-systems, and the order of precedence can be identified. Task sub-systems may be sequentially interdependent or reciprocally interdependent. Reciprocal interdependency occurs when two task sub-systems are mutually dependent on each other, for example, the architectural proposals for the external envelope of the building rely heavily upon the structural solution adopted, and vice versa. Thus, task sub-systems may be linked to one another in a variety of ways (interdependence is used here as defined by Thompson (5)).

Whilst the systems and sub-systems at primary and key decision level are differentiated by virtue of the decision points, task sub-systems are not only differentiated by operational decisions, but also on the basis of the personnel involved in them. They may be differentiated from each other on the basis of:

(a)   The type of skill demanded by the task
(b)   The geographical separation of the contributors
(c)   The sequence of the tasks (time)

Differentiation on any of these bases can be reinforced by the concept of sentience (6). A sentient group is a group to which individuals are prepared to commit themselves and upon which they depend for emotional support. In the building context, with substantial independence of contributing consultants, firms and professions, sentience can arise as both allegiance to a firm, and/or allegiance to a profession.

The description so far constitutes the operating system for actually carrying out the work required to progress the project. This is maintained, co-ordinated and kept going by the managing system, which is the source of the activities required to integrate the differentiation present in the operating system, together with the supervising and decision making activities. It is differentiated from the operating system on the basis of skill, i.e. management.

These concepts provide the abstract model which can be used for analysis of a building project organisation. The technique of Linear Responsibility Analysis is the vehicle for gathering, collating, presenting and analysing the data and uses the model for its basis. It is presented in graphical form and figure 1 shows part of an LRA to give some idea of how the information is mapped and presented.

The projects

The first case study undertaken in the course of this research was a £1.4m contract, at 1977 prices, for an

369

FIGURE 1

SAMPLE PORTION OF LINEAR RESPONSIBILITY ANALYSIS CHART

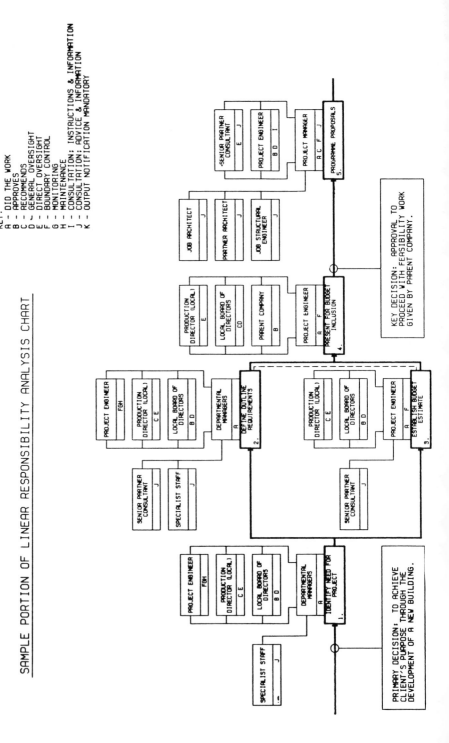

KEY:
A - DID THE WORK
B - APPROVES
C - RECOMMENDS
D - GENERAL OVERSIGHT
E - DIRECT OVERSIGHT
F - BOUNDARY CONTROL
G - MONITORING
H - MAINTENANCE
I - CONSULTATION; INSTRUCTIONS & INFORMATION
J - CONSULTATION; ADVICE & INFORMATION
K - OUTPUT NOTIFICATION MANDATORY

extension to existing buildings to provide a high rise warehouse, services block and packing line extension for a pharmaceutical manufacturer. The total floor area of the development was 5194 sq. m. A professional engineering consultancy firm with "in-house" expertise in structural and services engineering acted as professional advisers. In addition to providing these professional skills they also provided the project management on behalf of the client. They engaged an architect for architectural design aspects only. A quantity surveying firm were engaged directly by the client to work with the other professions. The contract was split into two phases, both of which were let by competitive tender.

Project number 2 was an engineering factory with offices and a link block. The total area of development was 4580 sq. m at a cost of £500,000 at 1977 prices. An industrial and commercial property consultancy firm provided all professional services "in-house". The contract was split into two separate contracts, one for the factory, the other for the offices. The former was let by competitive tender whereas the offices contract was negotiated on the basis of the factory tender.

The third project was an extension to existing buildings to provide butchery facilities for a large wholesale grocery business. The total floor area provided amounted to 1149 sq. m and cost £173,000 at 1979 prices. The contractual arrangements were conventional in that an architect acted as team leader with a private quantity surveyor, structural engineer and services engineer, each from small to medium sized practices. The contract was let by conventional competitive tender.

Primary and Key Decisions.

As stated above, the model proposes that the project can be broken down into three systems viz: Conception, Inception and Realization, and that these are defined by the discontinuities of the process which are caused by primary decisions. The primary decision to achieve the client's objective through the construction of a new building is the point at which the Project Realization process starts. As stated earlier, this paper is concerned with the organization and management of this process only.

The key decisions taken during the project realization process for each project are given in table 4. The decisions presented in the table bear no horizontal relationship to each other.

371

TABLE 4: KEY DECISIONS

PROJECT ONE	PROJECT TWO	PROJECT THREE
Key Decision no. 1 Approval to proceed with feasibility work given by parent co. Commitment to spend up to £25000 on fees.	Key Decision no. 1 Agreement to proceed to tender on factory but not on offices. Office design to be revised & tender negotiated.	Key Decision no. 1 Agreement to proceed on basis of sketch designs.
Key Decision no. 2 Permission by parent company to proceed with detailed design. Commitment to full design fees, based on approval of brief.	Key Decision no. 2 Acceptance of tender for factory. Agreement to construct. Commitment to constn costs and contractual obligations	Key Decision no. 2 Estimate too high, cost limit restated. Scale of acceptable reductions indicated. Proceed to tender on existing scheme subject to reductions.
Key Decision no. 3 Acceptance of tender. Commitment to construction costs & contractual obligations	Key Decision no. 3 Agreement on design of offices.	Key Decision no. 3 Acceptance of main tender subject to reduction. Agreement to construction. Commitment to contractual obligations.
Key Decision no. 4 Certificate of final completion. Contractual obligations discharged.	Key Decision no. 4 Agreement to further re-design of offices.	Key Decision no. 4 Acceptance of tender for refrigeration contract. Commitment to contractual obligations.
	Key Decision no. 5 Acceptance of office design. Agreement to instruct contractor to construct offices. Additional contractual obligations.	Key Decision no. 5 Certificate of final completion. Contractual obligations discharged.
	Key Decision no. 6 Certificate of final completion. Contractual obligations discharged.	

TABLE 5: Environmental Forces

	PROJECT ONE	PROJECT TWO	PROJECT THREE
Certainty/uncertainty at start of contract:	Outline of functional and technical requirements known. Project formed part of a planned expansion programme.	Reasonably clear outline functional requirements known for the factory but not for the offices. Recognition of need.	Outline functional and technical requirements known, and client's available funding although the two were not reconciled at the start.
Certainty/uncertainty as the project progressed:	During the process of realisation the programme was compressed for finacial and budgetary reasons. The requirement for early completion eased during construction.	Continuous hesitation by the client in pursuing the project because of the political and economic situation. Client had difficulty defining requirements.	During project realisation the decision had to be taken to reduce the size of the project to bring it to an acceptable cost.
Indication of the conflict identified within each project:	Change in services engineering manager. Change in contractor's site agent. Shortage of bricklayers.	Client felt consultants had over-committed workloads. Consultants felt client was unable to make decisions. Client felt that the consultsants had low standards. Shortage of bricklayers.	Client felt that the project was not as controlled as it could have been. Problems of reconciling the client's requirements with those of the planning authority.
Technical complexity; spatial	Tight constraints on the relationships of areas. High level of specialist equipment. Relationship to exisiting facilities important.	No severe restraints on location. Separation of classes of employees required. Functional requirements dominant.	Location pre-determined. Functional requirements dominant.
Technical complexity; structural	Structurally difficult site. Variety of structures.	Structurally straightforward solutions adopted. Site problems encountered.	Structurally straightforward.
Technical complexity; services	Complex provision of specialist equipment. Ventilation and temperature control important.	Services straightforward. Conventional solutions adopted.	Extensive refrigeration provided. Hygiene facilities extensive.
Aesthetic complexity	Match existing simple elevations	Functional efficiency more important than aesthetics.	Match existing simple elevations.

Environmental Forces

The three projects were subject to a variety of influences from their environments. These are summarised in table 5. Comparing the three projects, it is apparent that case study number 1 had the least uncertainty surrounding the project at the outset. All three projects were based on a clear recognition of need.

As projects 1 and 3 progressed their environment remained fairly stable level throughout the design stage, but project 2 was troubled by hesitation on the part of the client due to doubt arising from the political and economic situation.

Technically, the first project was highly constrained and this caused problems for the design team. This was not the case to the same extent for the other two projects.

TABLE 6:   Integration of the client with the process

Proj	Sub-sys.	No. of links	BETWEEN TASKS Good Integ. (%)	Partial Integ. (%)	No. of Tasks	WITHIN TASKS Good Integ. (%)	Part Integ (%)
One	1	5	0	0	4	0	0
	2	25	92	0	9	89	0
	3	24	63	0	8	75	25
	4	68	9	89	16	25	75
	Tot	122	36%	48%	37	48%	38%
Two	1	12	25	0	7	71	0
	2	35	80	0	8	100	0
	3	18	89	11	4	75	25
	4	34	100	0	4	100	0
	5	5	0	100	2	50	30
	6	22	100	0	4	100	0
	7	58	0	100	10	0	100
	Tot	184	60%	31%	39	64%	31%
Three	1	2	0	0	3	0	0
	2	10	20	0	5	60	0
	3	23	13	0	7	43	20
	4	5	20	0	5	40	0
	5	2	0	0	2	0	20
	6	35	0	91	11	0	0
	7	9	0	0	4	0	91
	Tot	86	7%	32%	37	22%	32%

Integration of the client with the process of building

Table 6 summarises the quantitative information distilled from the LRA about the degree to which the client was integrated with the process. This numeric data is based

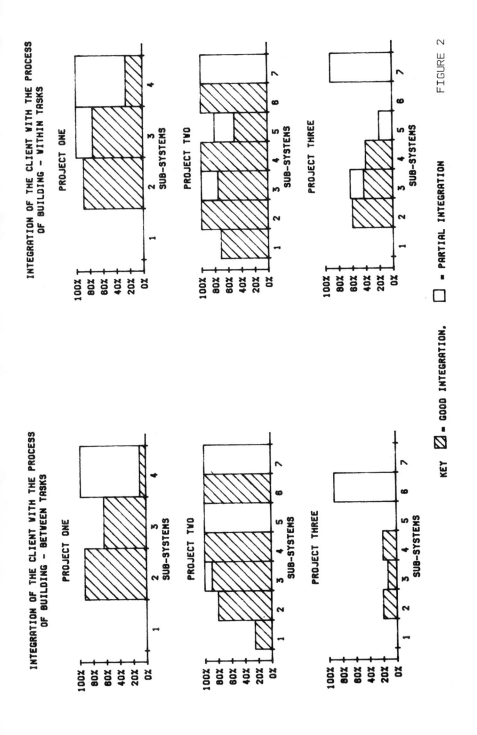

INTEGRATION OF THE CLIENT WITH THE PROCESS
OF BUILDING - WITHIN TASKS

PROJECT ONE

PROJECT TWO

PROJECT THREE

INTEGRRATION OF THE CLIENT WITH THE PROCESS
OF BUILDING - BETWEEN TASKS

PROJECT ONE

PROJECT TWO

PROJECT THREE

KEY ▨ = GOOD INTEGRATION.   ☐ = PARTIAL INTEGRATION

FIGURE 2

on the proposition that there are primarily two people responsible for integration or coordination on a building project; one from the client's organization, and one from the project team. For project number 1 the two primary integrators were identified as the Client's Project Engineer and the Project Manager. For Project number 2 they were the Group Chairman and the Project Manager, while for project number 3 they were the Managing Director and the Architect. Good integration of the client is said to exist when both primary integrators are in a managing role for each task, and partial integration is defined as one of them in a managing role with the other being consulted. If either or both are missing from a task sub-system then the client is not integrated with the project team for that particular sub-system.

By plotting this information graphically, as in figure 2, we can make direct comparisons between projects. Note that the overall width of the bar chart is the same for all three histograms, and that the individual bars on the charts are thinner where there are more sub-systems. This is so that we can view the overall area of shading on the chart as being a significant indicator of the comparative extent of integration for projects with differing numbers of subsystems.

All three projects have certain features in common. Firstly, there is little integration of the client with the team at the early stages of the projects. Projects 1 and 3 show no integration at all in their respective initial sub-systems. The third project shows the lowest client/project team integration.

Differentiation of the Operating System and the level of integration

Table 7 shows the data for this measure of project structure, and figure 3 shows the data in histogram form.

Projects 1 and 2 are well integrated whereas project 3 shows low integration. The measure of differentiation of the operating system can be seen as a measure of the need for integration, and unless this need is matched by a corresponding provision of integration then work is likely to be uncoordinated. This leads to the situation in project 3 where it can be seen that the project was lacked integration. Projects 1 and 2 were generally sound on this criterion, although it is evident from the graph that there was some shortfall in the provision of integration. On all of the projects, the majority of the different-iation was of the most complex type (i.e. under the code TABCD), differentiation within tasks was integrated slightly better than the differentiation between tasks, except on project 3 where the difference is more marked.

Table 7a: Differentiation of the operating system between tasks

Project	Sub-system	No. of Tasks	No. of Links	T	TB	TAB	TABCD	Level of Integration
One	1	4	5	50	50	-	-	100%
	2	9	25	20	-	14	66	95%
	3	8	24	7	14	7	72	95%
	4	16	68	3	3	-	94	100%
	Total	37	122	9	7	3	81	98%
Two	1	7	12	83	-	-	17	50%
	2	8	35	7	-	43	50	100%
	3	4	18	29	-	71	-	100%
	4	4	34	32	-	68	-	100%
	5	2	5	40	-	60	-	100%
	6	4	22	38	-	62	-	100%
	7	10	58	15	2	26	26	100%
	Total	39	184	26	1	46	27	93%
Three	1	3	2	50	-	-	50	-
	2	5	10	20	-	-	80	-
	3	7	23	14	-	-	86	-
	4	5	5	60	-	-	40	-
	5	2	2	100	-	-	-	-
	6	11	35	7	3	-	90	17%
	7	4	9	3	-	-	67	-
	Total	37	86	19	1	-	80	2%

Table 7b: Differentiation of the operating system within tasks

Project	Sub-system	No. of job position links	none	B	AB	ABCD	Level of Integration
One	1	4	-	50	-	50	100%
	2	34	-	-	50	50	100%
	3	22	-	4	14	82	89%
	4	74	-	3	28	69	100%
	Total	134	-	4	30	66	97%
Two	1	27	19	-	44	37	93%
	2	34	-	-	47	53	100%
	3	14	-	-	64	36	100%
	4	20	5	-	40	55	100%
	5	8	13	-	25	62	100%
	6	20	5	-	40	55	100%
	7	52	-	2	25	73	100%
	Total	175	4	1	39	56	99%
Three	1	5	-	100	-	-	100
	2	8	-	-	-	100	63
	3	8	-	-	-	100	-
	4	3	-	-	-	100	-
	5	-	-	-	-	-	-
	6	33	-	3	-	97	39
	7	11	-	-	-	100	9
	Total	68	-	9	-	91	30

*Key: T = Differentiation due to time
A = Differentiation due to company
B = Differentiation due to skill type
C = Differentiation due to location
D = Differentiation due to profession

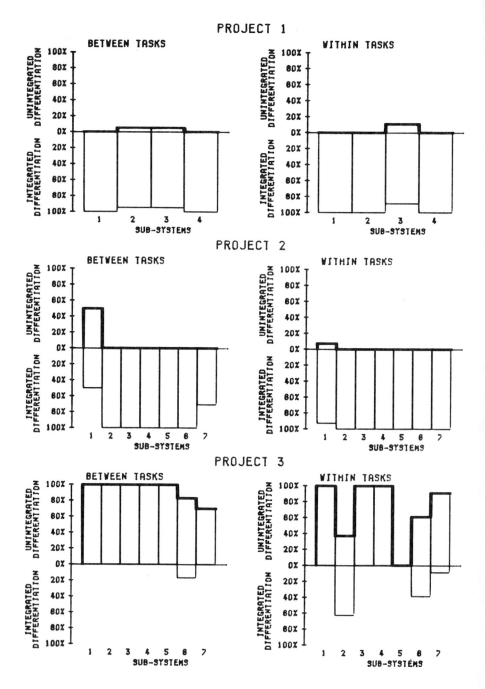

FIGURE 3

Differentiation of managing system and operating system

Table 8 summarises the data for this aspect of project
structure. It shows the degree to which someone who was a
manager of the project actually appeared in a task box on
the LRA as undertaking a professional (i.e. operational)
function, as opposed to a purely managerial function.
Such a situation fudges the boundary between "doing the
job" and "managing the job" and provides potential for
conflict in the process.

TABLE 8: Differentiation of managing and operating systems
% of task boxes (operating system)
occupied by a managing job position

Sub-system	Project 1	Project 2	Project 3
1	50	57	67
2	33	25	60
3	25	25	29
4	13	0	20
5		0	0
6		0	27
7		10	0
Total	24	20	30

Differentiation of the Managing System itself

The differentiation of the managing system itself is
quantified in table 9, and shown in histogram format in
figure 4.
    In order to act as an integrative mechanism and provide
continuity at decision points, ideally the managing system
should itself be undifferentiated, although this is not
usually practical. The best practical solution is to have
normally two people in the managing system, one from the
client's organization and one from the project team. The
more people are involved in the managing system the more
complex it becomes so that the managing system itself may
need integration to the detriment of its activities in
integrating the operating system.
    The measures used here indicate the frequency of a
number of people exercising management functions in
relation to a task (table 9a). In LRA terms this means
the number of different job positions in each control loop
which appear above the task boxes (see figure 1, e.g. in
sub-system 1 of project 1 there were three task sub-
systems which had three people exercising management
functions, and one task sub-system which had two.)
    The first project involved more people in the functions
of management in the control loops than the other two, the
lowest being project 3. Although project 3 shows no
duplication of integration effort, this is not surprising

Table 9a: Differentiation of the Managing system within tasks

Project	Sub-system	No. of control loops	No. of Job positions in control loops						Duplicated co-ord.
			0	1	2	3	4	5	
One	1	4	-	-	1	3	-	-	-
	2	9	-	-	1	3	5	-	-
	3	9	-	1	1	3	4	-	-
	4	16	-	2	4	5	5	-	31%
	Total	38	-	3	7	14	14	-	12%
Two	1	7	2	3	2	-	-	-	-
	2	8	-	2	6	-	-	-	-
	3	4	-	2	2	-	-	-	-
	4	4	-	-	4	-	-	-	-
	5	2	-	1	-	1	-	-	50%
	6	4	-	-	4	-	-	-	-
	7	10	-	-	5	2	1	2	100%
	Total	39	2	8	23	3	1	2	28%
Three	1	3	-	2	1	-	-	-	-
	2	5	-	4	1	-	-	-	-
	3	7	2	4	1	-	-	-	-
	4	5	2	1	2	-	-	-	-
	5	2	1	-	1	-	-	-	-
	6	11	4	3	2	2	-	-	-
	7	4	2	-	2	-	-	-	-
	Total	37	11	14	10	2	-	-	-

TABLE 9b: Differentiation of the Managing system between tasks

Project	Discontinuity as a percentage of all tasks	Discontinuity as a percentage of tasks at decision points
One	5%	75%
Two	0%	0%
Three	93%	100%

DIFFERENTIATION OF THE MANAGING SYSTEM WITHIN TASKS

PROJECT ONE

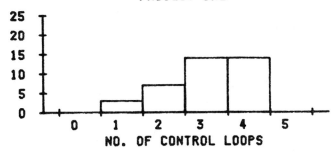

PROJECT TWO

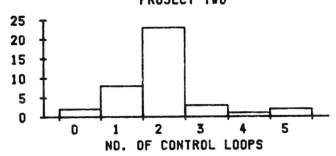

PROJECT THREE

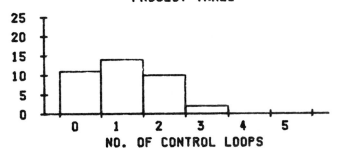

FIGURE 4

since few managing job positions are involved in the project. The other two projects show duplication of integration effort in the latter stages of the project, but this is to be expected because there are personnel involved with integrating work for the contractor as well as those involved with integrating the design team. The validity of accepting this duality of integration may be questioned, but it is a phenomena that is common to contracts let in competition after the design is substantially complete.

The differentiation of the managing system between tasks (table 9b) shows how discontinuities arose in the managing system. On a well managed project the number of discontinuous managing links should ideally be zero, especially across decision points because these form the major regions of control as the project passes from one sub-system to the next. This criterion shows that project 2 was ideal in this respect, while the others are less satisfactory, particularly project 3.

Conclusions

A paper of this kind which is addressed to a comparison of organization systems and their effect on project outcome must, to be containable, concentrate on the major issues. As a result much of the detail has had to be omitted (for a more detailed analysis of one of the case studies see Walker & Hughes, 1984 (7)). Nevertheless it can be seen that the organization and management of a project does have a strong effect upon the client's satisfaction with the project outcome.

TABLE 10: Comparison of the project teams' performance

Aspect	Project		
	1	2	3
1. Integration of the client with the process of building			
a) between tasks	2	1	3
b) within tasks	2	1	3
2. Match of differentiation with integration			
a) between tasks	1	1	3
b) within tasks	1	1	3
3. Differentiation of the managing from the operating system	2	1	3
4. Differentiation of the managing system itself			
a) between tasks	2	1	3
b) within tasks	2	1	3
Total	12	7	21
Client satisfaction	1	1	3

If the various aspects analysed are ranked on a simple scale of 1 to 3 with 1 representing the best performance for the three projects the pattern shown in table 10 emerges.

Whilst this is a crude method of comparison it does indicate that the quality of the aspects of organization and management identified do reflect clients' final satisfaction with their projects.

Whilst project 2 shows the best result for the aspects examined, it is only equal to project 1 in terms of client satisfaction. It is interesting to note that the environment in which it was carried out was rather less stable than for project 1. The additional quality of the aspects of organization and management on project 2 would appear therefore to have overcome the less stable environmental forces in producing the project outcome.

Project 3 scores lowest for client satisfaction and performs consistently worse than both other projects for each aspect of organization and management examined.

It is to be expected that if the aspects identified are carried out effectively then the project stands a good chance of providing the client with a satisfactory outcome.

The model and the LRA technique have been shown to be a very useful tool for uncovering the major organizational reasons for client dissatisfaction with their projects. They highlight the responsibilities of individuals and groups within the various organizations which make up building project teams. The quantitative assessments can be directly compared to similar analyses from different building projects, thus providing retrospective data which will help in the future design of organizational structures.

The LRA gives great visibility to project organizations and allows a range of studies of the structure to be undertaken far in excess of what is illustrated in this paper. One of the limitations of the approach is that it relies upon comparison of the client's satisfaction with the outcome of the project in order to verify the findings.

Work upon this and further development and refinement of the LRA and the quantitative data derived from it is continuing under the aegis of the Science and Engineering Research Council at Liverpool Polytechnic, Department of Surveying. As a major part of this coninuing research, further case studies are being undertaken. Many of the points in this paper are discussed further in "Project Management in Construction" by A. Walker, published by Granada (1984).

383

References

1. Walker, A. (1980) A model for the design of project management structures for building clients, Ph.D. thesis, Liverpool Polytechnic.
2. Walker, A. & Wilson, A.J. (1983) 'An approach to the measurement and performance of the project management process', Land Management Research Conference Proceedings, Leicester Polytechnic.
3. Walker, A. (1982) 'Linear Responsibility Analysis', Chartered Quantity Surveyor, 4(8), pp.228-230.
4. Walker, A. & Wilson, A.J. (1981) 'A model for the design of project management structures', The Quantity Surveyor, 37(4), pp.66-71.
5. Thompson, J.D. (1967) Organizations in action, McGraw-Hill.
6. Miller, E.J. & Rice, A.K. (1967) Sytems of organization, Tavistock Publications, London.
7. Walker, A. & Hughes, W.P. (1984) 'Private industrial project management: a systems-bases case study', Construction Management & Economics (in press).

SOME REFLECTIONS ON LIFE CYCLE COSTING

DONALD BISHOP
L.M.B.A. Professor of Building
The Bartlett School of Architecture and Planning
University College, London

INTRODUCTION

Despite the scale of industry's work few case studies have been made
of the considerations influencing clients' decisions to build. In
part this might be explained by the former dominance of public sector
programmes which had their roots in assessments of need strongly
influenced by considerations of public policy and by the requirements
of public accountability, and in part by the confidentiality surrounding most development decisions in the private sector. There is, of
course, much general guidance, a substantial literature on the assessment of proposals for property development, and vast experience of
cost planning techniques. These latter set out to achieve budgetary
control by enforcing the principle of "more for this implies less of
that", a tactic which often leads to design decisions offering better
value for money. Design guides and budgetary control systems have
featured in most public sector programmes whilst development appraisal
techniques evolved to inform with private sector decisions were
applied to some public sector programmes. For example cost benefit
analysis was widely used to appraise (and especially to rank) highway
programmes and to assess the viability of infrastructure investments.
In the main little regard was paid to the long term financial consequences of building new projects, other than the debts created by
borrowing, presumably because of an expectation that differences
between the running, maintenance and replacement costs of well-
designed projects would be small.
    At the time cost planning techniques were becoming well-established,
Stone amongst others launched the principle of costs-in-use and argued
doggedly for their relevance to design decisions. (1, 2, 3) In the
event the notion was then adopted neither by the industry's clients
nor by the professions although cost in - use - considerations
influenced product selections made by consortia and others. Interest
in the topic - perhaps stimulated by the energy crisis and by inflation - may be charted by the relevant publications on both sides of the
Atlantic; e.g. from the energy conservation programme in the USA (4)
and the publications of the Committee for Terotechnology of the
Department of Industry (5), both in the mid- to late-1970's, and the
practice guides e.g. by Haviland (6) and the admirably clear and

comprehensive handbook by Flanagan and Norman (7). Clearly life
cycle costing has come to age and has a place in the range of professional services offered to clients.

Clearly decision-makers should "examine the total cost implications
of any decision" in the expectation this will "facilitate an effective
choice between alternative methods of achieving a stated objective"*;
this is not in debate. However the technique involves four - at
least - assumptions which merit examination because they have different effects in different circumstances. This is not to aver, of
course, that they have been neglected in the development of the topic
but only that they are of sufficient importance to justify discussion.

FOUR ASSUMPTIONS IMPLICIT IN LIFE CYCLE COSTING

The four assumptions will be briefly examined before their consequences are considered. They are :-

(i)   the various ways of achieving design goals differ only in their
      life cycle costs;
(ii) the capital costs input into an analysis are the estimated cost
      outlays rather than the opportunity costs;
(iii) resources funding the initial investment are not constrained;
      and
(iv) analyses based on an examination of the estimated life cycle
      costs of the principal elements** will exhaust the possibilit-
      ies and lead to an efficient allocation of resources.

(i) ... differ only in their life cycle costs

Economic efficiency is the rational for life cycle costing, the
expectation being that decisions informed by life cycle costing will
achieve any given output at minimum total cost thus avoiding both
waste and technological inefficiency.*** This goal would be achieved
without additional analysis if the various ways of meeting design
objectives were neutral in all respects other than their life cycle
costs. That is, if the possible choices made no difference to the
satisfaction, pleasure, or fulfilment derived from "consuming a
quantity of a good"***. Whilst this may be conceptually possible, in
practice it is improbable.

Two of the examples quoted in ref. 7 will be used to illustrate the
point: floor coverings (example 4) and heating systems (example 6).
In addition to differences in life cycle costs some types of floor
coverings will be more appropriate to some usages than others (unless
they are virtually identical e.g. different grades of carpet tiles).
The difference in running costs (Table 4D) indicates this is not the
case. Hence the three methods will differ in their capacity to with-
stand concentrated wear, or to tolerate spillages, or in their

---

  * These quotations are not juxtaposed in reference (7).
  ** elements in a cost planning sense.
*** definitions are drawn from the Penguin Dictionary of Economics.

'warmth' and appearance, especially their appearance towards the end
of any replacement cycle (which may last for several years). Again
heating systems are compared with respect to energy, maintenance and
replacement costs. In practice, of course, some heating systems will
be more appropriate for some uses than others and their appropriateness varies with other characteristics of the building in question
especially with its thermal capacity and the benefits (or disbenefits)
from solar heating. Moreover, consumer preferences ought to be taken
into account.

One response to these arguments must be, "Of course, such considerations enter into any decision". If so, how might this be achieved?
One tactic might be to produce data for differences in performance
and for ranked consumer preferences (which would change with different circumstances). The difficulties of attributing prices either to
varying degrees of performance standards or to the revealed user
preferences are likely to frustrate any attempt to be numerate in the
sense that life cycle costing is numerate. Therefore these factors
would be incommensurate with life cycle costing and would have to
influence design decisions subjectively. To this extent life cycle
costing is a partial input decision procedure because two of the
major components which ought to be taken into account cannot be
measured. Therefore, life cycle costing information needs to be
complemented by parallel assessments of the relative utilities of
design decisions and these would greatly extend and complicate any
analysis.

(ii) ... estimated cost outlays rather than opportunity costs

Capital investment decisions are not made in a vacuum because any
decision to invest necessarily involves a decision to forego the
value of other possibilities - including consumption. Opportunity
costs analysis (which is not without its problems) attempts to
contrast the estimated returns from a course of action with the real
cost of undertaking it. This consideration bites on life cycle
costing (LCC) in two ways; first with respect to decisions within
the context of a given project, secondly with respect to decisions
clients might make outside a particular building project.

Consider, for example, the three floor coverings types A, B, and
C discussed above (Example 4, Ref. 7). In this their estimated
initial costs were £10,000, £11,200 and £13,600 and their present
values based on estimated running costs were   £8,216, £8,080 and
£8,231 respectively. Suppose they offered an identical performance
in other respects. Then a decision to spend £11,200 (type B) rather
than £10,000 (type A) implies no better trade-off exists within the
project i.e. LCC informed decisions when made element-by-element risk
sub-optimisation. This could be avoided by a decision procedure
which

(a) identified a set of feasible elements and for each calculated
the LCC and any constraints which would prevent one element being
juxtaposed with another - from performance or assembly or maintenance
considerations;
(b) identified the least present cost solution for each technic-

ally admissible element, evaluating the preference on performance and consumer preference grounds, and combined the two assessments - how? - to give a final selection of preferred elements.

Clearly such a procedure would be feasible. It would necessarily involve very much more effort by the whole design team both to identify and estimate the LCC for each of the possibilities for each element and to contribute to the many sequential decisions to be made jointly on LCC and utility grounds - each involving a different kind of subjective choice.

This is not the end of the matter. Every decision to invest more in order to achieve a lower LCC for a project foregoes the opportunity to consume or to invest the additional investment in some other way. Because the majority of clients need a building as a necessary adjunct to some other operational purpose e.g. to educate, or to manufacture, or to manage a commercial enterprise, relevant decisions must take these primary objectives into account. For example a manufacturer wishing to catch a market may prefer a solution that enables production to start at the earliest possible time, whatever the LCC consequences, another may prefer to invest in more efficient machinery rather than a building which would be efficient in LCC terms; another may opt for investing in a building entailing high maintenance costs but which will be sufficiently striking to act as a company logo. Attitudes to design decisions based on LCC considerations may be influenced also by the volume and scale of a programme which may emerge as a conflict of interest; for example the community may be served best by LCC solutions (because the community shoulders these costs in the long run) whilst a client's interests (in both the public and private sectors) may point to a least initial cost solution in order to achieve a programme within a restricted budget.

(iii) ... resources for the initial investment are not constrained

In many but not all instances each most economic LCC solution will entail higher initial costs than solutions which would have an adequate performance but higher life cycle costs. Therefore the sum of the initial costs justified by LCC considerations is likely to exceed a budget devised on the basis of initial costs. In these instances the already lengthy procedure outlined at ii(a) and (b) above would have to be further extended to reduce the sum of the initial costs of the least present cost solutions to the budget, presumably by replacing the least present cost by least initial cost solutions element-by-element, probably ranked in the order of the ratio of

$$\frac{IC \text{ (for LPC)} - LIC}{LPC - PC \text{ (for LIC)}} * \qquad \text{(without regard to signs)}$$

unless other design considerations dictated another order.

---

* LPC = least present cost   PC = present cost
  LIC = least initial cost   IC = initial cost

(iv) Analyses based on LCC will exhaust the possibilities and lead
      to an efficient allocation of resources

Life cycle costing literature makes great play of the benefits to be
obtained from better information and recognises the limitations
imposed by the paucity of current data. There is no questioning the
helpfulness of historic data to identify the relationships between
the various components of running, maintenance, and replacement costs,
their variability and the ways these change with time, and the differences between building types and categories of building usage.
Such data obviously make a useful contribution to any understanding
of the subject and will often form the basis of analysis. Unfortunately the very convenience and apparent usefulness of historic data
may disguise their weaknesses which could undermine attempts to
broaden the basis of cost advice. Two will be examined.
    Historic data when marshalled and replete with footnotes make a
convincing impression of authenticity - and durability. They involve,
inevitably, the assumption, "tomorrow will be like yesterday". For
some situations this may be sufficiently true. For buildings it is
not because the ways buildings are used and their occupational costs
respond to users' needs and to economic forces. Clients' expectations
(thermal efficiency and energy management for example), building
technology (high performance composites as claddings, the resurgence
of steel framed construction), new needs (security, information
technology) and changes in the relation of cost and quality of different solutions, all influence the way in which buildings are used and
hence the future stream of costs and benefits. To this extent,
historic costs are an unreliable guide to future costs especially if
applied in any form of any standard calculation which tends to reinforce the "reliability" of data by the apparent comprehensiveness of
the analyses. In this sense LCC has some of the inherent defects of
cost analyses based on price data which do not (and cannot) capture
the operational circumstances of any project. The answer may lie in
an operational approach to the principal components of LCC, expressed
as a number of related models (see also para. 5.8 to 5.11 inclusive
Ref. 7).
    Such an approach would have several advantages:

- attention would be focussed on the specific requirements of the
project in question, rather than on the generality of broadly similar
projects;
- a client could be involved in defining the expected operation mode,
a task that might lead to a better understanding of the implications
of the project and hence to an improved design brief;
- the design team would be brought into closer contact because few
if any of the issues fall within the domain of any one specialism; and
- the practical consequences of any unusual future cleaning, or
maintenance, or renewal requirements, or energy consumption are likely
to be detected during design (at the stage when they can be remedied
cheaply) rather than exposed after commissioning.

Modelling techniques should also overcome the second disadvantage inherent in historic costs. These imply a class of decision problem in which probability distributions can be derived for the outcomes of each action, that is a *decision problem under risk*. In the short term this assumption is probably valid for many building projects. In the longer term it is questionable especially in the light of the many factors now at work in society and in technology. Decisions by clients will usually have to be made in the absence of information about the possibilities for the building's use more distant than, say, 15 years hence. Such decisions - *decision making under uncertainty* - normally entail tackling ill-structured problems i.e. those for which no suitable analytic is available. For these brainstorming or Delphic techniques might identify objectives which would not otherwise be present in a client's brief. For example, analysis of the range of possibilities foreseen, say, two decades hence may indicate the advantage in terms of potentially more lettable space of designing for imposed loads in excess of those demanded by CP 3 Chapter V and for service routes far exceeding current requirements. Such considerations reinforce the potential advantages of formulating the analysis of each project anew rather than following any standard procedure and of making as much use as is feasible of operationally based models.

APPLICATIONS OF LIFE CYCLE COSTING

The previous section indicates there is scope for the further development of life cycle costing techniques, especially their application to particular circumstances. This is not to challenge the technique as an important addition to methods for informing clients about the consequences of decisions they might take or for obtaining an efficient allocation of a project's resources. Still less is this paper intended to launch another - and fruitless - debate about "the true meaning of cost". The important issue is that answers should be thought through from first principles to questions that have been formulated clearly in the first instance. This concluding section therefore considers the application of life cycle costing techniques (modified in the light of the examination of the four implicit assumptions) to four broadly defined circumstances, before turning to another and possibly more direct tactic for exploiting the principles of life cycle costing.

The four circumstances will be derived from the constraints on resources and programmes thus:

PROGRAMME	NOT CONSTRAINED	CONSTRAINED
NOT CONSTRAINED	Exceptional! e.g. programmes of prestige projects such as embassies; single prestige projects e.g. headquarters of multi-national organisations	Typically programmes of projects which may be adjusted to satisfy budgets e.g. bank or store refurbishment, highway or prison programmes
CONSTRAINED	Exceptional programmes such as defence projects in times of crisis	Most single projects and many programmes e.g. private housing, public sector programmes to meet statutory requirements (the former schools programmes), most industrial and commercial development

Clearly the category where neither the resources nor the programme are constrained may seek and obtain the best of all possible worlds because it may be assumed the design team will have the time (and the fees) to follow the iterative procedure outlined at above to achieve an efficient allocation of resources in the long term and the client will ignore opportunity costs. In these (unusual) circumstances life cycle costing techniques could have full sway and there should be ample scope to use methods to model the maintenance and servicing regimes and to employ brainstorming or similar techniques better to provide for more distant eventualities.

The next category, when programmes but not the resources are constrained is so exceptional that it need not long detain us. In these circumstances time is likely to be of the essence so that there may not be opportunities to explore the lengthy procedures of life cycle costing. If contracts for the various stages of a single major project were let sequentially, life cycle costing techniques could be used to select solutions for stages late in the sequence or to discriminate between bids by sub-contractors for stages involving a substantial degree of contractor design. Similarly life cycle costing could be used to improve the economic efficiency of projects late in a programme.

At first sight the flexibility offered by some construction programmes should enable budget constraints to be escaped by reducing the number of contracts let in any financial year when the higher initial costs dictated by life cycle costing considerations would otherwise exceed the budget for that programme. In practice flexibility might be illusionary because of clients' opportunities to use the funds released by a delayed programme for some other purpose. That is LCC - essentially long term arguments - would have to compete with the possibly more attractive short term possibilities revealed by opportunity costing. Therefore the flexibility offered by e.g. refurbishment or highway programmes may be the undoing of decisions

391

based on life cycle costing.

With the fourth (and most typical) category of projects both the budget and the programme are constrained. In these circumstances the four factors (value, opportunity costs, constrained budgets, and decision making under uncertainty) will all bite. Comparisons will need to take account of aspects of performance other than life cycle costing, comparisons will need to be between elements as well as between different solutions for the same element, sacrifices will have to be made to contain the initial costs within budgets, and resort should be made to modelling and brainstorming techniques to identify preferred strategies for anticipating longer term possibilities. Undoubtably such procedures are possible. Are they likely to be feasible in the great majority of projects given the normal pressures on time - and fees? Probably not. What then might be the way forward to insert life cycle costing techniques into decisions made by design teams and by their clients?

CONCLUDING DISCUSSION

The following are offered for debate, no more.

Whilst the uniqueness of all building projects is self-evidently true, for many purpose projects fall into convenient use groups which allow studies made at some level of generality usefully to indicate relationships and to identify potentially worthwhile options whilst rejecting those shown to be inefficient. Life cycle costing is a prime candidate for such comparative studies. Reference (7) concludes with a worked example which estimates the life cycle cost for a 240 place primary school. As an illustration of the technique it is good but not helpful in that it indicates neither the relative efficiency of the design nor ways in which the efficiency could be improved (it is appreciated it was not its purpose). Clearly there would be advantage if strategic and tactical issues for different building types were examined by repeated applications of life cycle costing (complemented by operational models) to variations in the basic design to explore e.g.

- the effect of the number of pupil places on the different components of life cycle costs;
- the effect of building form or configuration on life cycle costs;
- the sensitivity of life cycle costs to the principal design options e.g.

total area; net/gross area;
number of storeys
claddings: type
roof: flat or pitched, and the principal materials for each
heating and environmental control systems
finishes
landscaping; and so on.

The task of making these comparative studies would be considerable. It would be reduced by studies of individual elements such as those made by Stone many years ago. Whatever the effort entailed such studies could provide clients, design teams and specialist sub-contractors with the insight necessary to identify options which would have the greatest potential for development. In this way many of the benefits of life cycle costing might be realised without a detailed examination of all the possibilities for each project, which may be too demanding to be practical.

The contention "There must be a recognition that decisions made today carry cost implications for the future" is so self-evident it is astonishing that clients have not always set budgets for these expenditures as a central feature of any design brief and demanded matching estimates from their design teams.* At first sight there is no reason why such a tactic might not be implemented immediately, albeit in outline because the data for budgets and estimates would be rather sketchy. However action on these lines would generate a market for data much as clients' requirements for cost plans stimulated the development of the Building Cost Information Service.

What might be the consequences? Clients' demands make the industry address issues more promptly than any amount of exhortation. (8) Clients would need specifically to determine their policies for servicing, maintenance and renewal hence indicate the budget limits that would be acceptable. It is certain that many clients have neglected to do this in any detail. A duty to respond to a brief which dealt with these issues by a life cycle cost plan would focus the attention of a design team on servicing, maintenance, renewal and energy costs to an extent not always realised currently. The commissioning stage of projects would then assume an importance they now lack. Finally, there might well be a requirement for an independent post-commissioning audit to assess the extent to which expectations in the pre-tender report had been realised. This would have to be even-handed between the parties because the causes for a deficient or too costly performance (or both) might lie with a client's management as much as with the building team.

The arguments in this paper have led life cycle considerations away from a role as a technique for providing information on which to base economically efficient design decisions to a management tool for focussing the attention of clients, design teams and contractors on the performance of buildings as operational entities. This must be in the long term interests of the industry's clients and users of buildings. It would be regrettable if it also generated yet another ground for negligence actions arising for "lack of fitness-for-purpose", a possibility which must exist. Nevertheless this extension of the procedures described in the practice section of ref. 7 is probably the most direct way of making effective the potential of life cycle costing. The consequences would be far-reaching and consistent with the widely shared intention to improve both the quality of building and the service the industry affords to its clients.

---

* such a tactic stems directly from the procedure outlined at para 3.38 of ref. 7.

REFERENCES

1. Stone, P.A. (1960), 'The Economics of Building Designs', J. Roy. Statist. Soc., A 123 (3).
2. The Economies of Factory Buildings (1962), Factory Building Studies No. 12, HMSO.
3. Stone, P.A. (1980), Building Design Evaluation, 3rd Ed., Spon, London.
4. Marshall, H.E. and Ruegg, R.T. (1980), Energy Conservation in Buildings; an Economic Guidebook for Investment Decisions, National Bureau of Standards, Washington D.C.
5. Committee for Industrial Technologies, Dept. of Industry (1975), The Total Cost of Ownership; Terotechnology Concept and Practice, HMSO.
6. Haviland, D.S. (1977), Life Cycle Cost Analysis - a Guide for Architects, American Institute of Architects.
7. Flanagan, R. and Norman, G. (1983), Life Cycle Costing for Construction, RICS, London.
8. The Manual of the BPF System for Building Design and Construction (1983), British Property Federation, London.

EFFECTIVE, EFFICIENT AND PROFITABLE BUILDING PERFORMANCE: THE CASE
OF ENERGY EFFICIENT DESIGN AND COST EFFECTIVE HOUSES

PROFESSOR PATRICK O'SULLIVAN, University of Wales Institute of
Science and Technology
DOCTOR ROBIN WENSLEY, London Business School

In this paper, we wish to use the case of the marketability of energy
efficient house design as an example of the problems and issues
raised when one considers effectiveness and efficiency within a market
system;   that is within a system in which in general actions of the
various parties involved are strongly influenced by their financial
interest and therefore the profitability of the options.

   In this context we wish to distinguish between what might be termed
three types of cost effectiveness: those from the perspective of the
user, the supplier and the policy maker. We will argue that in many
cases within the building sector it is finally user cost effectiveness
which matters.   This is both because it is concepts of user cost
effectiveness which actually impact on user behaviour, and also
because most suppliers particularly speculative building developers
are themselves strongly influenced by their perceptions of user appeal
and benefit.

THE CASE OF ENERGY EFFICIENT DESIGN

We have previously argued (O'Sullivan and Wensley 1983) that attempts to influence actual energy performance of buildings by focussing exclusively on either statutory regulations and/or user attitudes is an ineffective approach to energy conservation policy, Vanhallen and Raaji (1981) suggested that the key factors which influenced actual household energy use behaviour itself and the characteristics of the house.

Other work by O'Sullivan and McGeevor (1982) implies that the inter-relationship between these two factors can be strong. For instance their study suggests that an understanding of actual energy use requires a more comprehensive knowledge of the meaning of such use patterns in both their social and physical context (McGregor 1981). Such an approach, for instance, leads to a hypothesis that in the absence of accurate knowledge, thermostat settings will be related to the warm up time of the principal living room.

Much of the design work on low energy buildings exhibit similar problems. It is, for instance, often based on a set of performance standards which despite their apparent logic do not conform well to the crucial concerns of users. Indeed even when attempts are made to register opinions, such as in the Bedford scale of perceived room temperatures, such a measure may still not properly reflect the underlying variable of comfort level. We now have extensive evidence that many designs can fail to achieve the designed energy savings often by a wide margin.

Responses to this problem have varied. At one extreme there is the temptation to remove the people variable from the equation altogether by developing more automatic systems with limited if any opportunity for over-ride. Except in rare instances however it appears that such actions will be self-defeating: they will only encourage even more substantial by-passing of the overall system. To have a real impact we need to recognise that certain designs are likely to be very sensitive to misuse by participants whilst others will be much more robust. Indeed the problem is actually rather more fundamental: if buildings are to be designed that are actually pleasant to live and work in, then design criteria must be based on what people actually want.

From more recent consumer research conducted on various house designs (Wensley 1984), it is clear that the consumer or user is often very well aware of what they want. For instance they want light airy living spaces but not at the cost of a feeling of lack of privacy because of large exposed glazing areas.

The building industry on the other hand is very confused. This is partly because customers themselves are different and therefore it is always easier to know what you want than to estimate what they want either individually or collectively. Indeed, in some of the research

it is clear that the industry experts at various levels be they
architects or developers are effectively responding to the design
choices much more as individual customers than as careful observers
of trends either in terms of market preference or actual performance.

We would suggest, however, that the reasons for the confusion
exhibited in the building industry in response to such design options
are in fact also more deep seated and structural. We need to
recognise that in domestic house design, many in the professional
groups are for various reasons ill informed and indeed sometimes
wrongly informed on technical design options. We believe the basic
source of confusion lies in the structure of the building industry
itself, with its high degree of sub-contracting and well defined
professional groupings. Such groups have strong vested interests not
only in defining problems in a manner which is containable within
their own competences but also to undertake work in a form in which
unanticipated risks are minimised.

The net result of such a structure is often that the final user,
lacking adequate guarantees becomes the bearer of the innovative
risks taken by others at earlier stages in the design and building
process. All will by now be aware of the continuing saga of failed,
or at best only partially successful building innovations, resulting
in uninhabitable dwellings and considerable loss of both comfort and
capital to those who were left holding the baby when the music
stopped: that is when the defects began to appear. Admittedly, we
have begun to recognise some of these problems in the development of
the NHBC 10 year guarantee although as some have pointed out, this is
actually a remarkably short period in the life of the building fabric.

THE BUILDING SECTOR

One of the important characteristics of buildings themselves and
hence the sector which finally makes such products available to the
ultimate customer is the extent to which the performance of individual
elements in the design is intimately effected by the overall structure
and system. Such effects occur not only at the initial design stage
but also later in terms of, first, on-site installation performance
and then both use patterns and maintenance schedules.

The actual performance over time of a particular technology depends
on the effective interactions between a number of both professional
and trade groups: from architects and mechanical engineers to
plumbers and electricians. To understand the likely process of
technology transfer in the building sector we need to recognise two
distinct and interacting effects:

i.   the difficulties posed by requiring close cooperation between
      professional and trade groups: for instance the recent Martin
      Centre report for ETSU on daylighting opportunities in commercial
      buildings and its impact on actual design practice illustrates
      clearly that an attempt to introduce such measures will only
      prove successful if it recognises not just the need for

appropriate information in different forms for the various
groups but also for, in many cases, different working relationships between the groups themselves.

ii. the nature of the building sector itself with various groups
     pursuing their own commercial and professional interests. This
     suggests that measures which show significant benefit to at
     least one group will be likely to be pursued and equally where
     the potential benefits are high we may expect to see a re-
     configuration on the supply side to provide a new mechanism for
     delivery of the desired "product".

We need to recognise, however, that the process whereby the clear
signals are developed to encourage a supply side response which
involves substantial re-configuration, is itself unclear. It is a
very important limitation of the market, that is the customers, can
only respond by making choices between the available offerings. Any
signals to indicate a preference outside the currently available
option set must either be inferred from existing choices or from
tentative market research. In such areas, market research can only
be tentative for the simple reason that the link between what can be
researched, in general attitudes, and actual purchase behaviour at a
later date is necessarily a tenuous one, both in general (Pickering
and Isherwood 1974) and in the specific area of energy conservation
measures (Philips, Mills and Nelson 1977).

One way of encouraging greater responsiveness and , indeed, in the
economist's terms, more entrepreneurial behaviour is by developing
more decentralised supply systems. For instance a number of volume
builders in the domestic sector, such as Barratts, allow their
regional organisations to exploit a considerable degree of local
autonomy in terms of their mix of designs and sites. It remains,
however, a key fact that even in such cases central control is exerted
through such mechanisms as the overall design catalogue. Interestingly enough, it would appear that the centralised strength of certain
professional groups, although important for other reasons, can substantially discourage such local initiatives as, for instance, in the
rather uncomfortable relationship between professional architects and
"designers" working for major builder/developers.

The recent BERU report (Leopold and Bishop 1983) on design
philosophy and practice in speculative housebuilding provides further
illustration of what might be dubbed this clash of cultures. Such a
clash involves both around the much greater commercial concern of the
builders and developers with "buildability", which, roughly translated
means the delivery of consistent product on time and cost, compared
with innovation design. Again this distinction must not be laboured
too far: the commercial housebuilding sector has had its own problems
with buildability and on-site quality control particularly when
managing a wide range of sub-contractors, but it is true that the
concerns about such problems are often central in the debate about
whether, for instance, new designs should be adopted.

Equally the "lack" of concern with innovative design should not be exaggerated. It is true that builder/developers tend to see the average customer as relatively conservative in their attitudes towards innovative design although they are also aware of the need to move the image of a modern domestic development away from that of an "estate". Their solution is, however, often to rely on relatively superficial design additions to overcome this problem. It is here that there does seem to be an opportunity for further joint develop-ments. Pilot work at the London Business School on this issue in the context of low energy, passive solar domestic housing suggests that the actual form in which a particular measure or measures are presented, broadly the aesthetics of both appearance and layout, does have a very significant effect on market acceptability. It would appear therefore that there is a significant opportunity for co-operation between Architects and builder/developers on the devel-opment of suitable designs which incorporate elements which are integral rather than merely add-ons.

The general issue remains, however, that the building sector supply side is dominated, for other good reasons, by strong professional groups, substantial specific supply interests, and indeed informal coalitions of builder/developers themselves who have a vested interest in restricting the range of choice available for the good reason that otherwise they could encounter substantial and unanticipated problems in both delivery and performance. In such circumstances the nature of most energy saving measures is such that compared with current standards such as 1982 Building Regulations for domestic housing, the additional benefits particularly in economic terms, are not large enough to dominate market choices, we must therefore consider instead the ways available to make energy efficient design more marketable.

MARKETABILITY ASSESSMENT

The assessment of the marketability of any particular energy conserv-ation approach is therefore crucial once we recognise that adoption will not be inevitable, and total once the economic cost/benefit evaluation achieves a particular single criterion such as in the UK crossing the Rubicon of the Treasury Test Discount Rate of 5% real, or indeed vice-versa. Whilst part of the explanation lies in the very diversity of the market for any of these measures so that it is almost inevitable that individual economic evaluations will exhibit a wide spread, we have also suggested above that the market structure itself and the nature of customer response may well be crucial.

Hence, assessing the marketability of any specific options goes wider than this concern with the inevitable spread of economic assessment. First, there needs to be an evaluation of the likely response from the various supply side groups. They are quite likely to assess a particular option in a significantly different way from either policy makers or the ultimate consumer. This evaluation must consider not only the differing attitudes of the groups themselves

399

but also their own degree of influence and power within the overall supply side. For instance, in the domestic sector, it may well be true that certain supply groups see a substantial benefit from the introduction of energy saving measures in new starter homes, but if such measures increase the basic cost and building societies are not willing to increase their valuations on the basis of lower anticipated running costs, then it is very likely that despite whatever pressure is put on them, most relevant builder/developers will not adopt the new measures because they believe, probably very correctly in many cases, that the key determinant of purchase behaviour in this market segment is affordability which is directly related to the building society valuation.

A marketability study also needs to assess the extent to which the ultimate customer would be likely to respond positively to a particular option when and if it were available. Inevitably such an assessment, as in any attempt to use market research to inform us about likely future market behaviour, will be subject to error but as we have suggested above, pilot work in this area does suggest that current customer opinions are often well enough defined and stable to allow reasoned extrapolations. It is also apparent that there are a number of design areas, such as open planning, south facing gardens and large windows, where the previous experience of customers living in houses wich such features has a very significant effect on their attitude and importance in the purchase decision for the next house. In particular there is evidence that individuals' experience of passive solar designed houses, such as those in Milton Keynes, has significantly increased the importance they would not attach to such features in their next purchase decision.

Finally, from a policy point of view, a marketability study needs to assess the extent to which such evidence of consumer acceptability, if presented in the right form, will actually influence the behaviour of the key intermediaries on the supply side. Although such a wide ranging marketability study has yet to be conducted, in the broad area of energy conservation measures for buildings as a whole, we would suggest that on the basis of current evidence, a number of tentative conclusions can be suggested:

i.   beneficial measures which can be incorporated in some form or other which avoids any direct capital cost increase are quite likely to be adopted.

ii.   if an increase in initial capital costs is inevitable, then one general problem arises in that energy costs are only one component, and often a minor one, of the total running costs and therefore not high on many people's priority list.

iii. the minor importance attached to energy running costs suggests that improvements in energy efficiency which do involve additional capital costs will only come about if the financial trade-offs are achieved through simple financing packages from the energy utilities or the energy saving features are an integral part of a more marketable package.

An example in the commercial office sector may involve an integrated design incorporating flexible and efficient air conditioning to cope with the changing heating and cooling demands of local office electronic work stations along with comprehensive wiring networks to cope with the distributed data and voice facilities required in the modern office environment. Despite the fact there there is reasonable evidence that air conditioning as such does not result in a reduced space heating load, in such circumstances, it may well be more realistic to design for an attractive energy efficient office environment rather than one which merely attempts to minimize energy consumption.

Similar arguments can be applied in the case of domestic housing, where we now have a substantial body of evidence that the benefits of a more energy efficient fabric are taken partly in reduced fuel consumption but also in increased comfort levels. It should, however, be recognised that at least within certain wide limits, the fact that part of the benefit is taken in increased comfort levels does not automatically exclude that portion as a measure of the total welfare value. We must be careful to distinguish between social welfare effects and net effects on energy demand.

Again, therefore, we need to consider the most marketable forms in which energy efficient designs might be developed for domestic housing. In general, it would appear that such designs would combine some of the features of the low infiltration/mechanical ventilation approached with some of the more aesthetically appealing features of incidental solar gain housing which generates a "light and airy" feeling for the living quarters.

POLICY ISSUES

In the light of the above analysis we wish to consider two broad areas of policy concern. First, there is the nature of the policy concerned with the design and use of buildings. Second, the problems of any policy based on adoption within a market system.

Design and Use of Buildings

It is widely recognised that the building stock, at various levels of analysis from the individual unit to the built environment represents a crucial social resource which can have either a severe constraining or facilitating effect on desirable social change. In the long run a concern with the energy efficiency of the building stock is more than just an attempt to minimize the energy costs. Properly designed an energy efficient building stock provides greater flexibility in terms of the detailed timing of supply provision and also proves more robust and hence insensitive to short term supply interruptions.

The problem with such a macro view of the built environment is the extent to which, at least at higher levels of aggregation, the maintenance and replacement cycles are of extremely long duration.

A realistic policy response to such problems must recognise the limited leverage points in the system related to the time cycles of the various components. Such leverage points include particularly new build but also significant retrofit work which can be related to the appropriate major cycle of refurbishment, such as in the commercial sector with current 1930's building stock. With the even longer time cycles of cities themselves we can probably only recognise the limited amount of modification that can be expected with an old and expensive infra-structure and rely on major significant developments either of the Docklands nature within London or on new "green field" sites such as Milton Keynes. This sense of the limitations of incremental changes also suggests that in some instances the stepwise, add-on approach may merely reinforce the limitations of the current structure despite the economic logic in each individual measure.

We also need to consider, from both a policy and a professional view, how the issues of risk might be better handled to aid the process of innovation. On the other hand, we must recognise the inevitable long time cycles in testing new technologies for buildings. It would seem we might learn something from the testing and approval procedures in other areas where long term effects can be critical such as the drug industry.

In terms of the allocation of professional risks, the cynic might well believe that it is the fact that many groups can avoid the risks and pass them on with fair impunity to the end-user, that also explains the apparent lack of concern to establish such mechanisms. Indeed one is sometimes forced to the rather strange conclusion that building professionals work in the name of one client (the public) but are paid by another (the builder) and are pretty unaccountable to both. This might be compared with, at one end, the role of the advertising agency, clearly both paid for and accountable to his commercial client, and at the other, the doctor, directly or indirectly, paid for and accountable to the customer (at least that is the theory!).

The Adoption Process within the Market

The market based qualitification is itself important in that adoption depends not purely on policy injunctions but on the choices of various independent agents operating in the marketplace. The key elements of such an adoption process include the importance of building examples "on the ground" the existence of exemplar designs in terms of fully evaluated "paper" studies, the need for adequate documents and professional reference books as well as professional articles and finally a need for a process of interaction and feedback between market evaluation and the renewed priorities of Research and Development. This focus on exemplars also implies a more appropriate model of innovation diffusion, are based on mutation of those seen both as leaders and also as trustworthy. Such considerations imply fairly rigorous criteria in approaching and assessing suitable initial contacts. Then practitioners must be seen as underwriting some, if not all, of the development risks and need to be compensated accordingly.

The importance of actual buildings "on the ground" has been widely recognised and is supported by programmes such as the Energy Conservation Demonstration Programme Scheme. There does, however, still appear to be some confusion about the role of monitoring and evaluation, at least from a marketability viewpoint. It would seem that the key concerns from an adoption perspective are that it can be done in practice, that it works and is affordable. Given the nature of these concerns it would seem that rather more effort might in some instances be applied to ensuring that standard rather than above average building practices are adopted and somewhat less effort might be expended on the detailed monitoring of performance. This should not be taken as an argument for reduced monitoring in total. Valid arguments have been presented that the duration of many monitoring programmes is too short to measure fully the impact of user behaviour based on experience of the system. From a different perspective there may also be a justification for more detailed monitoring to help refine our understanding of the underlying heat flow processes involved although it must be recognised that given the level of sophistication of the current available computer modelling systems, such analysis is hardly likely to validate such models except at the level of generating surprise results. In which case we return to the fact that the significant surprise results will probably also show through in a less detailed monitoring programme.

The issue of exemplar designs relates both to the system wide nature of many energy efficient designs as well as the fact that design practice will be much more likely to be influenced if such designs are widely available and fully evaluated. The current work by the Building Research Establishment (BRE) within the passive solar design studies project for domestic buildings managed by ETSU for the Department of Energy provides an example of the type of work involved.

The development of both professional articles and suitable reference material relates to the problem of both developing professional interest in the topic and then providing the technical support to encourage the interest. As an example of the first stage, the recent RIBA Journal was particularly focussed on the passive solar programme. There remains the issue of the most effective means of supporting this interest in terms of technical advice. One option which is worth further consideration, is the development of an energy advice service, using and supporting external energy consultants, for individual commercial users, along lines similer to a number of individual programmes within the successful Manufacturers Advice Scheme developed by the Department of Trade and Industry.

A focus on the issues of technology transfer helps us to recognise three particular issues. First, there are limits to how far we can argue that R & D is completed when the cost (in)effectiveness of a particular measure is established simply because in a market context it can never be unambiguously valued before the technology transfer itself has been attempted. Second, the process of attempting the tec@nology transfer itself is one of learning, not only about the technology in practice but also about the market response. As we learn more about the market it becomes appropriate to redefine both the R & D requirements and also the interrelated priorities. Third, the whole concept to technology transfer should not be seen as a "one-way" process. It often involves technical improvements in one particular sub-sector and as such can also produce anomolies and undesirable side-effects. The overall process becomes a way of evaluating both these effects and the risks involved.

In the energy conservation sector this problem is particularly acute because of the very specific nature of particular technological advances yet the wide systemic nature of the costs and benefits.

Finally, a crucial focus for development and encouragement involves multiple fora for debate discussion and learning between legitimate interests to build informal links and reinforce the feedback of market evaluation into technical priorities.

REFERENCES

1.   Leopold, Ellen & Donald Bishop (1983) 'Design philosophy and and practice in speculative housebuilding: Part 2' Construction Management and Economics, 1, pp 233-268

2.   Phillips, Nicolas, Pamela Mills & Elizabeth Nelson (1977) 'Domestic Energy Conservation: Development & Evaluation Programme

3.   Pickering, J.F. and Isherwood, B.S., 'Purchase Probabilities and Consumer Durable B ying Behaviour' (1974) Journal of Market Research Society, Vol. 16, No. 3, pp 203.

4.   Woodward, Nicholas, (Winter 1983) 'Dimensions of Complexity' London Business School Journal, Vol. 8, No. 2.

Index

407

409

411